T0335693

Topics in Grammatical Inference

Jeffrey Heinz · José M. Sempere
Editors

Topics in Grammatical Inference

Editors
Jeffrey Heinz
Department of Linguistics and Cognitive Science
University of Delaware
Newark, DE
USA

José M. Sempere
Universidad Politécnica de Valencia
Valencia
Spain

ISBN 978-3-662-48393-0 ISBN 978-3-662-48395-4 (eBook)
DOI 10.1007/978-3-662-48395-4

Library of Congress Control Number: 2015952765

Printed on acid-free paper

This Springer imprint is published by Springer Nature
The registered company is Springer-Verlag GmbH Berlin Heidelberg

José M. Sempere dedicates this book to Amparo, Noa and Rubén (his positive data). And Jeffrey Heinz dedicates this book to earthlings' outstanding abilities to infer grammars, especially those of Mika, Emma, and Maya.

Preface

Grammatical inference is one part of theoretical computer science that addresses the problem of how computers can learn from experience. In this way, it is like the well-known field of machine learning. Perhaps what most distinguishes grammatical inference from standard machine learning approaches is its focus on learning the structure which underlies the concept to be learned, i.e., in identifying the *nature* of the *target* concept.

A common ingredient that one can find in all the grammatical inference works is that the target concepts are formal objects like sets of strings or sets of trees, which can be represented either by accepting devices (for example, a finite automata, a neural network, etc.) or by generating devices (mainly, formal grammars which can be inserted into a formal framework such as the *Chomsky hierarchy*). Hence, it is common for the results and algorithms developed in the area of grammatical inference to have implications in other research areas such as computability and complexity theory, formal languages and applications, artificial intelligence, etc. Similarly, such areas frequently guide research in grammatical inference. So, grammatical inference and other research areas are continuously influencing each other.

Grammatical inference has been developed largely in the last 30 years. We can establish E.M. Gold's work *Language Identification in the Limit* (1967) as the seminal work where all the basic problems, concepts, and rules of the game were established for grammatical inference. Since then, the grammatical inference research area has been continuously in progress, especially since the landmark work in the 1980s by Dana Angluin. So, it is a well-developed area with a long tradition.

There is a well-established community of researchers who have a presence in the scientific scene through different publications and conferences. The main conference on grammatical inference is international in scale, and has been celebrated every 2 years since 1993 (the ICGI conference series). Its proceedings were published as part of the Springer LNCS or LNAI series until 2010, and since 2012 they have been published as part of the open-access JMLR Workshop and Conference

Proceedings series.[1] These conferences are directed by an international steering committee (of which we are current members). Additionally, the website http://www.grammarlearning.org/ provides online information about conferences, competitions, software, and other resources relating to the field of grammatical inference. Lastly, the 2010 book *Grammatical Inference* by Colin de la Higuera, published by Cambridge University Press, collects in one place the main body of results of this field and presents them from the ground up in a uniform fashion.

This book provides advanced treatments of topics in grammatical inference. In this way, this book complements de la Higuera's book, by addressing topics that are either unmentioned or only introduced there.

The topics included in this book are largely drawn from tutorials that were presented at the ICGI 2010 and 2012 conferences. They were selected for the following reasons: (1) the topic has reached a certain level of scientific maturity, so a reasonable number of (positive and negative) results, algorithms, and conclusions have been obtained; and (2) the topic is of fundamental interest to grammatical inference, so it attracts a significant number of researchers to study the many facets of the problems that it exhibits.

The first three chapters of the book deal with issues regarding the theoretical learning framework. So, John Case's chapter, Gold-Style Learning Theory, discusses different learning paradigms, relationships among them, and learning power associated with them more generally. Rémi Eyraud, Jeffrey Heinz, and Ryo Yoshinaka's chapter, *Efficiency in the Identification in the Limit Learning Paradigm,* pays special attention to the complexity issues associated with *identification in the limit* learning criteria. Then, a chapter by Colin de la Higuera, *Learning Grammars and Automata with Queries*, focuses on the main results of learning by changing the information source to what is called *active learning*, wherein the learner can ask an oracle for information about the target. The next part of the book focuses on the main classes of formal languages according to Chomsky's hierarchy: the regular languages and the context-free languages. With respect to the regular languages, two chapters deal with finite-state automata. First, the chapter by Damián López and Pedro Garca, *On the Inference of Finite State Automata from Positive and Negative Data,* shows the main aspects of learning regular languages from examples and counterexamples. The chapter by Jorge Castro and Ricard Gavaldà, *Learning Probability Distributions Generated by Finite-State Machines,* approaches the learning of stochastic regular languages in a probabilistic manner, with a special focus on spectral learning. The chapter *Distributional Learning of Context-Free and Multiple Context-Free Grammars,* by Alexander Clark and Ryo Yoshinaka, focuses on an algebraic approach to learning some subclasses of the context-sensitive languages, which include significant classes of context-free languages. The next chapter, by Johanna Björklund and Henning Fernau, *Learning Tree Languages*, largely deals with the learning of regular sets of tree languages. The relation between tree languages and context-free

[1] http://jmlr.csail.mit.edu/proceedings/.

languages of strings was established as an alternative approach to learn in what is named *learning from structural data*. Hence, most of the results of this chapter are relevant to learning context-free languages. Finally, the chapter by François Coste, *Learning the Language of Biological Sequences,* shows an area of application that has recently been approached by grammatical inference: the processing of biosequences.

One decision we made early in regards to this book was to give the authors a high level of autonomy in preparing their chapters. There are advantages and disadvantages to this approach. One disadvantage is that some overlap inevitably exists between the chapters. We have cross-referenced other chapters where appropriate. Another disadvantage is that the notation used in each chapter differs. However, we believe the advantages outweigh these disadvantages. Each chapter in this book stands on its own, and includes the concepts and references necessary to help the understanding of the results by the reader. Thus it is not necessary for the reader to approach the contents of this book in order. Additionally, as a consequence of this approach, this book is oriented to an audience with basic knowledge of mathematics, computer science, and formal language theory. It could be (under)graduate students, or computer scientists, linguistics researchers, cognitive psychologists, or other readers interested in the nature of learning and its relation to computer science, artificial intelligence, and a significant number of related areas.

The editing of this book has been a long process where we have been helped by many different persons. We would like to thank them all for the support that they have provided during this process. First, we would like to thank all the contributors and authors of this book for the patience that they have exhibited during all stages of the book's production, especially during the reviewing process. We specially thank Colin de la Higuera for his support and for giving us the idea of editing this book. We thank all the people from Springer for their support, specially Ronan Nugent for all his understanding about the delays and problems that we encountered during the preparation of this book. Last but not least, we give thanks to our family members and friends for all their support. This book has been a good conversation topic during all this time, although they avoided asking much about it on some occasions. To all of them, thank you very much.

Newark, DE, USA
Valencia, Spain
December 2014

Jeffrey Heinz
José M. Sempere

Contents

Contributors

Johanna Björklund Department of Computing Science, Universitet Umeå, Umeå, Sweden

John Case Computer and Information Sciences Department, University of Delaware, Newark, DE, USA

Jorge Castro LARCA Research Group, Universitat Politècnica de Catalunya-BarcelonaTech, Barcelona, Spain

Alexander Clark Department of Philosophy, King's College London, London, UK

François Coste Inria Rennes—Bretagne Atlantique, Rennes Cedex, France

Colin de la Higuera LINA, University of Nantes, Nantes, France

Rémi Eyraud QARMA Team, Laboratoire d'Informatique Fondamentale, Marseille, France

Henning Fernau FB IV—Abteilung Informatikwissenschaften, Universität Trier, Trier, Germany

Pedro García Departamento de Sistemas Informáticos y Computación, Universidad Politécnica de Valencia, Valencia, Spain

Ricard Gavaldà LARCA Research Group, Universitat Politècnica de Catalunya-BarcelonaTech, Barcelona, Spain

Jeffrey Heinz University of Delaware, Newark, DE, USA

Damián López Departamento de Sistemas Informáticos y Computación, Universidad Politécnica de Valencia, Valencia, Spain

Ryo Yoshinaka Graduate School of Informatics, Kyoto University, Kyoto, Japan

Chapter 1
Gold-Style Learning Theory

A Selection of Highlights Since Gold

John Case

Abstract This chapter is a tutorial addressed to, among other audiences, the grammatical inference community. It is on the computational learning theory initiated by E. Mark Gold's seminal 1967 paper. One of Gold's important motivations was to present a formal model of child language learning. This chapter introduces Gold's model and also presents an introduction to some selected highlights of the theory appearing since Gold 1967. Since both Gold and the present author had or have cognitive science motivations, many of the results discussed herein also have implications for, and were motivated by, questions regarding the human cognitive ability to learn. For this chapter some prior knowledge of computability theory would be helpful to the reader, as would some prior acquaintance with Gold's identification/learning in the limit concept. The first section concentrates on results which are independent of computational complexity considerations. Covered are: Gold's model of child language learning (including some reasonable, post-Gold criteria of successful learning and some critical discussion); some severe constraints on successful learning; some characterization results and important examples of successful learning; discussion and results on potential child insensitivities to data presentation order; and empirical U-shaped learning with some corresponding formal results interpreted for their cognitive science significance. The second section concentrates on relevant complexity issues. The issues considered are: large database size and corresponding *memory-limited learners*; unavoidable cases of complexity and information deficiencies *of the programs learned*; and the complexity of *learner updates*. In this section a few, seemingly difficult, open mathematical questions are indicated. Some of them are important for cognitive science.

J. Case (✉)
Computer and Information Sciences Department, University of Delaware,
Newark, DE 19716, USA
e-mail: case@udel.edu

J. Heinz and J.M. Sempere (eds.), *Topics in Grammatical Inference*,
DOI 10.1007/978-3-662-48395-4_1

1.1 Language Learnability: Gold 1967 and Beyond

In this section, covered are: Gold's model of child language learning (Sect. 1.1.1); criteria of success (Sect. 1.1.2); some severe constraints on successful learning (Sect. 1.1.3); some characterization results and important examples of successful learning (Sect. 1.1.4); discussion and results on potential child insensitivities to data presentation order (Sect. 1.1.5); and empirical U-shaped learning (Sect. 1.1.6) with some corresponding formal results interpreted for their cognitive science significance (Sect. 1.1.7).

1.1.1 Gold's 1967 Model of Child Language Learning

M, below in Fig. 1.1, is an algorithmic device (machine), and, for $t = 0, 1, 2, \ldots$, at "time" t, M reacts only to the finite sequence of utterances $u_0, u_1, \ldots, u_t$, with utterance u_i arriving from, for example, the Mother, at time i; with the (formal language) L to be learned $= \{u_0, u_1, \ldots\}$; and with g_t the child's t-th internal grammar *computed* by its M on input utterance sequence $u_0, u_1, \ldots, u_t$.

This leaves open for the moment what constitutes *successful* learning of a formal language L. In the next section (Sect. 1.1.2) we begin to take up this interesting topic. Further below, in the second main section entitled Complexity Considerations, i.e., Sect. 1.2, additional criteria of success are explored.

As for the plausibility of Gold's model of child language learning above, Gold [37] argued from the psycholinguistic literature (e.g., [47]) that children react to and need only *positive* data regarding languages L they learn—and that they really need *no* data regarding $\overline{L}$.

While this model of Gold's is clearly deficient regarding inputs to the child of semantic/denotational information and social reinforcers, I'll, nonetheless present below what I consider to be some resultant insights for cognitive science.

My view, argued in [19] (and to some extent in [18]) is that reality, *including cognitive reality*, has algorithmic *expected* behavior.

Hence, restricting learners to being algorithmic *from this point of view* is perfectly reasonable. Furthermore, it is a standard assumption of the field of cognitive science that cognition is algorithmic.

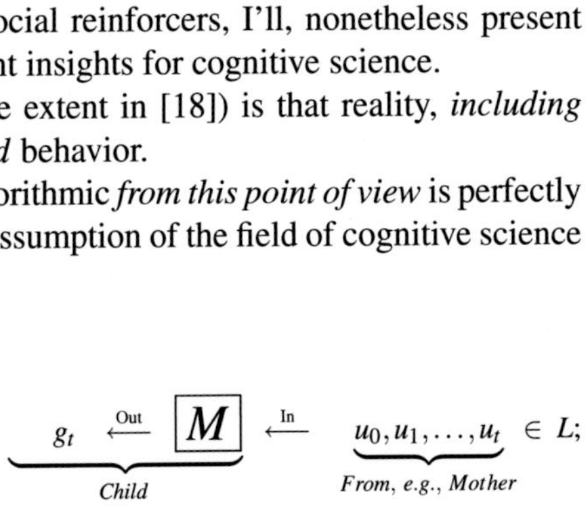

Fig. 1.1 Gold's model

$$\underbrace{g_t \xleftarrow{\text{Out}} \boxed{M}}_{\textit{Child}} \xleftarrow{\text{In}} \underbrace{u_0, u_1, \ldots, u_t}_{\textit{From, e.g., Mother}} \in L;$$

1.1.2 Some Criteria of Successful Learning

Definition 1.1

- T is a *text* for $L \overset{\text{def}}{\Leftrightarrow} \{T(0), T(1), \ldots\} = L$.[1]
- Depicted just below is the I/O behavior of learning machine M receiving successive text elements and outputting corresponding successive grammars.

$$g_0, g_1, \ldots, g_t, \ldots \overset{\text{Out}}{\longleftarrow} M \overset{\text{In}}{\longleftarrow} T(0), T(1), \ldots, T(t), \ldots.$$

- Below are criteria from [17, 27, 37, 53] for some machine M *successfully* learning *every* language L in a class of languages $\mathscr{L}$ (see also [40])—where the g_is are from the previous bullet.
  Suppose: $\mathbb{N} = \{0, 1, 2, \ldots\}$; $\mathbb{N}^+ = \{1, 2, \ldots\}$; $b \in (\mathbb{N}^+ \cup \{*\})$; and $x \leq *$ means $x < \infty$.
  - $\mathscr{L} \in \mathbf{TxtFex}_b \Leftrightarrow (\exists$ suitably clever $M)(\forall L \in \mathscr{L})(\forall T$ for $L)(\exists t)$ $[g_t, g_{t+1}, \ldots$ each generates $L \wedge \text{card}(\{g_t, g_{t+1}, \ldots\}) \leq b]$.
    We say M $\mathbf{TxtFex}_b$*-learns* (each) $L \in \mathscr{L}$. $\mathbf{TxtEx} \overset{\text{def}}{=} \mathbf{TxtFex}_1$, *Gold's criterion*! For example, the class $\mathscr{F}$ of all finite languages $\in$ **TxtEx** [37].
  - $\mathscr{L} \in \mathbf{TxtBc} \Leftrightarrow (\exists M)(\forall L \in \mathscr{L})(\forall T$ for $L)(\exists t)$ $[g_t, g_{t+1}, \ldots$ each generates $L]$.
    We say M **TxtBc**-*learns* (each) $L \in \mathscr{L}$. For example, $\mathscr{K} = \{K \cup \{x\} \mid x \in \mathbb{N}\} \in (\mathbf{TxtBc} - \mathbf{TxtFex}_*)$ [17], where K is the diagonal halting problem from [58].

1.1.2.1 $\mathbf{TxtFex}_b$-Hierarchy

Definition 1.2 Let $W_g \overset{\text{def}}{=}$ the language generated or enumerated by grammar or program g [58]. Informally: W_g can be thought of as the [summary of the] *behavior* of g.

I like the use of machine self-reference arguments [16], and in the next theorem (Theorem 1.3), the classes mentioned for witnessing that one $\mathbf{TxtFex}_b$ criterion has more learning power than another are each (finitarily) self-referential. The proof that they work is omitted but can be found in [17].

Theorem 1.3 [17] *Let $\langle \cdot, \cdot \rangle$ computably map $\mathbb{N} \times \mathbb{N}$ 1–1 onto $\mathbb{N}$ [58]. Here and below $\overset{\infty}{\forall} z$ means: for all but finitely many $z \in \mathbb{N}$ [9]. Suppose $n \in \mathbb{N}^+$. Let $\mathscr{L}_n =$ the set of all* infinite *L such that*

$$(\exists e_1, \ldots, e_n)[W_{e_1} = \cdots = W_{e_n} = L \wedge$$

$$(\overset{\infty}{\forall} \langle x, y \rangle \in L)[y \in \{e_1, \ldots, e_n\}]].$$

[1] Technically, a text T for L is a mapping from $\mathbb{N}$ onto L. A text for L is then a sequence of utterances of all and only the elements of L.

Let $\mathscr{L}_* = \bigcup_{n \in \mathbb{N}^+} \mathscr{L}_n$. *Then* $\mathscr{L}_{n+1} \in (\mathbf{TxtFex}_{n+1} - \mathbf{TxtFex}_n) \wedge \mathscr{L}_* \in (\mathbf{TxtFex}_* - \bigcup_{n \in \mathbb{N}^+} \mathbf{TxtFex}_n)$.

Does, say, Noam Chomsky, for each natural language he learns, eventually vacillate between up to 42 correct grammars in his head (but for some languages no fewer)? I.e., does he need $\mathbf{TxtFex}_{42}$-learning? The problem is that we can*not* (yet) see grammars inside peoples' heads. So, we don't know if humans employ **TxtEx**-learning, $\mathbf{TxtFex}_2$-learning, $\mathbf{TxtFex}_3$-learning, ... Hence, for now, $\mathbf{TxtFex}_n$-learning, for non-astronomical $n \in \mathbb{N}$, are all not unreasonable for modeling human behavior.

1.1.2.2 Discussion

In [49] a well-documented body of *experimental* evidence indicates Mothers' utterances to young children are *not* calibrated by increasing syntactic complexity to teach children gradually (but, instead, to fit the limited attention span and processing powers of children).

These considerations are *partly* formally mirrored in the success criteria by the requirement that the learners eventually correctly learn a language L *no matter in what order the input is presented*, i.e., ($\forall T$ for L)—as long as that input contains all and only the correct sentences of the language. At this point it is important to note that ($\forall T$ for L) includes both computable and also *un*computable texts T for L! It is noted in [52] that, since the utterances of children's caretakers depend heavily on external environmental events, such influences might introduce a random component into naturally occurring texts. Whence comes the interest at all in non-computable texts. As we shall see below in Sect. 1.1.5.2, *for many important criteria of success*, learning power is nicely *un*affected by whether we allow or disallow uncomputable texts!

Children *may* be *in*sensitive to some aspects of the *order* of data presentation. In Sect. 1.1.5 several *possibilities* are considered, and some corresponding theoretical results (for some learning criteria) will be presented.

First, though, in the next section (Sect. 1.1.3), the topic of constraints on successful learning are considered.

1.1.3 Constraints on Learnability

Angluin [2] introduced the following Subset Principle for **TxtEx**-learning. It places a severe constraint on the learnable.

Theorem 1.4 (Subset Principle [2, 17]) *Suppose* $\mathbf{C} \in \{\mathbf{TxtFex}_b, \mathbf{TxtBc}\}$ *and* M $\mathbf{C}$*-learns* L.
Then ($\exists$ finite $S \subseteq L)(\forall L' \subset L \mid S \subseteq L'))[M$ *does* not $\mathbf{C}$*-identify* $L']$.

N.B. The above Subset Principle *and* its depressing corollary just below (Corollary 1.5) hold *even if* M *is not algorithmic*! The proofs essentially depend on Baire Category (or Banach-Mazur Games) from Topology [51, 52] and *not* on algorithmicity! This is possible *only because* M in general can*not* infer data about $\overline{L}$. Case and Kötzing [26] studies which learning theory results depend only on topology, which on algorithmicity.

Angluin connected the Subset Principle to the machine learning problem of *overgeneralization*. From the proof: M *must* overgeneralize on (some Ts for) L', but it does not overgeneralize on L itself.

See [41, 68] for discussion regarding the possible connection between this subset principle and a more traditionally linguistically oriented one in [45].

Corollary 1.5 [37, 53] *If* $\mathscr{L}$ *contains an* infinite *language together with* all *its* finite *sublanguages, then* $\mathscr{L} \notin$ **TxtBc**.
Hence, for example, the class of regular languages $\notin$ **TxtBc**!

Definition 1.6 Consider the variant of **TxtBc**-learning called **TxtBc***-learning. **TxtBc***-learning is just like **TxtBc**-learning *except* the final grammars or programs, instead of each being perfectly correct, may each make *finitely many mistakes*.

Actually:

Remark 1.1 [53] While **TxtBc*** learners can learn more than **TxtBc**-learners, the class of regular languages $\notin$ **TxtBc***!

In [17] there are variants of Theorems 1.3 and 1.4 for criteria which allow a few mistakes in final programs. In [5] there are variants of Theorem 1.7 and Remark 1.2 for criteria which allow a few mistakes in final programs. Success criteria allowing a few mistakes in the final programs are also considered below in Sect. 1.2.2.

1.1.4 Characterizations and Pattern Languages

In this section presented are some characterization results (Sect. 1.1.4.1) and important positive learnability results based on pattern languages (Sect. 1.1.4.2).

1.1.4.1 Characterizations

Program p taking two inputs i, x is a *uniform decision procedure* for a class $\mathscr{U}$ of computably decidable languages iff

$$\mathscr{U} = \{U_i \mid (\forall x)[p \text{ on input } i, x \text{ decides whether or not } x \in U_i]\}.$$

Such a $\mathscr{U}$ is called *uniformly decidable*. Important examples are all the Chomsky Hierarchy classes, Regular, ..., Context-Sensitive [39], and, in the next section

(Sect. 1.1.4.2), Angluin's important class of Pattern Languages [1]. The following important characterization of Angluin *extends* the Subset Principle for the cases of *uniformly decidable* classes.

Theorem 1.7 (Angluin [2]) *Suppose* $\mathscr{U} = \{U_i \mid i \in \mathbb{N}\}$ *is uniformly decidable as above. Then* $\mathscr{U} \in$ **TxtEx** *iff there is a* computably enumerable sequence of enumerating programs *for* finite *sets* $S_0, S_1, \ldots$ *(*tell tales*) such that* $(\forall i)[S_i \subseteq U_i \wedge (\forall j \mid S_i \subseteq U_j)[U_j \not\subset U_i]]$. *The output programs of a witnessing M* can *be decision procedures.*

Remark 1.2 From [5], uniformly decidable $\mathscr{U}$ as just above is $\in$ **TxtBc** iff the associated tell tales $S_0, S_1, \ldots$ exist *but* do *not* have to be computably enumerable. In this context, the output programs of a witnessing M can*not*, in general, be decision procedures.

Remark 1.3 Angluin [2] exhibited a *uniformly decidable* $\mathscr{U}$ *with* tell tales but with *no* computably enumerable tell tales. Hence, from [5], her $\mathscr{U}$ is a *uniformly decidable class* $\in$ (**TxtBc** − **TxtEx**).

1.1.4.2 Pattern Languages

Next is an ostensive definition of Angluin's important class of *Pattern Languages*.

Definition 1.8 (*Angluin* [1]) A *pattern language* is one generated by all and only the *positive* length substitutions for variables (in upper case letter alphabet) of strings (over a lower case letter alphabet) in a pattern, such as, for example, `abXYcbbZXa`.

Angluin [1] showed the class of pattern languages to be **TxtEx**-learnable, and through further papers we have the following.

Theorem 1.9 [1, 63, 71] *For each* $n \in \mathbb{N}^+$, *the uniformly decidable class of unions of n pattern languages* $\in$ **TxtEx**!

These classes are not rendered unlearnable by the severe constraint of Theorem 1.4 as are the classes in the Chomsky Hierarchy (Corollary 1.5). This, in part, is because they *crosscut* the classes in the Chomsky Hierarchy. Perhaps the (somewhat ill-defined) class of natural languages is like that too. For example, most linguists consider each natural language to be infinite.

Applications of unions of n pattern languages, ranging from learning in molecular biology to more general machine learning, appear in, for example, [3, 4, 11, 50, 61, 64].

1.1.5 Insensitivities to Presentation Order

In this section we first consider some possible child insensitivities to order of presentation (Sect. 1.1.5.1); then corresponding formal results are presented (Sect. 1.1.5.2).

1.1.5.1 Order Insensitivities' Definitions

Children *may* be sensitive to the order or timing of data presentation (texts). First we present two *local* and, then, two *global* (formal) *in*sensitivities.

Definition 1.10

- M is *partly set-driven* [34, 35, 60] iff, on *sequence* $u_0, \ldots, u_t$, it reacts only to the *set* $\{u_0, u_1, \ldots, u_t\}$ of utterances *and* the *length* $t + 1$ of the utterance sequence—*not* to the *order* of the *sequence*. In effect, M reacts a little to timing (but not to order) of utterances.
- M is *set-driven* [69] iff, when shown utterance *sequence* $u_0, u_1, \ldots, u_t$, it reacts only to the corresponding *set* $\{u_0, u_1, \ldots, u_t\}$— *not* to the sequence's order and length.
- M is *weakly b-ary order independent* [17] iff, for each language L on which, for *some* T for L, M converges in the limit to a finite set of grammars, there is, corresponding to L, a finite set of grammars G of cardinality $\leq b$ such that M converges to a *subset* of this *same* G for *each* T for L.
- M is *b-ary order independent* [17] iff M is weakly so, *but*, instead of converging to a *subset* of G, it converges to *exactly* G. For $b = 1$, these two notions coincide and are essentially from [8, 52].

1.1.5.2 Order Insensitivities' Results

Results regarding (partly) set-driveness for **TxtEx** (the $b = 1$ case of $\mathbf{TxtFex}_b$) are from [34, 35, 60]. For example, *set-driveness strictly limits learning power for* **TxtEx**. That, for **TxtEx**, (weakly) 1-ary order independence is without loss of learning power is essentially from [8, 34, 35]. I found the $b > 1$ cases harder to prove than the $b = 1$ case.

Theorem 1.11 [17] *Any M can be algorithmically transformed into an M' so that M' is both* partly set-driven and weakly b-ary independent *and M' $\mathbf{TxtFex}_b$-learns* all *the languages M does (*even if *M only learns for* computable *texts).*

As noted above in Sect. 1.1.2.2, [52] argues that, since the utterances of children's caretakers depend heavily on external environmental events, such influences might introduce a random component into naturally occurring texts. This is whence comes the interest at all in non-computable texts. The just above theorem and the results below in this section (Sect. 1.1.5.2) imply that *for the important criteria considered in this section*, learning power is nicely *un*affected by whether texts are or are not allowed to be uncomputable.

Remark 1.4 [17] The preceding theorem holds with partly set-driven *replaced by* set-driven *but* with $\mathbf{TxtFex}_b$-identification *restricted to only infinite languages.*

Theorem 1.12 [17] *Any M can be algorithmically transformed into an M' so that M' is b*-ary order independent *and M'* **TxtFex**$_b$*-learns* all *the languages M does (*even if *M only learns for* computable *texts).*

It is not known (it's hard to tease out experimentally) whether children exhibit any of these insensitivities, but the formal results tell us something of how they affect learning power. It has not been investigated how the results in this section would be affected if complexity considerations were taken into account.

1.1.6 Empirical U-Shaped Learning

U-Shaped Learning follows the sequence *Learn, Unlearn, Relearn.* It occurs in child development [10, 46, 66, 67], e.g., verb regularization and understanding of various (Piaget-like) conservation principles, such as temperature and weight conservation and interaction between object tracking and object permanence.

Here is an example of U-shaped learning from irregular English past tense verbs. A child first uses *spoke*, the correct past tense of the irregular verb *to speak.* Then the child ostensibly overregularizes, incorrectly using *speaked.* Lastly, the child returns to using *spoke.* The major concern of the prior cognitive science literature on U-shaped learning is in *how* one *models* U-shaped learning. For example, for language learning, by general rules or tables of exceptions [10, 46, 54, 55]? With neural nets [59] and statistical regularities or statistical irregularities?

My own concern regarding U-shaped learning is whether it is an *un*necessary and harmless accident of human evolution *or* whether U-shaped learning is advantageous in that some classes of tasks *can* be learned in the U-shaped way, but *not* otherwise?

1.1.7 Formal U-Shaped Learning

In the interest of studying whether U shapes are necessary for full learning power, it is mathematically useful to define alternatives to success criteria, including those above but in which U shapes are *forbidden* on the way to success.

Definition 1.13

- Depicted just below is the I/O behavior of learning machine M receiving successive text elements and outputting corresponding successive grammars.

$$g_0, g_1, \ldots, g_t, \ldots \xleftarrow{\text{Out}} M \xleftarrow{\text{In}} T(0), T(1), \ldots, T(t), \ldots .$$

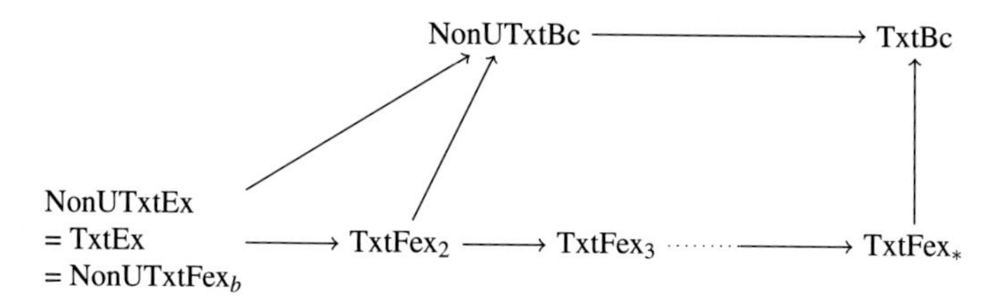

Fig. 1.2 Diagram of results

- Suppose **C** is a learning success criterion, for example, one $\in$ $\{\mathbf{TxtFex}_b, \mathbf{TxtBc}\}$. Then, where the g_is are from the first bullet of this definition, $\mathscr{L} \in \mathbf{NonUC} \Leftrightarrow (\exists M$ witnessing $\mathscr{L} \in \mathbf{C})(\forall L \in \mathscr{L})(\forall T$ for $L)(\forall i, j, k \mid i < j < k)[W_{g_i} = W_{g_k} = L \Rightarrow W_{g_j} = L]$. *Non-U-shaped* learners of $\mathscr{L}$ *never abandon correct behaviors output on texts for* $L \in \mathscr{L}$ *and subsequently return to them.*

1.1.8 Results/Question Regarding U-Shaped Learning

The results are presented first as a diagram (Sect. 1.1.8.1) and, then, as a verbal summary of key points—with an important cognitive science question at the end (Sect. 1.1.8.2).

1.1.8.1 Diagram of Some Results

The arrow $\longrightarrow$ above denotes class *inclusion*. The transitive closure of the inclusions in Fig. 1.2 below hold *and* no other inclusions hold [6, 12, 14, 15]. For *example*, from Fig. 1.2, we have $\mathbf{TxtFex}_2$ included in $\mathbf{TxtFex}_3$ *and* properly so—regarding the latter, the transitive closure of the inclusions of Fig. 1.2 has *no* arrow from $\mathbf{TxtFex}_3$ to $\mathbf{TxtFex}_2$.

For example, from the above, there is some $\mathscr{L} \in (\mathbf{TxtFex}_3 - \mathbf{NonUTxtBc})$! This same $\mathscr{L}$ then can*not* be $\in \mathbf{NonUTxtFex}_*$—else, it would, then, be in **NonUTxtBc**. The proof regarding this $\mathscr{L}$ *does* employ an interplay between *general rules* and (finite) sets of *exceptions* [12, 14]. As mentioned above in Sect. 1.1.6, some cognitive scientists, e.g., [10, 46, 54, 55], believe this interplay underpins human U-shaped learning.

1.1.8.2 Main Results and a Question

In the present section we summarize key results from the previous section (Sect. 1.1.8.1) and, at the end, pose and discuss a difficult question for cognitive science and the evolution of human cognition.

Remark 1.5

- *Main results*:
  - From **NonUTxtBc** $\longrightarrow$ **TxtBc**, U-shaped learning *is* needed for some class in **TxtBc**.
  - From **NonUTxtEx** = **TxtEx**, U-shaped learning is *not* needed for **TxtEx** learning, i.e., for learning *one* successful grammar in the limit.
  - From $\mathbf{NonUTxtFex}_*$ $\longrightarrow$ $\mathbf{TxtFex}_2$, U-shaped learning *is* needed for some class in $\mathbf{TxtFex}_2$ even if finitely many ($*$) grammars are allowed in the limit *but*, from $\mathbf{TxtFex}_2$ $\longrightarrow$ **NonUTxtBc**, it is *not* needed if we allow infinitely many grammars in the limit.
  - From the reasoning after the previous section's (Sect. 1.1.8.1's) diagram, there *exists* $\mathscr{L} \in (\mathbf{TxtFex}_3 - (\mathbf{NonUTxtFex}_* \cup \mathbf{NonUTxtBc}))$; in particular, U-shaped learning *is* needed for this $\mathscr{L} \in \mathbf{TxtFex}_3$—even if allow infinitely many grammars in the limit!

- *Question*: Does the class of tasks humans must learn to be competitive in the genetic marketplace, like this latter $\mathscr{L}$, *necessitate* U-shaped learning?
  Of course we have not yet modeled human cognition and its evolution sufficiently to answer this question. As pointed out in [12], on the *formal* modeling level, the pattern emerges that, for parameterized, cognitively relevant learning criteria, beyond very few initial parameter values, U shapes are *necessary* for full learning power! This is seen in the results just above as well as in Sect. 1.2.1.3. This latter section has, though, some important open questions very relevant to the emerging pattern.

1.2 Complexity Considerations

This section concentrates on computational complexity issues. Considered are: large database size and corresponding memory-limited learners (Sect. 1.2.1); unavoidable cases of complexity and information deficiencies of the programs learned (Sect. 1.2.2); and the complexity of learner updates (Sect. 1.2.3). In the present section a few, seemingly difficult, open questions are indicated. Some of them are important for cognitive science (Sect. 1.2.1.3); one pertains to Turing machine complexity (Sect. 1.2.3.6).

1.2.1 Hulking Databases and Memory-Limited Learners

The *Database (DB)* of utterances, $u_0, u_1, \ldots, u_{t-1}$, *prior* to time t, can, for *large* t, become an *Incredible Hulk (IH) un*pleasant to handle and query. I think of this

Fig. 1.3 The incredible hulk (IH) (©2013 Marvel and Subs)

$g_t \xleftarrow{\text{Out}}$ $\boxed{M}$ $\xleftarrow{\text{In}} u_t \in L$

$\downarrow\uparrow$

M remembers $\leq k$ elements (from time $t-1$)
$\{u_{i_1}, u_{i_2}, \ldots, u_{i_k}\}$ (and at t, can drop some u_{i_j}s and add u_t)
$\subseteq \{u_0, u_1, \ldots, u_{t-1}\} \subseteq L$;

Fig. 1.4 $\leq k$-bounded example-memory model

Marvel Comics character (Fig. 1.3) as metaphorically representing the problem of DBs that are too large.

Furthermore, humans (including children) may not remember, even subconsciously, every utterance they've ever heard.

In Sects. 1.2.1.1 and 1.2.1.2 we begin to introduce some *Memory-Limited* criteria of learning success to begin to deal with these *too large DB problems*.

Then Sect. 1.2.1.3 presents corresponding results. It indicates some results of relevance to cognitive science—as well as important, cognitive science-relevant, seemingly hard open questions.

1.2.1.1 One Kind of Memory-Limited Learning Criteria

The $\leq k$-Bounded Example-Memory Model [22, 36, 52] is *without access to the complete IH DB* and is depicted at time t below in Fig. 1.4.[2] At time t, M reacts only to: u_t, g_{t-1} and its memory from time $(t-1)$ of $\leq k$ of the prior utterances from L.

We say M **TxtBem**$_k$-learns L iff, for any text of all/only the utterances of L, at some time t, g_t above in Fig. 1.4 is a grammar for L and $g_t = g_{t+1} = g_{t+2} = g_{t+3} = \ldots$.

1.2.1.2 A Second Kind of Memory-Limited Learning Criteria

For **The $\leq k$-Queries Feedback Model** [22] ($k = 0, 1$ cases: [44, 70]) the learning machine M asks *little per hypothesis update of the complete IH DB*, and it is depicted at time t below in Fig. 1.5. In this model, at time t, M reacts only to: u_t, g_{t-1} and the answers to only $\leq k$ queries it generates regarding membership in the eventually large IH DB of the past inputs $\{u_0, u_1, \ldots, u_{t-1}\}$ up through time $(t-1)$.

---

[2] In Fig. 1.4, for $t = 0$, M's memory of prior data at time $(t-1)$ is empty; furthermore, in Fig. 1.4 and elsewhere below, for $t = 0$, g_{t-1} can be taken to be a fixed grammar, e.g., for the empty language.

Fig. 1.5 $\leq k$-queries feedback model

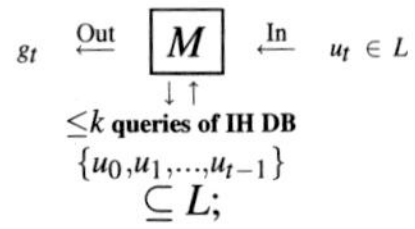

$g_t \xleftarrow{\text{Out}} \boxed{M} \xleftarrow{\text{In}} u_t \in L$

$\downarrow\uparrow$

$\leq k$ **queries of IH DB**

$\{u_0, u_1, \ldots, u_{t-1}\}$

$\subseteq L$;

M is said to **TxtFb**$_k$-learn L iff, for any text of all/only the utterances of L, at some time t, g_t above is a grammar for L and $g_t = g_{t+1} = g_{t+2} = g_{t+3} = \cdots$.

For $k = 0$ the two models coincide and the resultant learning criterion is called *iterative* or **TxtIt**-*learning* [69, 70].

1.2.1.3 Results About the Just Prior Two Models

Theorem 1.14 [22] *In* each *model, one can learn more with strictly larger k.*

Theorem 1.15 [22] *For* each $k > 0$, *for each model, there are language classes where that model is successful and the other is not!*

For each $k > 0$, for each of the two models, here is a *hard open problem*: does *forbidding* U shapes decrease learning power? For $k = 0$, the answer is *no* [28].

Even for $k = 0$, it is a hard open problem: does the case of adding the *counter t* to M's inputs u_t, g_{t-1} affect whether U shapes are necessary?

Remark 1.6 It is *known* [25], for example, for *each* $k \geq 3$, for the variant of **TxtBem**$_k$, where g_{t-1} is *not* available to M and where the memory of $\leq k$ utterances is *replaced* by memory of *any one of k-objects* (i.e., $\log_2(k)$-bits), that *forbidding* U shapes *decreases* learnability.

For $k = 1, 2$ here, U shapes are *ir*relevant [13].

Theorem 1.16 [22] *For each* $n > 0$, *the class of all* unions of n pattern languages $\in$ **TxtIt**!

1.2.2 Deficiencies of Programs Learned

Independently of other complexity issues is the problem of the reasonableness of the programs eventually successfully found by a learner. This is explored in the present section for the learning of programs for total characteristic/decision functions $\mathbb{N} \rightarrow \{0, 1\}$ from *complete* data about their graphs. In prior sections we explored the mathematically different vehicle of learning programs for generating formal languages from *positive information only about those languages*. For any *total* function, once one has an enumeration of its graph, one can algorithmically generate all that is and only what is in the complement of its graph. We restrict our attention in the present section to data presentations of such functions f in the order: $f(0), f(1), f(2), \ldots$—since from any presentation order of the graph of f (i.e., $\{(x, y) \mid f(x) = y\}$) one can algorithmically generate the presentation order $f(0), f(1), f(2), \ldots$.

In Sects. 1.2.2.1–1.2.2.3, the basic background material is presented. Then, in Sect. 1.2.2.4, results about inherent *computational complexity deficiencies* in programs learned are presented. Finally, in Sect. 1.2.2.5, some unexpected, inherent *informational deficiencies* of programs learned are laid out.

1.2.2.1 Successful Learning on Complete Data Regarding Total Functions

Suppose $\mathscr{S} \subseteq \mathscr{R}_{0,1}$, *the class of all (total) computable characteristic functions of subsets of* N. Defined just below are criteria of success analogs of **TxtEx** and **TxtBc** — but regarding learning programs for characteristic functions instead of grammars for languages.

Definition 1.17

- Depicted below is the I/O behavior of a learning machine M receiving successive values of a function f and outputting corresponding successive programs. Some later items in this definition refer to this depiction.

$$p_0, p_1, \ldots, p_t, \ldots \overset{\text{Out}}{\longleftarrow} M \overset{\text{In}}{\longleftarrow} f(0), f(1), \ldots, f(t), \ldots .$$

- Suppose $a \in \mathbb{N} \cup \{*\}$. a is for *a*nomaly count. For $a = *$, a stands for *finitely many.*[3]
- $\mathscr{S} \in \mathbf{Ex}^a \Leftrightarrow$ ($\exists$ suitably clever M)($\forall f \in \mathscr{S}$) [$M$, fed $f(0), f(1), \ldots$, outputs $p_0, p_1, \ldots \wedge (\exists t)[p_t = p_{t+1} = \cdots \wedge p_t$ *computes* f*—except at up to* a *inputs*]].
- $\mathscr{S} \in \mathbf{Bc}^a \Leftrightarrow (\exists M)(\forall f \in \mathscr{S})$ [M, fed $f(0), f(1), \ldots$, outputs $p_0, p_1, \ldots \wedge (\exists t)[p_t, p_{t+1}, \ldots$ each computes f—except at up to a inputs]].

1.2.2.2 Examples

Below are some important example classes together with some known results about them which will help us understand the results in Sects. 1.2.2.4 and 1.2.2.5 on the deficiencies of programs learned.

---

[3] Case and Smith [29, 30] motivate by anomalous dispersion from physical optics the presence of $a > 0$ anomalies in "successful" final programs. We omit herein details about that.

Remark 1.7

- For $k \geq 1$, let $\mathscr{P}^k$ = the class all 0–1 valued functions computable by (multi-tape) TMs in $O(n^k)$ time, with n = input length. Let $\mathscr{P} = \bigcup \mathscr{P}^k$, the class of polynomial time computable characteristic functions.
- Let slow be a *fixed* slow-growing unbounded function $\in \mathscr{P}^1$, e.g., $\leq$ Ackermann^{-1} [31]. Let $\mathscr{Q}^k$ = the class all $\{0, 1\}$-valued functions computable in $O(n^k \cdot \log(n) \cdot \text{slow}(n))$ time, again with n = input length. We have,

$$\underbrace{\mathscr{P}^k \subset \mathscr{Q}^k}_{\text{Tightest known separation [38, 39]}} \qquad \subset \mathscr{P}^{k+1}.$$

- $\mathscr{P} \in \mathbf{Ex}^0$. $\mathscr{P}^k \in \mathbf{Ex}^0$ too (*with each output conjecture running in k-degree poly time*); $\mathscr{CF}$, the class all *characteristic functions of* context-free languages, $\in \mathbf{Ex}^0$ [37].
- $\mathbf{Ex}^0 \subset \mathbf{Ex}^1 \subset \mathbf{Ex}^2 \subset \cdots \subset \mathbf{Ex}^* \subset \mathbf{Bc}^0 \subset \mathbf{Bc}^1 \subset \cdots \subset \mathbf{Bc}^*$ [29]. Hence, more anomalies tolerated in final successful programs entails strictly more learning power.
- Harrington in [30]: $\mathscr{R}_{0,1} \in \mathbf{Bc}^*$.

1.2.2.3 Basic Notation

Next is some basic terminology important to the statements of results in Sects. 1.2.2.4 and 1.2.2.5.

Definition 1.18

- $\mathscr{C}\text{of} = \{f \in \mathscr{R}_{0,1} \mid (\overset{\infty}{\forall} x)[f(x) = 1]\}$ ($\subset \mathscr{P}^1$ and $\mathscr{REG}$, the class all *characteristic functions of* regular languages). As will be seen, $\mathscr{C}$of is important as a nice example of a particularly trivial, easily learnable class of (characteristic) functions.
- φ_p^{TM} = the partial computable function $\mathbb{N} \to \mathbb{N}$ computed by (multi-tape) Turing machine program (number) p.
- $\Phi_p^{\text{TM}}(x)$ = the *runtime* of Turing machine program (number) p on input x, if p halts on x, and is undefined otherwise.
- $\Phi_p^{\text{WS}}(x)$ = the tape work space used by Turing machine program (number) p on input x, if p halts on x, and is undefined otherwise.
- $f[m]$ = the *sequence* $f(0), \ldots, f(m-1)$; for $m = 0$, $f[m]$ is the empty sequence.
- $M(f[m])$ = M's output based on $f[m]$. For the results herein, we may and will suppose without loss of generality that $M(f[m])$ is always defined.
- $M(f)$ denotes M's *final* output program on input f, *if any*; else, it is undefined.

1.2.2.4 Complexity Deficiencies of Programs Learned

First is a simple positive result which, in effect, is part of the third bullet in Remark 1.7 in Sect. 1.2.2.2. This result will contrast nicely with the theorem just after it.

Proposition 1.19 [20] *For each* $k \geq 1$, $\exists M$ *witnessing* $\mathscr{P}^k \in \mathbf{Ex}^0$ *such that* $(\forall f)(\forall m)$ $[\Phi^{\mathrm{TM}}_{M(f[m])}(x) \in O(|x|^k)]$.

Theorem 1.20 ([20], tightens [65]) *Suppose* $k \geq 1$ *and that* M *witnesses either* $\mathscr{Q}^k \in \mathbf{Ex}^*$ *or* $\mathscr{Q}^k \in \mathbf{Bc}^0$ (special case: *M witnesses* $\mathscr{Q}^k \in \mathbf{Ex}^0$).
Then: $(\exists f \in \mathscr{C}\mathrm{of})(\forall k\text{-degree polynomials } p)$
$(\overset{\infty}{\forall} n)(\overset{\infty}{\forall} x)[\Phi^{\mathrm{TM}}_{M(f[n])}(x) > p(|x|)]$.

Idea: if one ups the generality of a learner M from $\mathscr{P}^k$ to $\mathscr{Q}^k$, then the run-times of M's successful outputs on some trivial $f \in \mathscr{C}\mathrm{of}$ is *worse* than *any* preassigned k-degree polynomial bound, a complexity deficiency.

Of course, since $\mathscr{REG}$ is defined in terms of finite automata [32, 39], *no* Work Space is needed to compute the functions in $\mathscr{REG}$ (which includes the trivial functions in $\mathscr{C}\mathrm{of}$), but:

Theorem 1.21 [20] *For* $k \geq 1$ *and M witnessing* $\mathscr{CF} \in \mathbf{Ex}^*$ (special case: *M witnesses* $\mathscr{CF} \in \mathbf{Ex}^0$), $(\exists f \in \mathscr{C}\mathrm{of})(\exists x)[\Phi^{\mathrm{WS}}_{M(f)}(x) \geq k]$.

Idea: learning $\mathscr{CF}$ instead of $\mathscr{REG}$ produces some final programs on some trivial functions in $\mathscr{C}\mathrm{of}$ which *do* require some work space, another complexity deficiency.

1.2.2.5 Information Deficiencies of Programs Learned

First is a positive result which will contrast nicely with the theorem just after it.

Theorem 1.22 [20] $\exists M$ outputting only total poly-time conjectures *and witnessing* both $\mathscr{P} \in \mathbf{Bc}^*$ and $\mathscr{P}^1 \in \mathbf{Ex}^0$ *such that* $(\forall f \in \mathscr{P}^1)[\Phi^{\mathrm{TM}}_{M(f)}(x) \in O(|x|)]$.

In the next theorem f.o. **PA** is the version of Peano Arithmetic expressed within first order logic [48, 58]. **PA** is a well-known formal theory in which one can express and prove all the theorems in an *elementary* number theory book.

For a sentence E expressible in **PA**, $\ll E \gg$ is some natural translation of E into the language of **PA**.

In the next theorem, the example

$$E = \varphi^{\mathrm{TM}}_{M(f[m])} \text{ is computable}^* \text{ in } O(|x|^k)\text{time}$$

contains what would, without explanation, be a mysterious $*$. This E is meant to be an abbreviation of the clearer, but longer sentence: any total, finite variant of the function computed by the TM-program output by M on $f[m]$ runs in time $O(n^k)$, where $n =$ the length of this program's input.

Theorem 1.23 [20]
Suppose theory **T** *is any* true, computably axiomatized *extension of f.o.* **PA** *(so,* **T** *is a* safe, algorithmic *information extractor). Suppose $k \geq 1$ and M witnesses $\mathcal{Q}^k \in \mathbf{Bc}^*$. Then:* $(\exists f \in \mathcal{C}\text{of})(\overset{\infty}{\forall} m)[\mathbf{T} \nvdash \ll \varphi^{\text{TM}}_{M(f[m])}$ *is computable* in* $O(|x|^k)$ *time* $\gg]$.

It is particularly interesting to apply the preceding theorem (Theorem 1.23) to the case of a learner M from Theorem 1.22 earlier: while the programs learned by *this* M for functions in all of $\mathcal{C}$of are perfectly correct and *do* have excellent, linear-time run-times, some of these programs learned will be *informationally deficient*—since we cannot prove in **T** from *them* even considerably weaker upper bounds on the run-times of *any* total finite variants of the functions they compute!

1.2.3 Complexity of Learner Updates

This section begins with a general discussion of the important and well-investigated subject of learner update complexity (Sect. 1.2.3.1). Then an automata-theory-based paradigm is chosen for both illustration and since it may be new to the Grammatical Inference community as well as interesting to it (Sects. 1.2.3.2–1.2.3.6).

1.2.3.1 Complexity-Restricted Updates of Learners Generally

There has been tremendous interest in the Grammatical Inference community, e.g., [33], in *polynomial time updating of each learned output conjecture*, where the polynomials are typically in some measure of the size of data used to obtain the conjecture.

For example, the Pattern Languages are polytime **TxtIt**-learnable [43]—but at the interesting, apparently necessary cost of intermediate conjectures which do not generate some of the data on which they are based!

Pitt [56] notes that it's possible to *cheat* and always obtain (meaningless) polytime updates: Suppose q is a polynomial, and M on finite data sequence σ delivers its correspondingly conjectured program within time $q(|\sigma|)$. Pitt notes M can put off outputting conjectures based on σ until it has seen much larger data τ so that $q(|\tau|)$ is enough time for M to work on σ as long as it needs. He notes that this delaying trick is *un*fair—since it allows as much to be learned as in the case of no time bound on the learner updates.

Finding mathematical conditions to *guarantee no Pitt delaying tricks can be used* seems very difficult [24]. For a time I believed polytime $\mathbf{TxtFb}_k$-learning is fair—until Frank Stephan provided me a counterexample. I currently believe polytime $\mathbf{TxtBem}_k$-learning is reasonably fair—provided output conjectures are not padded up too much to carry over extra information from one update to the next. Reasonably fair such padding was employed for **TxtIt**-learning of some mildly context-sensitive classes in [7]. Most published polytime update learning algorithms I've seen seem *quite fair*, e.g., the **TxtEx**-learning of somewhat different mildly context-sensitive classes in [72] and, of course, the polytime **TxtIt**-learnability of the Pattern Languages mentioned above.

Next are presented, for example, some recent results about the use of learning functions whose graphs are regular (i.e., finite automata-acceptable). These are called *automatic* learning functions and are defined more formally later in Sect. 1.2.3.3.

Perhaps this material will be new to the Grammatical Inference community and of some interest to them. We believe finite automata are not smart enough to do Pitt tricks.

1.2.3.2 Automatic Structures

The study of automatic structures (defined below) began with the Program of Khoussainov and Nerode [42]: replace TMs by *finite automata* in computable model theory.

Below we provide some basic definitions regarding automatic structures which will lead, for example, to Remark 1.8 in Sect. 1.2.3.3 in which we are able to define automatic learning function and present a few results.

Definition 1.24

- The *automatic 1-ary relations* are the *regular (i.e., finite automata accepted) languages* $\subseteq \Sigma^*$, for some finite (non-empty) alphabet Σ.
- For our next defining *automatic binary relations* R over a finite alphabet $\Upsilon (\supseteq \Sigma)$, we need be able to *submit* a pair $(\alpha, \beta) \in (\Upsilon^* \times \Upsilon^*)$ to a finite automaton.
- *If* we feed α and, subsequently, β, the finite automaton will have trouble remembering much about α upon receiving β, *so* we use *convolution* to submit α and β *together*—details next.
- Suppose $\sqcup$ is a "blank" symbol $\notin \Upsilon$. We provide next an ostensive definition of $\text{conv}(\alpha, \beta)$.
  Example: $\text{conv}(ab, bba) = (a, b)(b, b)(\sqcup, a)$. Idea: these *pairs* are each new *single alphabet symbols* to be *sequentially* read by a finite automaton.
- We say a *binary* relation R is *automatic* iff $\{\text{conv}(\alpha, \beta) \mid R(\alpha, \beta)\}$ is *regular*—over the alphabet $((\Upsilon \cup \{\sqcup\}) \times (\Upsilon \cup \{\sqcup\}))$.
  The concept obviously generalizes to k-ary relations, and we consider it to be so generalized.

1.2.3.3 Automatic Classes and Functions

Remark 1.8

- $\mathscr{L}$ is said to be an *automatic class* iff each $L \in \mathscr{L}$ is a subset of Σ^*, and, for some *regular* index domain I and some *automatic* $S \subseteq (I \times \Sigma^*)$,
  $\mathscr{L} = \{L_\alpha \mid \alpha \in I\}$, where (the then regular) $L_\alpha = \{x \mid (\alpha, x) \in S\}$.
  Idea: such a $\mathscr{L}$ is *uniformly* regular.

- Automatic classes are the *automatic analog* of the *uniformly decidable classes*, e.g., from Angluin's characterization theorem, Theorem 1.7 from Sect. 1.1.4.1. The analog of the index domain above for the uniformly decidable classes is just $\mathbb{N}$.
- An *automatic function* M is an automatic, single-valued relation. I.e., a function M is *automatic* iff the relation $\{(\alpha, \beta) \mid M(\alpha) = \beta\}$ is automatic.
- In the later, learnability portion of this section, we are interested in modeling *learners* as *automatic and, also, as more general functions* M. Although more general functions can learn more, and
- Case et al. [21] notes there are *significant* automatic classes based on *erasing regular* patterns which are learnable by mere automatic learners.[4]

1.2.3.4 TM Input/Output of Automatic Functions

A finite automaton *accepting (the convolutions of) all/only the ordered pairs in the graph of an automatic function* M is *not* at all the same as *computing the value of* $M(x)$ *from each* x *in domain*(M). However:

Theorem 1.25 [23] *Suppose* M *is an* automatic function. *Then there is a* linear-time bounded, one-tape, deterministic TM *which* computes M *where each input* x *is given on the tape starting from the* marked left end *and output* $M(x)$ *starts from this* same left end.

This theorem has a strong converse as follows.

Theorem 1.26 [23] *Suppose a* linear-time bounded, one-tape, non-deterministic TM computes M *such that each input* x *is given on the tape starting from the* marked left end; *on each non-deterministic path the output is* $M(x)$ *or?, where at least one path has output* $M(x)$*; and outputs* $M(x)$ *or? start from this* same left end. *Then* M *is automatic. (Left end I/O is provably crucial.)*

1.2.3.5 Relevant General Learnability Definitions

We are next interested in the learnability (general or otherwise) of *automatic classes*. The next definition begins to explore how to handle this.

Definition 1.27

- As above, a *text* T for $L \subseteq \Sigma^*$ (in automatic class) $\mathscr{L}$ is a sequence of all and only the elements of L.
- A learner employs output *hypotheses* $\mathrm{hyp}_t \in I$ (I, the corresponding index set) and a sequence of long-term memories mem_t (each $\in \Gamma^*$).

---

[4] Above, in Sect. 1.1.4.2, we defined Angluin's pattern languages based on patterns, and *there* only *positive* length substitutions are allowed. The provably hard to learn *erasing* pattern languages [57] also allow empty substitutions. The *regular* ones [62] require that each variable in the associated defining pattern be present only once.

- A learner has initial long-term memory mem_0 (and initial hypothesis hyp_0). Think of each $t = 0, 1, \ldots$ as a time/cycle. Then *learner* $M : (T(t), \mathrm{mem}_t) \mapsto (hyp_{t+1}, \mathrm{mem}_{t+1})$.
- $\mathscr{L}$ is *learnable* by M: $(\forall L \in \mathscr{L})(\forall T \text{ for } L)(\exists t)(\forall t' > t)[\mathrm{hyp}_{t'} = \mathrm{hyp}_t \ \wedge\ \mathrm{hyp}_t$ is correct for $L]$.

Remark 1.9

- With *un*restricted memory and Ms *TM computable*, the preceding *learnability* criterion (*for automatic classes*) is equivalent to **TxtEx** (but with hypotheses restricted to I).
- Curiously: learnability *of automatic classes* does *not* change even if the Ms can *also* be *un*computable, and Angluin's Characterization can *in this case* omit computable enumerability of tell tales.

1.2.3.6 Linear Time Learners Suffice!

We explore next how efficiently the general learning *of automatic classes* can be done. First we introduce a relevant TM model.

Definition 1.28 Our $(k + 1)$-Tape TM Model:

- Tape 0 (base tape): At *the beginning of cycle* t, it contains $\mathrm{conv}(T(t), \mathrm{mem}_t)$ (with $|\mathrm{mem}_t| \leq$ longest text datum seen so far $+$ a constant).
  At *end of cycle* t, it contains $\mathrm{conv}(\mathrm{hyp}_{t+1}, \mathrm{mem}_{t+1})$ (with $|\mathrm{mem}_{t+1}| \leq$ longest text datum seen so far $+$ a constant). *Marked left end I/O* too.
- Additional Tapes $1, 2, \ldots, k$: *normal work-tapes*, with contents and head position *not* modified during *change* of cycle.
- *Each cycle* of the machine runs in the *linear-time* in the length of the longest text datum seen so far.

Theorem 1.29 [23] *Every* learnable automatic class *has* such *a* linear-time TM learner *employing only* $k =$ two *additional* work-tapes.
The two work tapes can be replaced by two stacks *or, instead,* one queue—a queue with non-overtaking, one-way heads to operate each end*!*

Open problem: does only $k =$ *one* additional *work tape* suffice?

1.3 Summary Including Open Problems

We briefly summarize what's been done in the present chapter.

As noted above, the chapter is about Gold's 1967 model of child language learning and selected highlights of the theory appearing since Gold's 1967 paper. It is divided into two major sections. The first concentrates on results that are independent of computational complexity considerations; the second concentrates on relevant complexity issues.

In the first major section are treated: Gold's Model; criteria of successful learning (with hierarchy results); reasonableness and weaknesses of the general model; provable severe constraints on learnability; related characterizations and the important example of Angluin's Pattern Languages; two local and two global formal insensitivities of learning to order of presentation of data with corresponding formal results, with no available conclusions about how these do or do not apply to human children; and the cognitive science empirically observed phenomenon of U-shaped learning with mathematically defined formal analogs, and corresponding interpreted formal results that suggest the U shape may be necessary for full human learning power. Also presented are some hard questions for the state of the art in empirical cognitive science: do some humans, after successfully learning at least some natural languages, vacillate between multiple grammars. Is U-shaped learning necessary for those humans who succeed in the genetic marketplace?

In the second major section on complexity considerations are treated: the problem of learning based on infeasibly large amounts of data, and presented are two formal models of data memory-limited learners, accompanied by corresponding formal hierarchy results; surprising results showing that some slight increases in learning generality inexorably lead to both complexity and information deficiencies in the programs that are learned for some very simple objects to be learned, but with no such Deficiencies appearing for the less general learning cases; and the important feasibly computable learner updates where it is noted that sometimes the feasibility of updates can be a cheat, and detailed results are given about both the use of finite automata accepted learning functions (where it is ostensibly difficult to cheat) and the learnability of uniformly finite automata accepted classes. Also presented are some seemingly difficult open mathematical questions (the first two are of relevance to cognitive science, the third to complexity theory): For each $k > 0$, for each of the two data memory-limited models of Sect. 1.2.1, does forbidding U shapes decrease learning power? For each $k \geq 0$, for these two models, does the case of supplying the counter t to M in addition to its inputs u_t, g_{t-1} affect whether U shapes are necessary? For Theorem 1.29 in Sect. 1.2.3.6, does only $k = 1$ additional work tape suffice?

Acknowledgments My thanks go to Sanjay Jain and Frank Stephan for catching a mistake in an earlier draft. I'm also grateful to an anonymous referee for excellent suggestions which I hope I've been able to follow sufficiently well in this chapter.

References

1. Angluin, D.: Finding patterns common to a set of strings. Journal of Computer and System Sciences **21**, 46–62 (1980)
2. Angluin, D.: Inductive inference of formal languages from positive data. Information and Control **45**, 117–135 (1980)
3. Arikawa, S., Miyano, S., Shinohara, A., Kuhara, S., Mukouchi, Y., Shinohara, T.: A machine discovery from amino-acid-sequences by decision trees over regular patterns. New Generation

Computing **11**, 361–375 (1993)
4. Arikawa, S., Shinohara, T., Yamamoto, A.: Learning elementary formal systems. Theoretical Computer Science **95**, 97–113 (1992)
5. Baliga, G., Case, J., Jain, S.: The synthesis of language learners. Information and Computation **152**, 16–43 (1999)
6. Baliga, G., Case, J., Merkle, W., Stephan, F., Wiehagen, W.: When unlearning helps. Information and Computation **206**, 694–709 (2008)
7. Becerra-Bonache, L., Case, J., Jain, S., Stephan, F.: Iterative learning of simple external contextual languages. Theoretical Computer Science **411**, 2741–2756 (2010). Special Issue for *ALT'08*
8. Blum, L., Blum, M.: Toward a mathematical theory of inductive inference. Information and Control **28**, 125–155 (1975)
9. Blum, M.: A machine independent theory of the complexity of recursive functions. Journal of the ACM **14**, 322–336 (1967)
10. Bower, M.: Starting to talk worse: Clues to language development from children's late speech errors. In: S. Strauss, R. Stavy (eds.) U-Shaped Behavioral Growth, Developmental Psychology Series. Academic Press, NY (1982)
11. Brazma, A., Jonassen, I., Eidhammer, I., Gilbert, D.: Approaches to the automatic discovery of patterns in biosequences. Journal of Computational Biology **5**(2), 279–305 (1998)
12. Carlucci, L., Case, J.: On the necessity of U-shaped learning. Topics in Cognitive Science (2012). Invited for Special Issue on Formal Learning Theory
13. Carlucci, L., Case, J., Jain, S., Stephan, F.: Memory-limited U-shaped learning. Information and Computation **205**, 1551–1573 (2007)
14. Carlucci, L., Case, J., Jain, S., Stephan, F.: Non U-shaped vacillatory and team learning. Journal of Computer and System Sciences **74**, 409–430 (2008). Special issue in memory of Carl Smith
15. Carlucci, L., Jain, S., Kinber, E., Stephan, F.: Variations on U-shaped learning. Information and Computation **204**(8), 1264–1294 (2006)
16. Case, J.: Infinitary self-reference in learning theory. Journal of Experimental and Theoretical Artificial Intelligence **6**, 3–16 (1994)
17. Case, J.: The power of vacillation in language learning. SIAM Journal on Computing **28**(6), 1941–1969 (1999)
18. Case, J.: Directions for computability theory beyond pure mathematical. In: D. Gabbay, S. Goncharov, M. Zakharyaschev (eds.) Mathematical Problems from Applied Logic II. New Logics for the XXIst Century, International Mathematical Series, Vol. 5, pp. 53–98. Springer (2007). Invited book chapter
19. Case, J.: Algorithmic scientific inference: Within our computable expected reality. International Journal of Unconventional Computing **8**(3), 192–206 (2012). Invited journal expansion of an invited talk and paper at the *3rd International Workshop on Physics and Computation 2010*
20. Case, J., Chen, K., Jain, S., Merkle, W., Royer, J.: Generality's price: Inescapable deficiencies in machine-learned programs. Annals of Pure and Applied Logic **139**, 303–326 (2006)
21. Case, J., Jain, S., , Le, T., Ong, Y., Semukhin, P., Stephan, F.: Automatic learning of subclasses of pattern languages. Information and Computation **218**, 17–35 (2012)
22. Case, J., Jain, S., Lange, S., Zeugmann, T.: Incremental concept learning for bounded data mining. Information and Computation **152**, 74–110 (1999)
23. Case, J., Jain, S., Seah, S., Stephan, F.: Automatic functions, linear time and learning. In: S. Cooper, A. Dawar, B. Löwe (eds.) How the World Computes - Turing Centenary Conference and Eighth Conference on Computability in Europe (CiE 2012), Proceedings, *Lecture Notes In Computer Science*, vol. 7318, pp. 96–106. Springer, Berlin (2012)
24. Case, J., Kötzing, T.: Difficulties in forcing fairness of polynomial time inductive inference. In: 20th International Conference on Algorithmic Learning Theory (ALT'09), *Lecture Notes in Artificial Intelligence*, vol. 5809, pp. 263–277 (2009)
25. Case, J., Kötzing, T.: Memory-limited non-U-shaped learning with solved open problems. Theoretical Computer Science (2012). In press online at doi:10.1016/j.tcs.2012.10.010, Special Issue for selected *ALT'10* papers

26. Case, J., Kötzing, T.: Topological separations in inductive inference. In: S. Jain, et al. (eds.) 24nd International Conference on Algorithmic Learning Theory (ALT'13), *Lecture Notes in Artificial Intelligence*, vol. 8139, pp. 128–142 (2013)
27. Case, J., Lynes, C.: Machine inductive inference and language identification. In: M. Nielsen, E. Schmidt (eds.) Proceedings of the 9th International Colloquium on Automata, Languages and Programming, *Lecture Notes in Computer Science*, vol. 140, pp. 107–115. Springer-Verlag, Berlin (1982)
28. Case, J., Moelius, S.: U-shaped, iterative, and iterative-with-counter learning. Machine Learning **72**, 63–88 (2008). Special issue for *COLT'07*
29. Case, J., Smith, C.: Anomaly hierarchies of mechanized inductive inference. In: Symposium on the Theory of Computation, pp. 314–319 (1978)
30. Case, J., Smith, C.: Comparison of identification criteria for machine inductive inference. Theoretical Computer Science **25**, 193–220 (1983)
31. Cormen, T., Leiserson, C., Rivest, R., Stein, C.: Introduction to Algorithms, second edn. MIT Press (2001)
32. Davis, M., Sigal, R., Weyuker, E.: Computability, Complexity, and Languages, second edn. Academic Press (1994)
33. de la Higuera, C.: Grammatical Inference: Learning Automata and Grammars. Cambridge University Press (2010)
34. Fulk, M.: A study of inductive inference machines. Ph.D. thesis, SUNY at Buffalo (1985)
35. Fulk, M.: Prudence and other conditions on formal language learning. Information and Computation **85**, 1–11 (1990)
36. Fulk, M., Jain, S., Osherson, D.: Open problems in Systems That Learn. Journal of Computer and System Sciences **49**(3), 589–604 (1994)
37. Gold, E.: Language identification in the limit. Information and Control **10**, 447–474 (1967)
38. Hennie, F., Stearns, R.: Two-tape simulation of multitape Turing machines. J. ACM **13**, 433–446 (1966)
39. Hopcroft, J., Ullman, J.: Introduction to Automata Theory, Languages, and Computation. Addison-Wesley Publishing Company (1979)
40. Jain, S., Osherson, D., Royer, J., Sharma, A.: Systems that Learn: An Introduction to Learning Theory, second edn. MIT Press, Cambridge, Mass. (1999)
41. Kapur, S., Lust, B., Harbert, W., Martohardjono, G.: Universal grammar and learnability theory: The case of binding domains and the 'subset principle'. In: E. Reuland, W. Abraham (eds.) Knowledge and Language, vol. I, pp. 185–216. Kluwer (1993)
42. Khoussainov, B., Nerode, A.: Automatic presentations of structures. In: Logical and Computational Complexity (International Workshop LCC 1994), *LNCS*, vol. 960, pp. 367–392. Springer (1995)
43. Lange, S., Wiehagen, R.: Polynomial time inference of arbitrary pattern languages. New Generation Computing **8**, 361–370 (1991)
44. Lange, S., Zeugmann, T.: Incremental learning from positive data. Journal of Computer and System Sciences **53**, 88–103 (1996)
45. Manzini, R., Wexler, K.: Parameters, binding theory and learnability. Linguistic Inquiry **18**, 413–444 (1987)
46. Marcus, G., Pinker, S., Ullman, M., Hollander, M., Rosen, T., Xu, F.: Overregularization in Language Acquisition. Monographs of the Society for Research in Child Development, vol. 57, no. 4. University of Chicago Press (1992). Includes commentary by H. Clahsen
47. McNeill, D.: Developmental psycholinguistics. In: F. Smith, G.A. Miller (eds.) The Genesis of Language, pp. 15–84. MIT Press (1966)
48. Mendelson, E.: Introduction to Mathematical Logic, fifth edn. Chapman & Hall, London (2009)
49. Newport, E., Gleitman, L., Gleitman, H.: Mother I'd rather do it myself: Some effects and noneffects of maternal speech style. In: C. Snow, C. Ferguson (eds.) Talking to children: Language input and acquisition, pp. 109–150. Cambridge University Press (1977)
50. Nix, R.: Editing by examples. Tech. Rep. 280, Department of Computer Science, Yale University, New Haven, CT, USA (1983)

51. Osherson, D., Stob, M., Weinstein, S.: Note on a central lemma of learning theory. Journal of Mathematical Psychology **27**, 86–92 (1983)
52. Osherson, D., Stob, M., Weinstein, S.: Systems that Learn: An Introduction to Learning Theory for Cognitive and Computer Scientists. MIT Press, Cambridge, Mass. (1986)
53. Osherson, D., Weinstein, S.: Criteria of language learning. Information and Control **52**, 123–138 (1982)
54. Pinker, S.: Language Learnability and Language Development. Harvard University Press (1984)
55. Pinker, S.: Rules of language. Science **253**, 530–535 (1991)
56. Pitt, L.: Inductive inference, DFAs, and computational complexity. In: Analogical and Inductive Inference, Proceedings of the Second International Workshop (AII'89), *Lecture Notes in Artificial Intelligence*, vol. 397, pp. 18–44. Springer-Verlag, Berlin (1989)
57. Reidenbach, D.: A negative result on inductive inference of extended pattern languages. In: N. Cesa-Bianchi, M. Numao (eds.) Proceedings of The 13th International Conference on Algorithmic Learning Theory (ALT'02), Lecture Notes in Artificial Intelligence, pp. 308–320. Springer (Lübeck, Germany, November, 2002)
58. Rogers, H.: Theory of Recursive Functions and Effective Computability. McGraw-Hill, New York (1967). Reprinted, MIT Press, 1987
59. Rogers, T.T., Rakinson, D.H., McClelland, J.L.: U-shaped curves in development: a PDP approach. Journal of Cognition and Development **5**(1), 137–145 (2004)
60. Schäfer-Richter, G.: Über eingabeabhängigkeit und komplexität von inferenzstrategien. Ph.D. thesis, RWTH Aachen (1984)
61. Shimozono, S., Shinohara, A., Shinohara, T., Miyano, S., Kuhara, S., Arikawa, S.: Knowledge acquisition from amino acid sequences by machine learning system BONSAI. Trans. Information Processing Society of Japan **35**, 2009–2018 (1994)
62. Shinohara, T.: Polynomial time inference of extended regular pattern languages. In: RIMS Symposia on Software Science and Engineering, Kyoto, Japan, *Lecture Notes in Computer Science*, vol. 147, pp. 115–127. Springer-Verlag (1982)
63. Shinohara, T.: Inferring unions of two pattern languages. Bulletin of Informatics and Cybernetics **20**, 83–88. (1983)
64. Shinohara, T., Arikawa, A.: Pattern inference. In: K.P. Jantke, S. Lange (eds.) Algorithmic Learning for Knowledge-Based Systems, *Lecture Notes in Artificial Intelligence*, vol. 961, pp. 259–291. Springer-Verlag (1995)
65. Sipser, M.: (1978). Private communication
66. Strauss, S., Stavy, R. (eds.): U-Shaped Behavioral Growth. Developmental Psychology Series. Academic Press, NY (1982)
67. Strauss, S., Stavy, R., Orpaz, N.: The child's development of the concept of temperature (1977). Unpublished manuscript, Tel-Aviv University
68. Wexler, K.: The subset principle is an intensional principle. In: E. Reuland, W. Abraham (eds.) Knowledge and Language, vol. I, pp. 217–239. Kluwer (1993)
69. Wexler, K., Culicover, P.: Formal Principles of Language Acquisition. MIT Press, Cambridge, Mass (1980)
70. Wiehagen, R.: Limes-Erkennung rekursiver Funktionen durch spezielle Strategien. Elektronische Informationverarbeitung und Kybernetik **12**, 93–99 (1976)
71. Wright, K.: Identification of unions of languages drawn from an identifiable class. In: R. Rivest, D. Haussler, M. Warmuth (eds.) Proceedings of the Second Annual Workshop on Computational Learning Theory, Santa Cruz, California, pp. 328–333. Morgan Kaufmann Publishers, Inc. (1989)
72. Yoshinaka, R.: Learning multiple context-free languages with multidimensional substitutability from positive data. Information and Computation **412**, 1821–1831 (2011)

Chapter 2
Efficiency in the Identification in the Limit Learning Paradigm

Rémi Eyraud, Jeffrey Heinz and Ryo Yoshinaka

Abstract The most widely used learning paradigm in Grammatical Inference was introduced in 1967 and is known as *identification in the limit*. An important issue that has been raised with respect to the original definition is the absence of efficiency bounds. Nearly fifty years after its introduction, it remains an open problem how to best incorporate a notion of efficiency and tractability into this framework. This chapter surveys the different refinements that have been developed and studied, and the challenges they face. Main results for each formalization, along with comparisons, are provided.

2.1 Introduction

2.1.1 The Importance of Efficiency in Learning

Gold [24] introduced in the 1960s a definition of learning called *identification in the limit*, which works as follows. An algorithm is fed with an infinite sequence of data exemplifying a target language. When a new element is given to the algorithm, it may output a hypothesis. The algorithm identifies the language in the limit if for any possible sequence of data for this language, there exists a moment from when the algorithm does not change its hypothesis, and this hypothesis is a correct representation of the target language. When a whole class of languages is considered, the algorithm identifies the class in the limit if it can identify all languages of the class.

R. Eyraud (✉)
QARMA Team, Laboratoire d'Informatique Fondamentale, Marseille, France
e-mail: remi.eyraud@lif.univ-mrs.fr

J. Heinz
University of Delaware, Newark, DE, USA
e-mail: heinz@udel.edu

R. Yoshinaka
Graduate School of Informatics, Kyoto University, Kyoto, Japan
e-mail: ry@i.kyoto-u.ac.jp

J. Heinz and J.M. Sempere (eds.), *Topics in Grammatical Inference*,
DOI 10.1007/978-3-662-48395-4_2

The fact that the convergence is required to hold whatever the sequence of data is what makes this paradigm adversarial [14]. This worst-case scenario principle strengthens the value of any algorithmic idea that yields an identification in the limit result for a class of languages [25].

However, Gold's formulation can be of little help for practical purposes, when one wants to study a learning idea with the aim of applying it to real-world data. This is mainly due to the fact that no efficiency property is required and thus one can assume infinite time and space. This is the reason why several refinements of Gold's model which add polynomial bounds to the requirements of the paradigm have been developed. The purpose of this chapter is to comprehensively review the proposed refinements and the challenges they face. Main results of each approach, along with comparisons, are provided.

Instead of augmenting the learning framework to incorporate a notion of efficiency, one response to this state of affairs could be to utilize a different learning framework altogether, preferably one which contains a built-in notion of efficiency, such as the Probably Approximately Correct framework [43]. Section 2.2 discusses some issues with PAC-learning of formal languages, which makes this option less attractive than it otherwise may appear at first.

Section 2.3 studies the limitations of the initial identification in the limit definition and previous attempts to overcome them. These include requirements based on the running time of the studied algorithm. Efficiency requirements depending on the incremental behavior of the algorithm, and a set-based refinement of Gold's paradigm are detailed in Sect. 2.4. Finally, Sect. 2.5 introduces two recent reformulations of the paradigm.

2.1.2 *Preliminary Definitions*

An *alphabet* Σ is a finite non-empty set of symbols called *letters*. A *string* w over Σ is a finite sequence $w = a_1 a_2 \dots a_n$ of letters. Let $|w|$ denote the length of w. Given a set of strings S, we denote $|S|$ its cardinality and $\|S\|$ its size, i.e. the sum of $|S|$ with the lengths of the strings S contains.[1] In the following, letters will be indicated by $a, b, c, \dots$, strings by $u, v, \dots, z$, and the empty string by λ. Let Σ^* be the set of all strings and Σ^+ the set $\Sigma^* \setminus \{\lambda\}$.

We assume a fixed but arbitrary total order $<$ on the letters of Σ. As usual, we extend $<$ to Σ^* by defining the *hierarchical order* [33], denoted by $\lhd$, as follows:

$$\forall w_1, w_2 \in \Sigma^*, w_1 \lhd w_2 \textit{ iff } \begin{cases} |w_1| < |w_2| \text{ or} \\ |w_1| = |w_2| \text{ and } \exists u, v_1, v_2 \in \Sigma^*, \exists a_1, a_2 \in \Sigma \\ s.t.\ w_1 = u a_1 v_1, w_2 = u a_2 v_2 \text{ and } a_1 < a_2. \end{cases}$$

[1] We define $\|S\| = |S| + \sum_{w \in S} |w|$ so that $\|\{a\}\| < \|\{\lambda, a\}\|$.

$\lhd$ is a total strict order over Σ^*, and if $\Sigma = \{a, b\}$ and $a < b$, then $\lambda \lhd a \lhd b \lhd aa \lhd ab \lhd ba \lhd bb \lhd aaa \lhd \ldots$

We extend this order to non-empty finite sets of strings: $S_1 \lhd S_2$ iff $\|S_1\| < \|S_2\|$ or $\|S_1\| = \|S_2\|$ and $\exists w \in S_1 \setminus S_2$ such that $\forall w' \in S_2 \setminus S_1$, $w \lhd w'$. For instance $\{a\} \lhd \{\lambda, a\}$ and $\{a, b\} \lhd \{aaa\}$.

By a language L over Σ we mean any set $L \subseteq \Sigma^*$. Many classes of languages were investigated in the literature. In general, the definition of a class $\mathbb{L}$ relies on a class $\mathbb{R}$ of abstract machines,[2] here called *representations*, that characterize all and only the languages of $\mathbb{L}$. The relationship is given by the *naming function* $\mathscr{L} : \mathbb{R} \to \mathbb{L}$ such that: (1) $\forall R \in \mathbb{R}, \mathscr{L}(R) \in \mathbb{L}$ and (2) $\forall L \in \mathbb{L}, \exists R \in \mathbb{R}$ such that $\mathscr{L}(R) = L$. Two representations R_1 and R_2 are *equivalent iff* $\mathscr{L}(R_1) = \mathscr{L}(R_2)$.

Many different classes of representations have been studied in the literature. It is beyond the scope of this chapter to exhaustively list them. However, we introduce the following definition, which is a generalization of some well-known classes of grammars. We will mainly focus on the classes of representations whose characterization can be done in this context.

Definition 2.1 (*Generative grammar*) $G = \langle \Sigma, N, P, I \rangle$ where Σ is the alphabet of the language, N is a set of variables usually called non-terminals, $P \subset (N \cup \Sigma)^+ \times (N \cup \Sigma)^*$ is the set of generative (production) rules, I is the finite set of axioms, which are elements of $(\Sigma \cup N)^*$.

A generative rule (α, β) is usually denoted $\alpha \to \beta$. It allows the rewriting of elements of $(\Sigma \cup N)^*$ into elements of $(\Sigma \cup N)^*$. Given $\gamma \in (\Sigma \cup N)^*$ we say that a production rule $\alpha \to \beta$ applied to γ if it exists $\eta, \delta \in (\Sigma \cup N)^*$ such that $\gamma = \eta\alpha\delta$. The result of applying this rule on γ is $\eta\beta\delta$. We write $\gamma \Rightarrow \eta\beta\delta$. $\Rightarrow^*$ is the reflexive and transitive closure of $\Rightarrow$, and $\Rightarrow_P^*$ is the reflexive and transitive closure of $\Rightarrow$ restricted to the production rules in P.

We define the size of a generative grammar to be the size of the set of its rules, plus the size of its set of axioms: $\|G\| = \|I\| + |P| + \sum_{\alpha \to \beta \in P}(|\alpha\beta| + 1)$.

Definition 2.2 (*Generated language*) Let $G = \langle \Sigma, N, P, I \rangle$ be a generative grammar. $\mathscr{L}(G) = \{w \in \Sigma^* : \exists \alpha \in I \ s.t. \ \alpha \Rightarrow_P^* w\}$.

Example 2.1 The usual classes of the Chomsky hierarchy are classes of generative grammars. Regular grammars correspond to the restriction $P \subset N \times (\Sigma N \cup \{\lambda\})$, or $P \subset N \times (N\Sigma \cup \{\lambda\})$ by symmetry. The context-free grammars are the ones where $P \subset N \times (\Sigma \cup N)^*$ while the context-sensitive grammars are the ones such that if $\alpha \to \beta \in P$ then $\exists (\gamma, \delta, \eta) \in (\Sigma \cup N)^*$, $A \in N$: $\alpha = \delta A \eta$ and $\beta = \delta\gamma\eta$. If no restrictions are imposed on the rules of the grammar, then the resulting class of representations corresponds to that of the unrestricted grammars. All of these classes were formerly defined with a set of axioms reduced to one element of N [11].

[2]This is not strictly necessary: for instance, the substitutable languages [13] have no grammatical characterization.

Example 2.2 String Rewriting Systems (SRS) [9] are generative devices where $N = \emptyset$. A rule corresponds to an element of Σ^* rewritten into an element of Σ^* and the set of axioms is made of elements of Σ^*. The language represented by an SRS is the set of strings that can be rewritten using the rules from an element of I.

Some classes of representations that have been studied in grammatical inference are not covered by Definition 2.1. This is the case for instance for multiple context-free grammars [39], patterns [2], tree [17] and graph [37] grammars, etc. While it is not difficult to generalize the definition in order to cover these classes, we conduct our discussion in the context of the above definition for concreteness and due to its familiarity.

2.2 PAC Learning and Other Learning Paradigms

2.2.1 PAC Paradigm

The best known paradigm in machine learning is certainly the Probably Approximately Correct (PAC) criterium [43] and its refinements [29, 30]. Unlike the identification in the limit paradigm, the PAC framework comes with built-in efficiency requirements so PAC-learners are efficient in important senses. A natural question then is: Why modify the identification in the limit paradigm when the PAC framework can be utilized instead? We argue that the PAC paradigm is not well-adapted to learning formal languages, as even very simple and well-characterized classes of languages are not PAC-learnable [4]. Several theoretical reasons explain this inadequacy, and each of them relates to aspects of the formal grammars used to describe formal languages.

One of the main reasons is that the VC-dimension of even the simplest models of language representations, namely the finite state automata, is not bounded [28] which make them not learnable in the PAC sense [8]. Indeed, not even the class of finite languages has finite VC-dimension. This is closely related to the fact that the learning principle of empirical risk minimization [44], inherent in most approaches studied under the PAC framework, is of little use when formal languages are considered. Indeed, the number of representations *consistent* to a given set of data of a target language, that is to say representations that correctly explain all the data, is often infinite. It is then useless to reduce the hypothesis space to the ones that minimize the error on a given set of data.

Similarly, consider the fact that the PAC paradigm does not suffer from the main drawback of identification in the limit of being asymptotic. Unlike PAC learning, in identification in the limit, there is no guarantee provided about the quality of the hypothesis before the (exact) convergence happens. But this drawback seems to be inherent to the kind of representations of the learning targets considered. Even if two generative grammars have all but one of their rules in common, the languages of these two grammars can be as far apart as one wishes. This problem is inherent to the nature of formal languages and their grammatical representations. This '*Gestalt*-like' property is unavoidable in the formalization of learning: the whole grammar is more than the sum of its rules. In our view, this mainly justifies the use of identification in the limit in the context of grammar learning.

Another reason is that a representation of a formal language is not only a classifier, that is to say a device that defines what is in the language and what is not, but it also gives *structural* information about the elements of the language.

Also, there are concerns that are more independent of the representations. Another particularity of language learning is that a lot of algorithms use only positive examples of a target concept, while the usual machine learning framework relies on labeled data.

Finally, the PAC paradigm is particularly pertinent in the case of statistical models, where the probability of making a mistake can be evaluated using the hypothesis. This particular attribute of the PAC paradigm is of less value when non-stochastic model learning is of interest. But even while grammatical inference is concerned with learning probability distributions over strings, the power of the considered models makes the paradigm inadequate: there are for instance infinitely many structurally different probabilistic context-free grammars that define the same set of distributions [26].

To be complete, some positive learning results exist in restrictive versions of the PAC-paradigm, mostly in the case where the target distribution is known to be drawn using a given class of stochastic grammars, and with additional restrictions that allow us to distinguish the different parts of the target from any sample (see [15, 36, 41] for examples).

2.2.2 Other Learning Paradigms

There are other less known learning frameworks which eschew identifiability in the limit in order to incorporate notions of computational efficiency. The aim here is not to give an exhaustive list of such paradigms: we just want to give pointers to the main ones.

The first that is worth mentioning is known as *query learning* in which the learner interacts with an oracle (see Chap. 3, *Learning Grammars and Automata with Queries*, de la Higuera). A wide range of types of queries have been investigated, from membership queries [31] where the oracle answers whether given strings belong to the language or not, to equivalence queries [3] that allow the learner to know if its current hypothesis is the target one, including correction queries [6] that correspond to membership queries where the oracle returns a 'close' element of the language if the submitted string is not part of the target (different definitions of string distance can be considered). In this approach, efficiency is measured by the number of queries the algorithm needs to converge to a hypothesis exactly equivalent to the target. Another learning paradigm derived from the former one requires access to a finite set of examples of the language and a membership oracle [16, 34].

Although these paradigms can be of practical interest (see the work on model checking for instance [27]), and though they can also be motivated by the study of first language acquisition [14], the need for an oracle clearly reduces the practical value of an algorithm investigated in this context.

Another learning paradigm that can be used to study algorithms in the context of grammatical inference is the one of *stochastic finite learning* [51]. In this framework, an algorithm is said to have learned a language if, from any infinite sequence of data of this language drawn from a probability distribution, it stops after having seen a finite number of elements and its hypothesis at that point is correct with high probability. The expected number of examples that the learner needs before convergence forms a measure of the algorithm's efficiency. This approach is similar in its aims to identification in the limit, but it can also be seen as a probably exactly correct paradigm. It is thus a tempting way to fill the gap between PAC-learning and identification in the limit. However, results in this paradigm are hard to obtain and even simple classes of languages are known to be not learnable. Many of the arguments of the previous section on the PAC-paradigm work can be used for this formalization. On the other hand, there are positive results for some classes of pattern languages [52].

We believe the reasons above, or some combination thereof, have led many scholars to seek a way to incorporate efficiency into the identification in the limit paradigm (as opposed to abandoning the paradigm altogether).

2.3 The Limits of Gold's Paradigm

2.3.1 Identification in the Limit

We now provide a detailed formalization of the identification in the limit paradigm.

A presentation P of a language L is an infinite sequence of data corresponding to L. We note $P[i]$ the ith element of P and P_i the set of the ith first elements of P. If the data contains only elements of L then the presentation is called a *text* of language L. A text T is a complete presentation of L iff for all $w \in L$ there exists $n \in \mathbb{N}$ such that $T[n] = w$. If data in the presentation are instead pairs (w, l), such that $w \in \Sigma^*$ and l is a Boolean valued TRUE if $w \in L$ and FALSE otherwise, then the presentation is called an *informant*. An informant I is a complete presentation of L iff for all $w \in \Sigma^*$ there exists $n \in \mathbb{N}$ such that $I[n] = (w, l)$. In the rest of the chapter, we will only consider complete presentations.

A learning algorithm in this context, sometimes called an *inductive inference machine*, is an algorithm that takes as input larger and larger initial segments of a presentation and outputs, after each input, a hypothesis from a pre-specified hypothesis space.

Definition 2.3 (*Identification in the limit* [24]) A class $\mathbb{L}$ of languages is *identifiable in the limit (IIL)* from text (resp. from informant) if and only if there exists a learning algorithm $\mathfrak{A}$ such that for all languages $L \in \mathbb{L}$, for all text T (resp. informant I) of L,

- there exists an index N such that $\forall n \geq N, \mathfrak{A}(T_n) = \mathfrak{A}(T_N)$ [resp. $\mathfrak{A}(I_n) = \mathfrak{A}(I_N)$]
- $\mathscr{L}(\mathfrak{A}(T_N)) = L$ [resp. $\mathscr{L}(\mathfrak{A}(I_N)) = L$]

Angluin [5] characterizes exactly those classes of languages that are identifiable in the limit from text. The central theorem in this work refers to the presence of 'telltale' finite subsets for each language in the class. Later, in Sect. 2.4.2, we will see an efficiency bound in terms of 'characteristic' finite subsets of languages (these are not exactly the same as Angluin's telltale subsets).

Gold [24] established three important results in this paradigm. The first is that the class of all finite languages is identifiable in the limit from text. The second is that no superfinite class of languages can be identified in the limit from text. Despite what the name may evoke, a class of languages is superfinite if it contains all finite languages and at least one infinite language (the class contains thus an infinite number of languages). The third is that any computably enumerable class whose uniform membership problem is decidable[3] is identifiable in the limit from an informant.

The proof of the second result relies on the fact that given a presentation of an infinite language L, there does not exist any index N from which a learner can distinguish the finite language made of the strings seen so far and the infinite language. If the algorithm converges to L on a complete text T for L at N then there is a text for the finite language containing all and only the strings in T_N for which the algorithm will *also* converge to L. Hence the algorithm fails to identify this finite language in the limit.

On the other hand, the learning algorithm for the third result (learning any computably enumerable class with informant) is really naive: it enumerates the elements of the class until it finds the first one consistent with the information so far. In other words, the algorithm always conjectures the first language in the enumeration that accepts all positive examples (labeled TRUE) and rejects all negative ones (labeled FALSE). If it is the correct hypothesis, the algorithm has converged. If not, then there will be an example later in the presentation that will be inconsistent with the current hypothesis and consequently the algorithm will move along down the enumeration to the next consistent language.

This third result, though of positive nature, is one of the main reasons that the identification in the limit paradigm needs to be refined to include a notion of tractability. 'Learning by enumeration' is clearly not tractable and thus is of little use. While it meets the letter of the definition of learning, it violates our intuitions of what learning should be like. At first glance, a natural way to exclude such learning 'solutions' is to add a tractability requirement to the definition in some way. However, as we now discuss, this is more difficult than it may initially appear.

[3]The uniform membership problem is the one where given a string and a representation one needs to determine whether the string belongs to the represented language.

For more on variations of Gold's original paradigm see Chap. 1, *Gold-Style Learning Theory* (Case).

2.3.2 Polynomial Time

Given the limitations of IIL shown in the previous section, designing requirements to add to the paradigm is needed to strengthen the validity of learning ideas. An intuitive way to deal with that is to constrain the time allowed for the algorithm to make its computations.

Limiting the overall running time appears inappropriate since languages may have infinite cardinality and concomitantly there is no bound on the length of the strings. Thus for any polynomial function p, infinite language L, and number n, there is a presentation P for L such that the first element of P is larger than $p(n)$. Stochastic finite learning [51] would be of great interest to readers concerned with this problem since it replaces this worst-case scenario with a learning framework that focuses on expected convergence (where presentations are drawn according to probability distributions).

A more consensual requirement is *update-time efficiency*. An algorithm is update-time efficient if it outputs a new hypothesis in time polynomial in the size of the data seen so far. This is reasonable as far it goes. Unfortunately, this requirement turns out to be no real restriction at all.

In a seminal paper [35], Leonard Pitt shows that update-efficiency is not sufficient to prove the validity of a learning approach. Indeed, using a method now known as Pitt's trick, he proves that any algorithm that can identify a class in the limit can be transformed into an algorithm that identifies the class in the limit and is update-time efficient.

Informally the proof relies on the fact that, given a presentation P, if a learner converges to a correct hypothesis on the initial sequence P_i, a variant can delay the computation of any interesting hypothesis until having seen P_j $(j > i)$ such that the computation time of the initial learner on P_i is polynomial in $\|P_j\|$. This variant of the learning algorithm then has an efficient update-time while also fulfilling the conditions of identification in the limit. Pitt's trick essentially trades time for data so that enforcing tractability in this way has no impact. The set of classes of languages identifiable in the limit without the update-time requirement is exactly the same as the set of classes of languages identifiable in the limit with it.

Pitt's trick reveals that algorithms may be able to efficiently output hypotheses, but convergence can only occur after non-reasonable amounts of *data* have been provided. This lessens the practical utility of the theoretical results when real data is taken into account.

One may wonder if one can prohibit Pitt's trick, which ignores the great part of the given data, by forcing a learner to respect all the given data. Case and Kötzing [10] show that apparently reasonable properties to force a learner to take all the examples into account are not strong enough to prevent Pitt's trick actually when learning from text.

2.3.3 *Identification of a Language and the Size of a Target Representation*

Despite the problem described in the previous section, the requirement to have polynomial update-time is still desirable. Efforts have been made to enrich the paradigm further such that Pitt-style delaying tricks are not possible.

Most additional requirements are based on the same method: they link the behavior of the algorithm to the size of a representation of the target language. Indeed, though the identification of the target language is required, a polynomial bound cannot be established with respect to the size of the language since non-trivial classes of languages often contain an infinite number of infinite languages. A representation of finite size of the target language is thus needed. Choosing a target representation also focuses the attention on the hypothesis space of the algorithm, which is relevant from a machine learning standpoint.

However, the choice of representations is not central at all in Gold's learning paradigm as a learner's hypotheses can converge to an arbitrary one among equivalent representations for the correct language. The apparent consequence is that the choice of a representation class for a target language class does matter when taking the representation size into account.

But this duality between the identification of a language and an efficiency bound on the size of a target representation has consequences that need to be handled carefully. For example, it is well known that a nondeterministic finite automaton can be exponentially smaller than the smallest deterministic finite automaton accepting the same language. A learning algorithm that behaves efficiently with respect to the size of deterministic finite automata may not be admitted as an efficient algorithm in terms of the size of nondeterministic finite automata. The reader is referred to Chap. 4, *On the Inference of Finite State Automata from Positive and Negative Data* (López and García), for details on this question.

In general, an inefficient learner can be seen as an efficient learner by choosing a class of redundant representations. Therefore, it is important to make clear under which class of representations the efficiency of a learner is discussed.

In principle, the choice of a representation class is arbitrary and seems hard to justify, but in practice there exist orthodox or natural representations for target language classes. For example, minimal deterministic (canonical) finite state automata are widely used to represent regular languages. Since they are uniquely determined based on an algebraic property of regular languages, there is no room to inflate the representation size.

An intuitive way to deal with the duality exposed above would be to define a paradigm where identification is on a target representation and not on a language. The formalization of this idea is known as *strong identification* [12]. However, this approach only makes sense for classes of representations where each language admits a unique representative: otherwise, it is impossible for any algorithm to distinguish between the different grammars generating the same language, and thus the identification cannot succeed. The use of canonical finite-state automata in the work on regular languages [33] is an example of such an approach.

2.4 First Refinements

2.4.1 Mind Changes and Implicit Errors of Prediction

One way to formalize the notion of convergence with a reasonable amount of data with respect to the size of the representation is to measure the number of *mind changes* [1, 7]. Another way is to measure the number of *implicit prediction errors* [35].

A *mind change* occurs when a learning algorithm replaces its current hypothesis with another. Then one adds to the identification in the limit paradigm the requirement that the number of mind changes made before convergence must be bounded by a polynomial function in the size of the representation.

However, Pitt [35] presents another trick where the algorithm postpones changing its mind solely to meet the requirements of the mind change bound. Consequently, the algorithm maintains untenable hypotheses (ones inconsistent with the data) until a sufficient amount data is seen so that a mind change can occur without violating the polynomial bound on the number of mind changes.

Measuring implicit predictions errors can get around this trick when learning from an informant. When the learner's current hypothesis is inconsistent with a new datum, it is called *an implicit error of prediction*. Then one adds to the identification in the limit paradigm the requirement that the number of times the current hypothesis is in contradiction with the new example has to be polynomial in the size of the target representation. More formally:

Definition 2.4 (*Identification in polynomial number of implicit errors*)

- Given a presentation P, an algorithm $\mathfrak{A}$ makes an implicit error of prediction at step n if the language of the hypothesized target $\mathfrak{A}(P_n)$ is in contradiction with $P[n+1]$.
- A class $\mathbb{G}$ of representations is polynomial-time identifiable in the limit in Pitt's sense if $\mathbb{G}$ admits a polynomial-time learning algorithm $\mathfrak{A}$ such that for any presentation of $\mathcal{L}(G)$ for $G \in \mathbb{G}$, $\mathfrak{A}$ makes implicit errors of prediction at most polynomial in $\|G\|$ [35].
- A class $\mathbb{G}$ of representations is polynomial-time identifiable in the limit in Yokomori's sense if $\mathbb{G}$ admits a polynomial-time learning algorithm $\mathfrak{A}$ such that for any presentation P of $\mathcal{L}(G)$ for $G \in \mathbb{G}$, for any natural number n, the number of implicit errors of prediction made by $\mathfrak{A}$ on the nth first examples is bounded by a polynomial in $m \cdot \|G\|$, where $m = \max\{|P[1]|, \ldots, |P[n]|\}$ [46].

Notice that Yokomori's formulation is a relaxed version of that of Pitt's.

However, if the presentation is a text, there is yet another unwanted Pitt-style delaying trick: the algorithm can output a representation for Σ^*, which will never be in contradiction with the data. It can then wait to see enough examples before returning a pertinent hypothesis.

On the other hand, if the presentation is an informant, then the additional requirement limiting the number of implicit prediction errors is significant because there is no language like Σ^* which is consistent with both the positive and negative examples. Consequently, it can be shown that not all classes of languages identifiable in the limit in polynomial update time are identifiable in the limit in Pitt's sense or in Yokomori's sense: in the former paradigm, an algorithm working in polynomial time can change its hypothesis an exponential number of times before convergence, while in the latter paradigms this is not allowed and cannot be circumvented as in the case of texts. Note this is different from the mind-change requirement, where the delaying trick there works in both kinds of presentations: in that case, the algorithm can choose to not update its hypothesis when a new example contradicts it.

Another property of these requirements is that they are mainly designed for incremental algorithms. Indeed, these paradigms give a lot of importance to the sequence of data, in particular as the parts of two sequences that contain the same elements in a different order might not correspond to the same number of implicit errors (or mind changes). This forces the complexity analysis to consider particularly malevolent sequences of data. However, in many practical frameworks, for instance in Natural Language Processing or Bio-informatics, we are interested in algorithms that work from a finite *set* of data, where the order of presentation is irrelevant. From this perspective, the (inadvertent) focus on an incremental process appears to be a drawback.

The main positive learning results using this approach concerns the class of very simple languages [47, 49]: an algorithm has been designed that fulfills the requirements of Yokomori's formulation of the paradigm. This class of languages is incomparable with the class of regular languages and contains context-free languages.

2.4.2 Characteristic Sample

The most widely used definition of data efficiency relies on the notion of *characteristic sample*. The characteristic sample is a finite set of data from a language L that ensures the correct convergence of the algorithm on any presentation of L as soon as it is included in the data seen so far. For some, these characteristic samples evoke Angluin's telltale subsets [5], also of finite size, which were central to characterizing the nature of classes of formal languages identifiable from text.

In this learning paradigm [18], it is required that the algorithm needs a characteristic sample whose size[4] is polynomial in the size of the target representation. Formally:

Definition 2.5 (*Identification in the limit in polynomial time and data*) A class of languages $\mathbb{L}$ is *identifiable in the limit in polynomial time and data* from a class $\mathbb{R}$ of representations iff there exist a learning algorithm $\mathfrak{A}$ and a polynomial $p()$ such that for any language $L \in \mathbb{L}$, for any representation $R \in \mathbb{R}$ of L:

- $\mathfrak{A}$ has a polynomial update-time,
- there exists a set of data CS, called a characteristic sample, of size at most $p(\|R\|)$ such that for any presentation P of L, if $CS \subseteq P_n$ then $\mathfrak{A}(P_n)$ is equivalent to R, and for all $N > n$, $\mathfrak{A}(P_N) = \mathfrak{A}(P_n)$.

The idea underlying the paradigm is that if the data available to the algorithm so far does not contain enough information to distinguish the target from other potential targets then it is impossible to learn. This complexity requirement diverges from update-time requirements above in that incremental learning algorithms no longer sit at the core of the paradigm. Indeed, limiting the complexity in terms of the characteristic sample makes possible the set-based definition that we are developing below.

Definition 2.6 Let $\mathbb{L}$ be a class of languages represented by some class $\mathbb{R}$ of representations.

1. A *sample* S for a language $L \in \mathbb{L}$ is a finite set of data consistent with L. A *positive sample* for L is made only of elements of L. A positive and negative sample for L is made of pairs (w, l), where l is a boolean such that $l =$ TRUE if $w \in L$ and $l =$ FALSE otherwise. The *size* of a sample S is the sum of the size of all its elements plus $|S|$.
2. An $(\mathbb{L}, \mathbb{R})$-learning algorithm $\mathfrak{A}$ is a program that takes as input a sample for a language $L \in \mathbb{L}$ and outputs a representation from $\mathbb{R}$.

We can now formalize the notion of characteristic sample in the set-based approach.

Definition 2.7 (*Characteristic sample*) Given an $(\mathbb{L}, \mathbb{R})$-learning algorithm $\mathfrak{A}$, we say that a sample CS is a *characteristic sample* of a language $L \in \mathbb{L}$ if for all samples S such that $CS \subseteq S$, $\mathfrak{A}$ returns a representation R such that $\mathscr{L}(R) = L$.

Hopefully it is evident that the class of representations is especially relevant in this paradigm.

The learning paradigm can now be defined as follows.

---

[4]The size of a sample is the sum of the length of its elements: it has been shown [35] that its cardinality is not a relevant feature when efficiency is considered, as it creates a risk of collusion: one can delay an exponential computation on a given sample of data and wait for a sufficient number of examples to run the computation on the former sample in polynomial time in the size of the latter.

Definition 2.8 (*Set-based identification in polynomial time and data* [18]) A class $\mathbb{L}$ of languages is *identifiable in polynomial time and data (IPTD)* from a class $\mathbb{R}$ of representations if and only if there exists an $(\mathbb{L}, \mathbb{R})$-learning algorithm $\mathfrak{A}$ and two polynomials $p()$ and $q()$ such that:

1. Given a sample S of size m for $L \in \mathbb{L}$, $\mathfrak{A}$ returns a consistent hypothesis $H \in \mathbb{R}$ in $\mathcal{O}(p(m))$ time.
2. For each representation R of size k of a language $L \in \mathbb{L}$, there exists a characteristic sample of L of size at most $\mathcal{O}(q(k))$.

Notice that the first item is a reformulation of the polynomial update time requirement, which is now in terms of the size of the sample. The second item corresponds to the additional requirement that the amount of data needed to converge is computationally reasonable. By forcing the algorithm to converge to a correct hypothesis whenever a characteristic sample of reasonable size has been seen, this paradigm tackles the risk of collusion by forbidding Pitt's delaying tricks.

The main reason this unusual way to formalize identification is chosen is because by formalizing learning when a set of data is available it corresponds to the most common framework when real-world data is considered.

Furthermore, the set-based approach encompasses the incremental approach since any algorithm studied in the latter can easily be cast into a set-based one. In other words, any algorithm that learns a class of languages in the sense of Definition 2.5 also learns the class in the sense of Definition 2.8.

However, it is not easy to cast set-based learners into incremental ones. Naively one may believe that for any algorithm $\mathfrak{A}$ satisfying Definition 2.8, there exists an incremental algorithm which satisfies Definition 2.5. The idea would be, for each new data, to run $\mathfrak{A}$ on the set of data seen so far. However, as shown in Appendix, this simple approach will not always work. There is an algorithm for learning the substitutable context-free languages which satisfies Definition 2.8 for which this incremental construction fails. In Appendix, it is shown that unless this incremental algorithm $\mathfrak{A}$ is conservative,[5] $\mathfrak{A}$ will not converge to a single grammar. However, if $\mathfrak{A}$ is conservative then there is a presentation at a point at which the characteristic set is seen but $\mathfrak{A}$ has not yet converged to the correct grammar. It remains to be seen whether for every set-based learner satisfying Definition 2.8, there is an incremental learner satisfying Definition 2.5.

Main Results Many learning algorithms have been studied in the context of IPTD. The main positive results concerns approaches that used positive and negative examples as input. In this context, regular languages are learnable [18] using deterministic finite state automata, and so are deterministic even linear languages as the question of inferring these grammars can be reduced to that of inferring deterministic finite state automata [40]. Another related class of languages that have been positively investigated in this context is the deterministic linear one [20]. The algorithms is fed with positive and negative examples and outputs a deterministic linear grammar.

[5] An incremental learner is *conservative* if it changes its current hypothesis H if and only if the next datum is inconsistent with H.

Context-free languages that are representable by delimited, almost non-overlapping string rewriting systems are also IPTD-learnable [22] from positive and negative examples. Comparisons of this class with the previous ones are difficult since they are not defined using the same kind of representation.

The whole class of context-free languages is learnable in the IPTD sense [19, 38] from structural positive examples, that is to say derivation trees with no information on the internal nodes. Given a positive integer k, the target class of representation is that of k-reversible context-free grammars [32] and the elements of the sample have to correspond to derivation trees of these grammars.

Limitations We have already discussed one drawback to measuring the complexity of the learning problem in terms of the size of the representation. It can be unclear what counts as a 'reasonable' representation. Consequently, it may be possible to artificially inflate representations to allow learning. This is another kind of trick since the algorithm would be efficient according to the letter of the definition but not its spirit.

Identification in polynomial time and data also suffers from the opposite kind of drawback. As we will see, for non-regular languages, there can be exponentially *compact* representations of languages. For such cases, IPTD-learning appears to give the wrong results: classes which intuitively ought to count as tractably learnable (because they return a very compact representation of the target language) can in fact be shown to not be IPTD-learnable. As IPTD was developed and studied in the context of learning regular languages, neither of these problems arose since minimal deterministic finite-state automata are considered to be reasonable representations of regular languages.

Example 2.3 illustrates the problem for the IPTD-learning of non-regular languages. It proves that context-free languages cannot be learned under this criterion using context-free grammars. Indeed, the characteristic sample of any grammar of the series has to contain the only string in the language, but the length of this string is exponentially greater than the size of the grammar.

Example 2.3 [18] let $\mathbb{G}_1 = \cup_{n>0}\{G_n\}$ be the class of context-free grammars such that for any n, the unique axiom of G_n is N_0 and its production rules are $N_i \rightarrow N_{i+1}N_{i+1}$, for $0 \leq i < n$, and $N_n \rightarrow a$. The language of G_n is the singleton $L(G_n) = \{a^{2^n}\}$.

The reason why this example is not learnable does not come from the hardness of the languages: they are made of only one string. But the use of any class of representations that contains $\mathbb{G}_1$ is not identifiable in the limit.

It seems that in this case the problem comes from the definition of what learning means, that is to say from the learning criterion, rather than the properties of the language. From an information theory point of view, it is obviously interesting to have an algorithm that is able to find a model explaining the data it is fed with that is exponentially smaller than the data. This is actually a desired property in many fields of machine learning (see [23] for instance). Hence, the trouble here comes from the learning paradigm.

2.5 Recent Refinements

In this section, we review two contemporary approaches that develop a definition of efficient learning which can be applied to non-regular classes of languages. They are both refinements of the identification with polynomial time and data.

2.5.1 *Structurally Complete Set*

We first introduce the following definition:

Definition 2.9 (*Structurally complete set*) Given a generative grammar G, a structurally complete set (SCS) for G is a set of data SC such that for each production $\alpha \rightarrow \beta$, there exists an element $x \in SC$, an element $\gamma \in I$ and two elements $\eta, \tau \in (\Sigma \cup N)^*$ such that $\gamma \Rightarrow^* \eta\alpha\tau \Rightarrow \eta\beta\tau \Rightarrow^* x$. The smallest structurally complete set (SSCS) S for a grammar G is the sample such that for all SCS S' for G, $S \trianglelefteq S'$.

A notion of structurally complete sample has already been defined in the context of regular language learning [21]. However, this former definition relied on a particular representation, namely the finite state automaton, and it considered only the case of positive and negative examples. Definition 2.9 is more general as it does not depend on a particular representation and does not consider a particular type of data. Definition 2.9 is a generalization of the notion of *representative sample* [42] that has been introduced in the context of learning from membership queries and a sample of positive examples of a subclass of context-free languages named simple deterministic languages.

Definition 2.10 (*Polynomial structurally complete identification*) A class $\mathbb{L}$ of languages is *identifiable in polynomial time and structurally complete data (IPTscD)* for a class $\mathbb{R}$ of representations if and only if there exists an algorithm $\mathfrak{A}$ and two polynomials $p()$ and $q()$ such that:

1. Given a sample S for $L \in \mathbb{L}$ of size m, $\mathfrak{A}$ returns a consistent hypothesis $H \in \mathbb{R}$ in $\mathcal{O}(p(m))$ time.
2. For each representation R of a language $L \in \mathbb{L}$, there exists a *characteristic sample* CS whose size is in $\mathcal{O}(q(k))$, where k is the size of the smallest structurally complete set for R.

Notice that in the case where negative data is also available, the size of the characteristic sample has to be polynomial in the size of a SCS which contains only positive examples. This implies that the amount of negative evidence has to be polynomially related to that of the positive evidence.

This paradigm shifts the perspective considerably: the efficiency does not rely anymore directly on the size of the representation but instead on the kind of strings it can generate. This move is anticipated, and pursued in part, in the approach by Ryo Yoshinaka [48], discussed in Sect. 2.5.2.

Comparison with IPTD Consider the class of languages $\mathbb{L}_2 = \bigcup_{n \in \mathbb{N}} \{\{a^i : 0 \leq i \leq 2^n\}\}$. This class is identifiable in polynomial time and data from positive data only using the class of representations $\mathbb{G}_2 = \bigcup_{n \in \mathbb{N}} \{\langle \{a\}, \{S, A\}, \{S \rightarrow A^{2^n}, A \rightarrow a|\lambda\}, \{S\} \rangle\}$. Indeed, given a target language, the simple algorithm that returns the only grammar consistent with a sample admits the characteristic sample $\{a^{2^n}\}$ which is linear in the size of the target. However, the smallest structurally complete set of any target grammar is $\{\lambda, a\}$ which is of size 2. As the size of the smallest SCS is constant and the class of languages infinite, $\mathbb{L}_2$ is not identifiable in polynomial time and structurally complete data.

On the other hand, let us consider the class of languages of Example 2.3: $\mathbb{L}_1 = \bigcup_{n \in \mathbb{N}} \{\{a^{2^n}\}\}$ and its class of representations $\mathbb{G}_1 = \bigcup_{n \in \mathbb{N}} \{\langle \{a\}, N_n, P_n, \{N_0\} \rangle\}$, with $P_n = \{N_n \rightarrow a\} \cup_{0 \leq i < n} \{N_i \rightarrow N_{i+1} N_{i+1}\}$. Given n, the characteristic sample is $\{a^{2^n}\}$ which is also the smallest structurally complete set for the target grammar. However, this sample is not polynomial in the size of the target grammar. Therefore $\mathbb{L}_1$ is identifiable in the limit in polynomial time and structurally complete data using $\mathbb{G}_1$ but not in polynomial time and data.

This shows that these two paradigms are thus non-comparable. However, most non-trivial language classes studied under the former paradigm are identifiable in polynomial time and structurally complete data. This is the case for instance for the regular languages from positive and negative examples and for all sub-regular classes studied in the context of grammatical inference: there is a linear link between the size of a regular grammar and what can be derived from any of its non-terminals.

2.5.2 *Thickness*

In a recent paper [48], Ryo Yoshinaka introduced the identification from a characteristic sample whose size is a polynomial in the size of the target grammar and of a measure called the thickness of the grammar.

Definition 2.11 (*Thickness*) Let $G = \langle \Sigma, N, P, I \rangle$ be a generative grammar. The *thickness* of G is $\tau_G = \max\{|\omega(\alpha)| : \exists \beta, \alpha \rightarrow \beta \in P\}$ where $\omega(\alpha) = \min_{\triangleleft}\{w \in \Sigma^* : \alpha \Rightarrow_G^* w\}$.

Informally, the thickness is the length of the longest string in the set of the smallest strings that can be generated from a left hand-side of a grammar rule.

This definition is an extended version of the one that was first introduced for context-free grammars in the context of model complexity [45]. Notice that it has nothing to do with the usual notion of thickness in learning theory.

Definition 2.12 (*Polynomial thick identification* [48]) A class $\mathbb{L}$ of languages is *identifiable in polynomial time and thick data (IPTtD)* for a class $\mathbb{R}$ of representations if and only if there exists an algorithm $\mathfrak{A}$ and two polynomials $p()$ and $q()$ such that:

1. Given a sample S for $L \in \mathbb{L}$ of size m, $\mathfrak{A}$ returns a consistent hypothesis $H \in \mathbb{R}$ in $\mathcal{O}(p(m))$ time.
2. For each representation R of a language $L \in \mathbb{L}$ of size k, there exists a *characteristic sample CS* whose size is in $\mathcal{O}(q(k, \tau_R))$.

IPTtD is clearly a refinement of IPTD since it simply adds the thickness as a parameter of the paradigm. It is however a fundamental move since it links the efficiency of the learning not only on the target representation but also to the kind of strings the grammar produces. This shift in perspective is a way to indirectly take into account the length of the strings in the language in the learning criterium. On the other hand, since it does not go so far as to remove the requirement that the characteristic sample be polynomial in the size of the grammar, it is still susceptible to inflation tricks.

Learnable Classes Since the size of the representation is used in Definition 2.12, it is clear that every class of languages that is IPTD is also IPTtD.

However, the converse is not true. Consider the grammars of Example 2.3: The thickness of any $G_n \in \mathbb{G}_1$ is 2^n.

More interesting examples are the classes of languages that have been investigated in the context of what is called distributional learning (see Chap. 6, *Distributed hearing of contest-free and multiple contest free Grammars*, Clark and Yoshinaka). For instance, a context-free language is *substitutable* if whenever two substrings appear once in the same context, then they always appear in the same context in the language [13].

There exists a polynomial-time algorithm that identifies the class of context-free substitutable languages from positive examples only, in the sense of Definition 2.8, but the exhibited characteristic sample might be of size exponential in the size of the target representation (this is the case for the languages of Example 2.3, which are substitutable). Thus, this algorithm is not IPTD. On the other hand, it is easy to see that this characteristic sample is polynomial in the size *and the thickness* of the target grammar, so the algorithm is IPTtD. This result can be extended to the more complex classes that have been studied in the context of distributional learning from positive examples only (see for instance [48, 50]).

2.5.3 *Comparison of the Two Refinements*

Since the IPTD and IPTscD classes are incomparable and every IPTD class is IPTtD, clearly there is an IPTtD class which is not IPTscD (this is the case for instance of the class $\mathbb{L}_2$ introduced at the end of Sect. 2.5.1). However, one can show that every

IPTtD class of unambiguous CFGs is IPTscD. Also, it is easy to see that every IPTscD class of context-free languages is IPTtD using the same class of representations.

The two refinements of polynomial identification share a basic idea – to measure the complexity by the size of the simplest strings that a grammar generates, rather than the description size of it. Indeed one can show that the size of the smallest SCS of G is polynomially bounded by $\tau_G \|G\|$. That is, if a language class is IPTscD for a class of context-free grammars, then it is also IPTtD.

However, the converse is not necessarily true. The following discussion illustrates a particularly difficult problem for IPTscD learning: ambiguity. Let G_n consists of the following rules:

$$P_n = \{ A \rightarrow a,\ A \rightarrow b,\ B \rightarrow b \} \cup \{ S \rightarrow X_1 \dots X_n \mid X_i \in \{A, B\} \},$$

which generates $L(G_n) = \{a, b\}^n$. Then the set $\{a^n, b^n\}$, whose size is $2n + 2$, is the smallest SCS for G_n. On the other hand, $\|G_n\| \in \mathcal{O}(n2^n)$ and $\tau_{G_n} = n$. When learning the class $\mathbb{L} = \bigcup_{n \in \mathbb{N}} \mathbb{L}_n$ where $\mathbb{L}_n = \{ L \subseteq \{a, b\}^n \} \setminus \{\emptyset\}$ with a positive sample, the only possible characteristic sample of $L(G_n)$ is $L(G_n)$ itself for any learning algorithm. Therefore, $\mathbb{L}$ is not IPTscD for any representation class. One can easily see that $\mathbb{L}$ is IPTtD for a reasonable class of grammars where G_n is the unique grammar for $\{a, b\}^n$.

The grammar G_n is very redundant and highly ambiguous—there are 2^n ways to derive b^n. If the redundancy is removed from G_n by deleting the nonterminal B and the rules involving B, the size of the grammar is now $O(n)$ and it is not IPTtD any more. In fact, one can show that $\tau_G \|G\|$ is polynomially bounded by the size of the smallest SCS when only unambiguous context-free grammars are considered.

2.6 Conclusion

The purpose of this chapter was to address the problem of efficiently learning formal languages and grammars. We argued that the PAC framework is not the best suited one even though its efficiency requirements are well-designed. On the other hand, we argued in favor of identification in the limit paradigms provided they are adequately modified to include efficiency requirements. This survey showed doing so is more challenging than anyone may have anticipated. We discussed the challenges that have been encountered by different attempts. For regular languages, de la Higuera's [18] solution is satisfactory due to the canonical representation given by the smallest deterministic acceptors. For non-regular languages, challenges remain. We discussed two promising paths forward to address efficient learning in the identification in the limit paradigm in the realm of non-regular languages. One was based on the notion of a structurally complete sample, and the other was based on the 'thickness' of strings generated by production rules. Both are measuring efficiency at least partly in terms of the size of particular strings generated by grammars. We believe further developments along these lines will help shape future directions in grammatical inference.

Appendix

Here we present an example that shows that a learning result in a set-based approach (that of IPTtD) may not yield to a learning result in the incremental approach.

A characteristic sample has been exhibited for a set-based polynomial-time learning algorithm[6] for the class of substitutable context-free languages [13]. The size of this characteristic sample is polynomial in the size of the target grammar and its thickness [48]. From any superset of this set, the algorithm returns a representation that generates the target language. Therefore, one can state that the algorithm learns the class of substitutable context-free languages in a set-based approach.

A particularity of this algorithm is that from two different supersets of the characteristic sample, it may returns two different equivalent grammars, and the number of such pairs of samples is infinite (this is due to the infinite number of congruence classes that a context-free language defines). Consider the incremental version of the algorithm that computes a new grammar for every new example. It therefore does not fit the conditions of identification in the limit since there does not exist a moment after which the algorithm always returns the same hypothesis, though there exists a point after which the generated language will always be the target one.[7]

An intuitive solution is then to make the algorithm conservative: the incremental version of the algorithm has to change its hypothesis only if the new example is not recognized. However, this is not working as is shown with the following example.

Consider the language $a(\{b, c\}\{b, c\})^*$, which is substitutable. It is also context-free as it can be generated by the grammar whose rules are $S \rightarrow a|SBB$ and $B \rightarrow b|c$, with S being the only axiom.

As defined in the previously cited papers, the characteristic sample is the following set: $CS = \{lur \in \Sigma^* : \exists N \rightarrow \alpha, (l, r) = C(A) \text{ and } u = \omega(\alpha)\}$, where $C(A)$ is the smallest context in which the non-terminal A can appear in a derivation, and $\omega(\alpha)$ is the smallest element of Σ^* that can be derived from α in the grammar.

If we assume $a < b < c$ and $(ab, \lambda) < (a, b)$, the characteristic sample is then $CS = \{a, abb, abc\}$.

Suppose the learner gets examples $a, abb, abbbc$ in this order. As the letter c is new, the conjecture has to be updated at this point. The new conjecture is then the string rewriting system $\{a \rightarrow abb, a \rightarrow abc, b \rightarrow bbc\}$ with a being the only axiom.

---

[6]Notice that the algorithm was originally presented in an incremental paradigm. However, its study was (mostly) done in a set-based framework and, as is shown in this appendix, the proofs are valid only in this context.

[7]This is known as *behaviorally correct* identification in the limit.

It generates every sentence in the characteristic sample.[8] However the hypothesis is not correct since for example acc is in the target language but not in the current one. Therefore, if the next example is the missing string of the characteristic sample, abc, the algorithm will not change its hypothesis: though all elements of the characteristic sample are available, the current hypothesis is not correct. Once an element of the language that is not generated by the hypothesis is seen, the hypothesis will be updated using a set containing a characteristic sample and thus the new conjecture will correspond to a correct representation of the language.

References

1. A. Ambainis, S. Jain, and A. Sharma. Ordinal mind change complexity of language identification. *Theoretical Computer Science*, pages 323–343, 1999.
2. D. Angluin. Finding patterns common to a set of strings. *Journal of Computer and System Sciences*, 21:46–62, 1980.
3. D. Angluin. Queries and concept learning. *Machine Learning*, 2(4):319–342, 1987.
4. D. Angluin, J. Aspnes, and A. Kontorovich. On the learnability of shuffle ideals. In *Proceedings of the Algorithmic Learning Theory Conference*, pages 111–123, 2012.
5. Dana Angluin. Inductive inference of formal languages from positive data. *Information and Control*, 45:117–135, 1980.
6. L. Becerra-Bonache, A. Dediu, and C. Tîrnăucă. Learning DFA from correction and equivalence queries. In *Proceedings of the International Colloquium on Grammatical Inference*, pages 281–292, 2006.
7. L. E. Blum and M. Blum. Toward a mathematical theory of inductive inference. *Information and Control*, 28(2):125–155, 1975.
8. A. Blumer, A. Ehrenfeucht, D. Haussler, and M. Warmuth. Learnability and the Vapnik-Chervonenkis dimension. *Journal of the ACM*, 36(4):929–965, 1989.
9. R. Book and F. Otto. *String-Rewriting Systems*. Springer Verlag, 1993.
10. J. Case and T. Kötzing. Difficulties in forcing fairness of polynomial time inductive inference. In *Proceedings of the Algorithmic Learning Theory Conference*, pages 263–277, 2009.
11. N. Chomsky. Three models for the description of language. *IRE Transactions on Information Theory*, 2:113–124, 1956.
12. A. Clark. Learning trees from strings: A strong learning algorithm for some context-free grammars. *Journal of Machine Learning Research*, 14:3537–3559, 2014.
13. A. Clark and R. Eyraud. Polynomial identification in the limit of substitutable context-free languages. *Journal of Machine Learning Research*, 8:1725–1745, 2007.
14. A. Clark and S. Lappin. *Linguistic Nativism and the Poverty of the Stimulus*. Wiley-Blackwell, 2011.
15. A. Clark and F. Thollard. PAC-learnability of probabilistic deterministic finite state automata. *Journal of Machine Learning Research*, 5:473–497, 2004.
16. A. Clark and R. Yoshinaka. Distributional learning of parallel multiple context-free grammars. *Machine Learning*, 96:5–31, 2014.
17. H. Comon, M. Dauchet, R. Gilleron, C. Löding, F. Jacquemard, D. Lugiez, S. Tison, and M. Tommasi. Tree automata techniques and applications. Available on: http://tata.gforge.inria.fr/, 2007.

[8]The algorithm is consistent which implies that the two first elements of the characteristic sample are in the conjectured language and we have the rule $a \rightarrow abc$ that generates the third element of CS from the axiom.

18. C. de la Higuera. Characteristic sets for polynomial grammatical inference. *Machine Learning*, 27:125–138, 1997.
19. C. de la Higuera. *Grammatical inference: learning automata and grammars*. Cambridge University Press, 2010.
20. C. de la Higuera and J. Oncina. Learning deterministic linear languages. In *Proceedings of Conference on Learning Theory*, pages 185–200, 2002.
21. P. Dupont, L. Miclet, and E. Vidal. What is the search space of the regular inference? In *Proceedings of the International Colloquium on Grammatical Inference*, pages 25–37, 1994.
22. R. Eyraud, C. de la Higuera, and J.-C. Janodet. LARS: A learning algorithm for rewriting systems. *Machine Learning*, 66(1):7–31, 2007.
23. F. Girosi. An equivalence between sparse approximation and support vector machines. *Neural Comput.*, 10(6):1455–1480, 1998.
24. E. M. Gold. Language identification in the limit. *Information and Control*, 10(5):447–474, 1967.
25. J. Heinz. Computational theories of learning and developmental psycholinguistics. In J. Lidz, W. Synder, and J. Pater, editors, *The Oxford Handbook of Developmental Linguistics*. Cambridge University Press, in press
26. D. Hsu, S. M. Kakade, and P. Liang. Identifiability and unmixing of latent parse trees. In *Advances in Neural Information Processing Systems (NIPS)*, pages 1520–1528, 2013.
27. M. Isberner, F. Howar, and B. Steffen. Learning register automata: from languages to program structures. *Machine Learning*, 96:65–98, 2014.
28. Y. Ishigami and S. Tani. VC-dimensions of finite automata and commutative finite automata with k letters and n states. *Discrete Applied Mathematics*, 74:123–134, 1997.
29. J. Langford. Tutorial on practical prediction theory for classification. *Journal of Machine Learning Research*, 6:273–306, December 2005.
30. M. Li and P. Vitanyi. Learning simple concepts under simple distributions. *SIAM Journal of Computing*, 20:911–935, 1991.
31. E. Moore. Gedanken-experiments on sequential machines. In Claude Shannon and John McCarthy, editors, *Automata Studies*, pages 129–153. Princeton University Press, 1956.
32. T. Oates, D. Desai, and V. Bhat. Learning k-reversible context-free grammars from positive structural examples. In *Proceedings of the International Conference in Machine Learning*, pages 459–465, 2002.
33. J. Oncina and P. García. Identifying regular languages in polynomial time. In *Advances in Structural and Syntactic Pattern Recognition*, volume 5 of *Series in Machine Perception and Artificial Intelligence*, pages 99–108. 1992.
34. T.-W. Pao and J. Carr III. A solution of the syntactical induction-inference problem for regular languages. *Computer Languages*, 3(1):53 – 64, 1978.
35. L. Pitt. Inductive inference, DFA's, and computational complexity. In *Analogical and Inductive Inference*, number 397 in LNAI, pages 18–44. Springer-Verlag, 1989.
36. D. Ron, Y. Singer, and N. Tishby. On the learnability and usage of acyclic probabilistic finite automata. In *Proceedings of the Conference on Learning Theory*, pages 31–40, 1995.
37. G. Rozenberg, editor. *Handbook of Graph Grammars and Computing by Graph Transformation: Volume I. Foundations*. World Scientific, 1997.
38. Y. Sakakibara. Efficient learning of context-free grammars from positive structural examples. *Information and Computation*, 97:23–60, 1992.
39. Hiroyuki Seki, Takashi Matsumura, Mamoru Fujii, and Tadao Kasami. On multiple context-free grammars. *Theoretical Computer Science*, 88(2):191–229, 1991.
40. J. M. Sempere and P. García. A characterization of even linear languages and its application to the learning problem. In *Proceedings of the International Colloquium in Grammatical Inference*, pages 38–44, 1994.
41. C. Shibata and R. Yoshinaka. PAC-learning of some subclasses of context-free grammars with basic distributional properties from positive data. In *Proceedings of the Algorithmic Learning Theory conference*, pages 143–157, 2013.

42. Y. Tajima, E. Tomita, M. Wakatsuki, and M. Terada. Polynomial time learning of simple deterministic languages via queries and a representative sample. *Theoretical Computer Science*, 329(1-3):203 – 221, 2004.
43. L. G. Valiant. A theory of the learnable. *Communications of the Association for Computing Machinery*, 27(11):1134–1142, 1984.
44. V. Vapnik. *The nature of statistical learning theory*. Springer, 1995.
45. M. Wakatsuki and E. Tomita. A fast algorithm for checking the inclusion for very simple deterministic pushdown automata. *IEICE TRANSACTIONS on Information and Systems*, VE76-D(10):1224–1233, 1993.
46. T. Yokomori. On polynomial-time learnability in the limit of strictly deterministic automata. *Machine Learning*, 19:153–179, 1995.
47. T. Yokomori. Polynomial-time identification of very simple grammars from positive data. *Theoretical Computer Science*, 1(298):179–206, 2003.
48. R. Yoshinaka. Identification in the limit of k, l-substitutable context-free languages. In *Proceedings of the International Colloquium in Grammatical Inference*, pages 266–279, 2008.
49. R. Yoshinaka. Learning efficiency of very simple grammars from positive data. *Theoretical Computer Science*, 410(19):1807–1825, 2009.
50. R. Yoshinaka. Efficient learning of multiple context-free languages with multidimensional substitutability from positive data. *Theoretical Computer Science*, 412:1821–1831, 2011.
51. T. Zeugmann. Can learning in the limit be done efficiently? In *Proceedings of the Algorithmic Learning Theory conference*, pages 17–38, 2003.
52. T. Zeugmann. From learning in the limit to stochastic finite learning. *Theoretical Computer Science*, 364(1):77–97, 2006.

Chapter 3
Learning Grammars and Automata with Queries

Colin de la Higuera

Abstract When learning languages or grammars, an attractive alternative to using a large corpus is to learn by interacting with the environment. This can allow us to deal with situations where data is scarce or expensive, but testing or experimenting is possible. The situation, which arises in a number of fields, is formalised in a setting called active learning or query learning. By controlling better the information to which one has access, this setting provides us with a better understanding of the hardness of learning tasks. But the setting also allows us to solve practical learning situations, for which new algorithms are needed.

3.1 Introduction

Grammatical inference deals with the question of learning grammatical models, such as automata, grammars or transducers, given information about a language [32]. The most general setting is that of learning from examples over which the learner has no say, often allowing us to reformulate the learning problem as a combinatorial question: find the smallest automaton such that some condition is met. An alternative introduced by Angluin is to allow the learning algorithm to have access to its environment, and question it about the language for which a grammar is sought [29, 44, 50].

Query learning [5] can be described as a game where the learner can ask questions (queries) to an Oracle (teacher) about the target language. The game ends when the learner guesses the target. Of course, the learning results strongly depend on the sort of queries that the learner is allowed to make.

This work was partially supported by the IST Programme of the European Community, under the PASCAL 2 Network of Excellence, IST–2007-216886. The author also acknowledges support by the Région Pays de Loire. This publication only reflects the author's views.

C. de la Higuera (✉)
LINA, University of Nantes, Nantes, France
e-mail: cdlh@univ-nantes.fr

J. Heinz and J.M. Sempere (eds.), *Topics in Grammatical Inference*,
DOI 10.1007/978-3-662-48395-4_3

If grammatical inference concerns the learning of grammars for strings, but also, trees and graphs, below we will only consider string languages, and point in the conclusion to further resources where more complex types are used.

Typically, the formalism involves the presence of a learner (**he**) and an Oracle (**she**). There has to be an agreement on the type of queries one can make. Some typical queries are:

- Membership queries: the learner presents a string to the Oracle and receives as answer the label, i.e., a Boolean indicating if the string belongs or not to the language.
- Strong equivalence queries: the learner presents a machine or a grammar and receives as answer TRUE (whenever the machine is equivalent to the target) or a counterexample which is a string misclassified by the hypothesis.
- Weak equivalence queries: the learner presents a machine and receives as answer TRUE or FALSE.
- Correction queries: the learner presents an example and receives as answer TRUE or a close element from the target language ('close' implies that a topology is defined).

There are a number of ways of describing or understanding the Oracle. She can be probabilistic (and randomly draw a permissible answer), follow a worst-case policy (in which case, the worst-possible legal answer will be returned), helpful (and then the Oracle may be seen as a teacher). Settings can be described where the Oracle may make noisy answers [43].

The type of Oracle one uses will entirely depend on the type of learning situation one has to model; but in all cases the task is to learn a grammatical representation of a language, while querying the Oracle.

A number of negative results have been described in the literature, used to prove that some class of languages/grammars could not be efficiently learned by using queries of one type or another. Deterministic finite automata (DFA) were thoroughly investigated in this setting: as negative results, it was proved that they could not be learned from just a polynomial number of membership queries [4] nor from just a polynomial number of strong equivalence queries [7].

On the other hand, algorithm LSTAR is able to learn DFA from a polynomial number of strong equivalence queries and membership queries [6].

Correction queries offer new possibilities [12, 13, 46]. The idea behind these queries is that the learner produces an example, submits it to the Oracle and obtains either a positive answer or a correction of the suggested example. There can be various types of corrections: some obey language-theoretic considerations, and some are intended to fit with real-life applications.

One may argue that the query learning setting is purely formal and ill-adapted to practical situations. When considering most of the early papers in the field this would be true: the papers were mostly published in theoretical conferences and journals, and no mention of applications seemed necessary. Furthermore with more and more data available why rely on interactive learning?

On the other hand, the fact of being able to choose what data needs labelling, during the actual learning process, offers many algorithmic advantages and justifies the current renewed interest in these techniques.

Indeed, active learning is used nowadays to check a learnt model. A typical example is an online translation system, which allows the user to undo a proposed translation: in this case, the query is the suggested translation and the Oracle is the end user.

The mathematical definitions are given in Sect. 3.2. In Sect. 3.3 we detail some applications in which these questions are interesting and have found use. In Sect. 3.4 the learning model is mathematically scrutinised: we present many types of queries and identify paradigms in which they can be used, Sects. 3.4.3 and 3.4.4 deal with identification in the limit and PAC learning. Negative results are recalled in Sect. 3.5: these are important in order to understand that being allowed to query does not suddenly open all the doors. Positive results and applications depend heavily on algorithm LSTAR; this algorithm is described and analysed in Sect. 3.6, and some of the corresponding implementation issues are discussed in Sect. 3.6.4. Extensions of LSTAR and alternative active learning results and challenges are surveyed (briefly) in Sect. 3.7, with an up-to-date presentation of the state of the art, specially following the Zulu competition held in 2010 (Sect. 3.7.1).

3.2 Definitions and Notations

$\mathbb{N}$ is the set of positive or null integers; $\mathbb{Q}$ is the set of all rational numbers; $[n] = \{1, \ldots, n\}$ (and for $n = 0$, $[0] = \emptyset$).

An *alphabet* Σ is a finite nonempty set of symbols called *letters*. A *string* $w = a_1 \ldots a_n$ is any finite sequence of letters. We write λ for the empty string and $|w|$ for the length of w. Let $\Sigma^\star$ denote the set of all strings over Σ. We say that u is a *subsequence* of v, denoted $u \preceq v$, if $u = a_1 \ldots a_n$ and there exist $u_0, \ldots, u_n \in \Sigma^\star$ such that $v = u_0 a_1 u_1 \ldots a_n u_n$. We suppose there is a predefined alphabetical order over the letters, denoted by $\leq_{lex}$; we introduce the *hierarchical order*: $u \trianglelefteq v$ if $|u| < |v|$ or ($u = v$ and $u \leq_{lex} v$).

A *language* is any subset $L \subseteq \Sigma^\star$. Let $\mathbb{N}$ denote the set of non-negative integers. For all $k \in \mathbb{N}$, let $\Sigma^{\leq k} = \{w \in \Sigma^\star : |w| \leq k\}$ and $\Sigma^{>k} = \{w \in \Sigma^\star : |w| > k\}$. We define $A \oplus B = (A \setminus B) \cup (B \setminus A)$, where $A \setminus B = \{x \in \Sigma^\star : x \in A \wedge x \notin B\}$.

A *grammar* is a machine that allows us to generate, recognize or define strings.

Grammatical inference aims at learning the languages of a fixed class $\mathcal{L}$ represented by the grammars of a class $\mathcal{G}$. $\mathcal{L}$ and $\mathcal{G}$ are related by a naming function $\mathbb{L} : \mathcal{G} \to \mathcal{L}$ that is total ($\forall G \in \mathcal{G}, \mathbb{L}(G) \in \mathcal{L}$) and surjective ($\forall L \in \mathcal{L}, \exists G \in \mathcal{G}$ such that $\mathbb{L}(G) = L$). When considering a learning task, we will denote by T the target grammar and H the hypothesis. In each case the corresponding language is $\mathbb{L}(T)$ (or $\mathbb{L}(H)$).

For any string $w \in \Sigma^\star$ and language $L \in \mathcal{L}$, we shall write $\ell_L(w) = 1$ if $w \in L$. If $w \notin L$, $\ell_L(w) = 0$. The same notation will be used with grammars instead of languages, i.e., $\ell_G(w) = \ell_{\mathbb{L}(G)}(w)$

Concerning the grammars, they may be understood as any piece of information allowing a given parser to recognise the strings. $\|G\|$ will denote the size of the grammar G (e.g., the number of states in the case of DFA, for a fixed alphabet). Moreover, given a set X of strings, we will write $|X|$ for the cardinality of X and $\|X\|$ for the sum of the lengths of the strings in X.

A *deterministic finite automaton* (DFA) is a 5-tuple $A = \langle \Sigma, Q, q_0, F, \delta \rangle$ such that Q is a set of states, $q_0 \in Q$ is an initial state, $F \subseteq Q$ is a set of final states and $\delta : Q \times \Sigma \to Q$ is a transition function. Every DFA can be *completed* with one sink state such that δ is a total function. As usual, δ is extended to $\Sigma^\star$. The *language recognised by* A is $\mathbb{L}(A) = \{w \in \Sigma^\star : \delta(q_0, w) \in F\}$. The size of A is $|Q|$. We will write $\mathcal{DFA}(\Sigma)$ for the class of all DFA over the alphabet Σ. The class of all DFA of size at most n is denoted by $\mathcal{DFA}_n(\Sigma)$.

The *edit distance* $d(w, w')$ is the minimum number of *primitive edit operations* needed to transform w into w' [47]. The operation is either (1) a *deletion*: $w = uav$ and $w' = uv$, or (2) an *insertion*: $w = uv$ and $w' = uav$, or (3) a *substitution*: $w = uav$ and $w' = ubv$, where $u, v \in \Sigma^\star$, $a, b \in \Sigma$ and $a \neq b$. For example, $d(\texttt{abaa}, \texttt{aab}) = 2$ since $\texttt{a}\underline{\texttt{b}}\texttt{aa} \to \texttt{aa}\underline{\texttt{a}} \to \texttt{aab}$ and the rewriting of abaa into aab cannot be achieved with less than 2 steps. $d(w, w')$ can be computed in $\mathcal{O}\left(|w| \cdot |w'|\right)$ time by dynamic programming [59].

The edit distance is a metric, so topological balls over Σ can be introduced. The *ball of centre $o \in \Sigma^\star$ and radius $r \in \mathbb{N}$*, denoted $B_r(o)$, is the set of all strings whose distance is at most r from o: $B_r(o) = \{w \in \Sigma^\star : d(o, w) \leq r\}$. The size of a ball is therefore $|o| + \log r$.

For example, if $\Sigma = \{\texttt{a}, \texttt{b}\}$, then $B_1(\texttt{ba}) = \{\texttt{a,b,aa,ba,bb,aba,baa,bab, bba}\}$ and $B_r(\lambda) = \Sigma^{\leq r}$ for all $r \in \mathbb{N}$. We will write $\mathcal{BALL}(\Sigma)$ for the family of all the balls and $\mathcal{BALL}_n(\Sigma)$ for the family of all balls of size at most n.

3.3 Motivations and Applications

A number of typical situations in which Oracle learning makes sense are described in [30]. In most cases, algorithm LSTAR, or one of its variants, has been used. LSTAR has been designed to learn DFA using equivalence and membership queries. The earliest task addressed by LSTAR-inspired algorithms was that of map building in robotics: the (simplified) map is a graph and the outputs in each state are what the robot may encounter in a state. A particularity here is that the robot cannot be reset: the learning algorithm is to learn from just one very long string and all its prefixes [35, 52]. Another issue is that if the graph is not strongly connected, the robot may not even be able to explore all the graph and will therefore only learn part of the language/map.

A task somehow related is that of discovering the rational strategy followed by an adversary. The setting is that of a game played by two players. One of these follows a strategy described by a finite state machine. In order to beat this player, the algorithm should play against him, introducing (with a cost!) the moves that allow him to learn this strategy. This line was investigated in papers related to agent technologies [22, 23].

A task on which grammatical inference is proving to be particularly useful is that of wrapper induction: the idea is to build expressions (typically XPath) allowing us to find particular types of information in webpages sharing a similar structure. Since webpages will be encoded in HTML or XML, learning tree automata is a good approach (see Chap. 7, *Learning Tree Languages*, Björklund and Fernau). Furthermore the role of the Oracle is played by the human user [20, 21]. The system will interrogate the (human) user who will mark webpages. The marking will be used to learn a tree automaton (or transducer).

In different tasks linked with checking if the specifications of a system or a piece of hardware are met, the item to be checked is used as an Oracle. Queries are made in order to build a model of the item, and then the learnt model can be checked against the specifications. Learning test sets [18] and testing hardware [40] are other activities where there is an unsuspected Oracle: the actual electronic device we are testing can be physically tested, by entering a sequence. The device will then be able to answer a membership query. Note that in that setting equivalence queries will be usually simulated by sampling [16, 18, 51].

The World Wide Web can also be interpreted as an Oracle. The knowledge it contains is huge and cannot be sampled. It also may be inexact and can be queried in a number of ways. Interrogating the web in order to pick the useful information for learning is becoming an important task.

Interactive learning is becoming an interesting approach: the user is asked to correct a solution proposed by the computer, in this way the user plays the role of the Oracle.

One may also consider active learning in order to deal with concepts that have been learnt but may not be intelligible. Consider for instance neural networks [37]: an alternative approach to learning from sequences is to train recurrent neural networks [3]. But once these are learnt, we may choose to interpret these and one way to do this is to use the black box/neural network as an Oracle that can be interrogated. In some way, what is going to happen is that we are going to attempt to translate the neural network into an automaton.

3.4 Query Learning Models

In this section we detail the different mathematical elements we need: the Oracle, the queries and the learning paradigms.

3.4.1 The Oracle

The active learning paradigm is based on the existence of an Oracle which can be seen in principle as a device that

- knows the language and has to answer correctly,
- can only answer to queries from a given set of queries.

An Oracle (*she*) is generally supposed to be fair: she can answer any specific query of a predefined type. She can even answer queries that a concrete machine would not be able to cope with and therefore might even solve undecidable questions. The ability of the Oracle is determined by the learning setting.

In some cases the Oracle may have various possible answers. In this case she should be allowed to give any admissible answer. As our goal when studying Oracle learning is to consider worst case scenarios, then we will always have to suppose the Oracle is giving us the least informative of all possible answers.

3.4.2 The Different Types of Queries

In Fig. 3.1 we give a running example to illustrate the different sorts of queries we will present. We will suppose this DFA is the target. There are 3 states, two of which are final accepting states (q_0 and q_2, denoted by a double circle) and one is rejecting (q_1, denoted by a thick grey line).

3.4.2.1 Sampling Queries

The first type of queries one can use to learn is *sampling queries*: the learner just asks the Oracle for a *random* example. The Oracle then has to randomly draw a string from a distribution (fixed but usually unknown) over all possible strings. Let $\mathscr{D}$ denote this distribution.

The query of an example or a counterexample will be denoted $\text{EX}_{\mathscr{D}}()$ or EX() when the distribution is clear. When the Oracle is only queried for a positive example, we will write POS-EX(). And when the Oracle is only queried for strings of length $\leq m$, we will write $\text{EX}(\Sigma^{\leq m})$ and $\text{POS-EX}(\Sigma^{\leq m})$ respectively. Formally, the Oracle will then return a string drawn from $\mathscr{D}$, or $\mathscr{D}(\mathbb{L}(H))$, or $\mathscr{D}(\Sigma^{\leq m})$, or $\mathscr{D}(\mathbb{L}(H) \cap \Sigma^{\leq m})$,

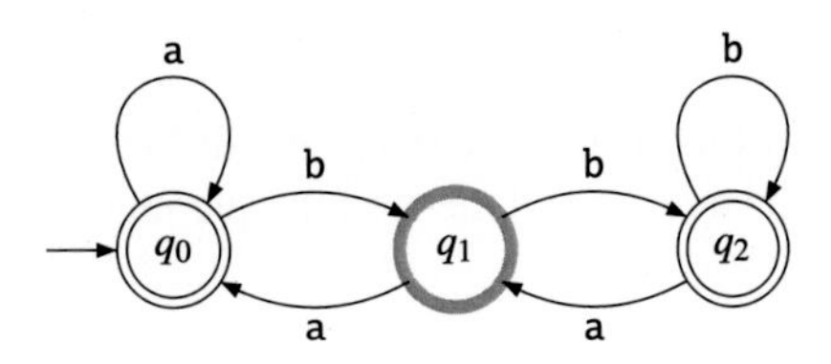

Fig. 3.1 A target automaton

respectively, where $\mathscr{D}(L)$ is the restriction of $\mathscr{D}$ to the strings of L: $Pr_{\mathscr{D}(L)}(x) = Pr_{\mathscr{D}}(x)/Pr_{\mathscr{D}}(L)$ if $x \in L$, 0 otherwise. $Pr_{\mathscr{D}(L)}(x)$ is not defined if $L = \emptyset$.

Example 3.1 Suppose that $\mathscr{D}$ is defined as follows: $\forall n > 0,\ Pr_{\mathscr{D}}(\mathtt{a}^n\mathtt{b}^n) = \frac{1}{2^n}$ and $Pr_{\mathscr{D}}(x) = 0$ for all the other strings. Then, for the running example (aabb, 1) will be drawn with probability $\frac{1}{4}$ when using $\text{EX}_{\mathscr{D}}()$, but aabb will be the answer in $\frac{1}{2}$ of the cases to POS-EX(), and will always be the answer to POS-EX($\Sigma^{\leq 5}$).

A *specific sampling query* is made by submitting a pattern: the Oracle draws a string that matches some chosen pattern sampled according to the distribution $\mathcal{D}$. Specific sampling queries are intended to fit the idea that the user can ask for examples matching some pattern he is interested in. We will denote such a query by EX(Π), where Π is a pattern, typically a regular expression.

For the running example, we may want to use EX(), which could return (aa, 1) or (ab, 0), POS-EX() which could return aa or NEG-EX() whose result could be ab. A specific sampling queries could be EX(Σ^3) (asking for a string of length 3) or EX($\mathtt{a}^*$) which would ask for a string, in the language, containing only occurrences of symbol a.

3.4.2.2 Membership Queries

The most studied queries are called *membership queries*. The learner presents a string to the Oracle and gets back the label of this string ($\ell_T(x)$). The Oracle is therefore playing the role of the characteristic function of the language. A membership query is made by proposing a string to the Oracle who answers TRUE if the string belongs to the language and FALSE if not. We will denote this formally by:

$$\text{MQ} : \Sigma^\star \rightarrow \{\text{TRUE, FALSE}\}$$

In the case of the target represented in Fig. 3.1, MQ(ba) returns TRUE.

3.4.2.3 Equivalence Queries

The idea behind the *equivalence queries* is that the learner can somehow build a representation of its hypothesis and submit it to the Oracle. We distinguish *weak* equivalence queries in which the Oracle just answers TRUE or FALSE from *strong* equivalence queries in which the Oracle will return a counterexample if the answer is FALSE.

Formally, a *weak equivalence queries* is denoted by WEQ. A weak equivalence query is made by proposing a grammar H to the Oracle. The Oracle answers TRUE if the grammar is equivalent to the target and FALSE if not:

$$WEQ : \mathcal{G} \rightarrow \{\text{TRUE, FALSE}\}$$

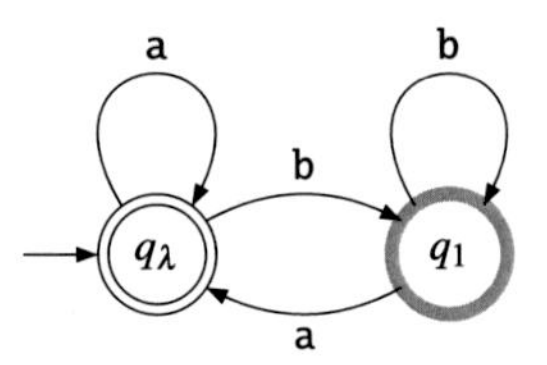

Fig. 3.2 A hypothesis automaton H

A *strong equivalence queries* (EQ) is made by proposing a grammar (or an automaton) H to the Oracle. The Oracle answers TRUE if the grammar is equivalent to the target and returns a string in the symmetrical difference between the target and $\mathbb{L}(H)$ if not:

$$\mathrm{EQ} : \mathcal{G} \rightarrow \{\mathrm{TRUE}\} \cup \Sigma^{\star}$$

In the case of the target represented in Fig. 3.1, WEQ(H) (for H represented in Fig. 3.2) returns FALSE whereas the answer to EQ(H) might be counterexample abb. But the counterexample could be any string of the type $\mathtt{ab}^n$ with $n > 1$.

In principle the Oracle is not computationally limited; in the case of more complex classes of grammars or automata, the Oracle may actually be solving an $\mathcal{NP}$-hard or even undecidable equivalence problem in order to answer the query.

3.4.2.4 Subset Queries

A *subset query* (SSQ) is made by proposing a grammar (or an automaton) H to the Oracle. The Oracle answers TRUE if $\mathbb{L}(H)$ is a subset of the target language and returns a string from $\mathbb{L}(H)$ that is not in the target language if not:

$$SSQ : \mathcal{G} \rightarrow \{\mathrm{TRUE}\} \cup \Sigma^{\star}$$

In the case of the target represented in Fig. 3.1, SSQ(H) (for H represented in Fig. 3.2) returns counterexample bba (but other counterexamples would have been possible, including $\mathtt{bb}^{1000000}\mathtt{a}$ or any valid arbitrarily long string).

Definition of superset queries is obtained in a similar—but opposite—manner.

3.4.2.5 Presentation Queries

A presentation of a language is an enumeration of all strings in $\Sigma^{\star}$, with a label indicating if a string belongs or not to $\mathbb{L}(T)$ (this is an *informed presentation*) or an enumeration of all strings in $\mathbb{L}(T)$ (this is a *text presentation*). A *presentation query* (PQ) is made by submitting an integer i: the Oracle then returns $f(i)$, the ith element of the presentation.

In a presentation there can be repetitions. If the presentation is a text presentation of the empty language, the Oracle just returns some special value #.

The presentation has to be fair (sometimes this is called complete): in a text presentation of language L, given any string w in L there exists at least one integer i such that $f(i) = w$; in an informed presentation, given any string w in $\Sigma^\star$, there exists at least one integer i such that $f(i) = (w, \ell_L(w))$.

A possible informed presentation of $\mathbb{L}(T)$ is $(\lambda, 1)$, $(\mathtt{ab}, 0)$, $(\mathtt{ab}, 0)$, $(\mathtt{bb}, 1)$, $(\mathtt{a}, 1)\ldots$ A possible text presentation is a, aaa, bb... In terms of notations, this would lead in the first case to $\mathrm{PQ}(0) = (\lambda, 1)$, $\mathrm{PQ}(1) = (\mathtt{ab}, 0)$,..., and in the second (text) case to $\mathrm{PQ}(0) = \mathtt{a}$, $\mathrm{PQ}(1) = \mathtt{aaa}\ldots$

Note that for a learner to use presentation queries, somehow the Oracle has to be able to select a specific presentation, and is supposed to stick to it.

3.4.2.6 Correction Queries

Correction queries (*CQ*) were introduced in [15] and studied since [13, 46, 56].

With the correction queries based on the edit distance [13] ($\mathrm{CQ}_{\mathrm{EDIT}}$), the learner submits a string w and the Oracle answers TRUE if $w \in L$, and any correction $z \in L$ at minimum edit distance of w otherwise.

With the corrections based on the shortest suffix ($\mathrm{CQ}_{\mathrm{SUFFIX}}$) [14], what is returned is a string wz in $\mathbb{L}(T)$ where z is the shortest possible string in length-lex order. If no correction allows us to be in the language, the Oracle just returns some value #.

A comparison of the power of these different types of correction queries can be found in [56].

Following with the example, $\mathrm{CQ}_{\mathrm{SUFFIX}}(\mathtt{bba})$ may receive as an answer bbaa, whereas for $\mathrm{CQ}_{\mathrm{EDIT}}(\mathtt{bba})$ the answer will be any string in L at edit distance one, *i.e.* any string from set {ba, bb, aba, baa, baba, bbab}.

3.4.2.7 Probabilistic Queries

If the target is a PFA or a probabilistic context-free grammar, one may want to to use probabilistic queries. With these, the learner submits a string and receives from the Oracle the probability of this string. In [34] the authors suggest to use *specific sampling queries* (see Sect. 3.4.2.1) to learn PFA: only a specific zone is sampled.

3.4.2.8 Translation Queries

Vilar [58] introduced *translation queries* in order to learn transducers. With these, an input string is submitted to the Oracle who returns the translation of this string. When learning transducers, these translation queries play the role that membership queries play with automata.

3.4.2.9 Combinations of Queries

In most cases, various types of queries are allowed to be used at the same time. We denote by QUER the types of queries that one may use for a specific task. A typical combination is QUER = {EQ, MQ} which is called MAT (*Minimally adequate teacher*, named by Angluin this way to emphasize the fact that with this particular combination DFA are learnable, whereas they are not with less).

3.4.3 *Query Learnability*

The query learning process supposes an interaction between the learner and the Oracle. Moreover, at some point, the learner is supposed to halt and return a solution. Ideally, the hypothesis is always correct (identification). Sometime, it may only be (probably) approximately correct (PAC).

This should be contrasted with *identification in the limit* [38], which was inspired by the cognitive process of a child that would acquire its native language by picking up the sentences that are broadcast in its environment. More formally, information keeps on arriving about a target language and the learner keeps making hypotheses. We say that convergence takes place if there is a moment where the process is stationary and the hypothesis is correct.

In the case of learning with queries, the learner is not only allowed to choose when and what to query, but he also has to decide when to halt.

The following definitions are based on works by [6, 11, 32].

Definition 3.1 A class $\mathcal{G}$ is *identifiable with queries* from QUER if there exists an algorithm $\mathfrak{A}$ such that given any grammar T in $\mathcal{G}$, $\mathfrak{A}$ makes a finite number of queries from QUER, returns a grammar H and halts. Furthermore H is equivalent to T.

It should be noted that Definition 3.1 corresponds to a strict definition in which the key point is that the algorithm must be able to decide when to halt or not. This makes the definition much more restrictive than identification in the limit. For example, the finite languages are not learnable from membership queries, whereas the same class can be identified in the limit from text.

In order to put some polynomial bounds on the learning process there are several difficulties:

- having a global limit is impossible: as many queries require an example (or a counterexample) to be given, it is difficult, if not impossible, to impose a limit on the length of what the Oracle may return;
- letting the complexity depend on the quantity of information received can also lead to unfair or collusive results: a learner can deliberately decide to postpone convergence until he has received enough information to justify the time he is taking to learn.

The above remarks are analysed in more detail in [8] and [24]. To take them into account, we need to introduce some extra notations depending on the actual runs of the learning algorithm:

Let us suppose that the information received from the environment by $\mathfrak{A}$ during some run ρ is stocked in a list $\mathbf{Info}(\rho)$ for which we can compute $l_{\mathbf{Info}}(\rho)$ indicating the length of this list and $max_{\mathbf{Info}}(\rho)$ the length of the longest element in this list (typically the longest counterexample returned by the Oracle at a certain point). On a particular run ρ, $\mathbf{Info}_n(\rho)$ corresponds to the information received from the Oracle as answers to the n first queries.

A first imposed condition concerns the time the learner is allowed to build his next hypothesis.

Definition 3.2 A query learning algorithm $\mathfrak{A}$ makes *polynomial updates* if there exists a polynomial $p(\cdot, \cdot)$ such that, given any run ρ, the time used by $\mathfrak{A}$ before (finishing) making the $n+1$st query is in $\mathcal{O}(p(max_{\mathbf{Info}_n}(\rho), l_{\mathbf{Info}_n}(\rho)))$.

In other words, in order to make his next query the learner can (only) use time polynomial in the amount of information received so far.

Definition 3.3 A class $\mathcal{G}$ is *polynomial update identifiable with queries* from QUER if there exists an algorithm $\mathfrak{A}$ making polynomial updates such that given any grammar T in $\mathcal{G}$, $\mathfrak{A}$ returns a grammar H equivalent to T and halts.

But this definition does not limit in any sense the total number of queries that can be made. For this, we add the following condition.

Definition 3.4 A class $\mathcal{G}$ is *polynomially identifiable with queries* from QUER if there exists a polynomial $q(\cdot, \cdot)$ and an algorithm $\mathfrak{A}$, such that given any grammar T in $\mathcal{G}$, $\mathfrak{A}$ makes polynomial updates and returns a grammar H equivalent to T before halting and for which the total number of queries is in $\mathcal{O}(q(\|T\|, max_{\mathbf{Info}}(\rho)))$.

The above definitions will not be able to adapt exactly to all possible types of queries. For example, they will need to be modified to take into account probabilistic queries. What is important to note, nevertheless, is that in certain cases, the Oracle may choose to return an unnecessarily long string, and just the parsing of this string would cost us more than what our polynomial bounds allow. Because of this, the length of the information received from the Oracle is both part of the data and of the result.

Another issue has been noted before: if the polynomial updates condition is not present, a learning algorithm may, in certain cases, take an exponential amount of time to build the correct hypothesis then interrogate the Oracle in such a way as to force the Oracle to return an extraordinarily long example, justifying a posteriori the time wasted.

3.4.4 PAC Query Learning

The second paradigm, PAC learning (Probably approximately correct) [57] is more pragmatic as it doesn't oblige the learner to exactly identify the target. In this setting, one assumes that there is a distribution $\mathcal{D}$ over the strings of the target language, which is used to sample learning and testing examples. Two parameters are introduced: ε is related to the error of the model (i.e., the probability of a string to be misclassified) and δ is related to the probability that the sample randomly drawn according to $\mathcal{D}$ is not representative of the target language.

The PAC paradigm [57] has been widely used in machine learning. It aims at building, with high confidence ($> 1 - \delta$), good approximations (error less than ε) of an unknown concept.

Definition 3.5 (ε*-good hypothesis*) Let T be the target grammar and H a hypothesis grammar. Let $\mathcal{D}$ be a distribution over $\Sigma^\star$ and $\varepsilon > 0$. We say that H is an ε*-good hypothesis w.r.t.* T if $Pr_{\mathcal{D}}(x \in \mathbb{L}(T) \oplus \mathbb{L}(H)) < \varepsilon$.

A learning algorithm is now asked to learn a grammar given a *confidence* parameter δ and an *error* parameter ε. The algorithm must also be given an upper bound n on the size of the target grammar and an upper bound m on the length of the examples it is going to get (perhaps computed using Lemma 3.1). The algorithm can query an Oracle for examples or labels, depending on QUER.

Definition 3.6 (*Polynomial PAC-learnability*) Let $\mathcal{G}$ be a class of grammars and QUER a combination of queries. $\mathcal{G}$ is PAC-learnable from QUER if there exists an algorithm $\mathfrak{A}$ such that $\forall\varepsilon, \delta > 0$, for any distribution $\mathcal{D}$ over $\Sigma^\star$, $\forall n \in \mathbb{N}$, $\forall T \in \mathcal{G}$ of size at most n, for any upper bound $m \in \mathbb{N}$ on the size of the examples returned by the Oracle, if $\mathfrak{A}$ has access to ε, δ, n and m, then with probability higher than $1 - \delta$, $\mathfrak{A}$ returns an ε-good hypothesis *w.r.t.* T. If $\mathfrak{A}$ runs in time polynomial in $\frac{1}{\varepsilon}$, $\frac{1}{\delta}$, n and m, we say that $\mathcal{G}$ is *polynomially* PAC-*learnable*.

Again, the above definition depends strongly on the sort of queries the learner is allowed to make. If it has no control over the length of the counterexamples (for instance) then the definition should take this into account.

The PAC-learnability of grammars from strings of unbounded size poses specific technical questions [44, 45, 60]. Indeed, with the standard definition, a PAC-learner can ask an Oracle to return a sample randomly drawn according to the distribution $\mathcal{D}$. However, in the case of strings, there is always the risk (albeit small) to sample a string too long to account for in polynomial time. In order to avoid this problem, we can impose that we sample from a distribution restricted to strings shorter than a specific value. A particular value ν can be built by unrestricted sampling using the following lemma.

Lemma 3.1 ([33]) *Let $\mathcal{D}$ be a distribution over $\Sigma^\star$. Then for all $\varepsilon, \delta > 0$, with probability larger than $1 - \delta$, if one draws a sample S of at least $\frac{1}{\varepsilon} \ln \frac{1}{\delta}$ strings following $\mathcal{D}$, the probability for any new string x to be longer than all the strings of S is less than ε. Formally, let $\nu_S = \max\{|y| : y \in S\}$, then $Pr_{\mathcal{D}}(|x| > \nu_S) < \varepsilon$.*

3.5 Some Important Hardness Results

It may sound too easy: if one can ask anything one likes, learning should always be feasible. But obviously the actual power of the different queries is limited.

3.5.1 Learning with Membership Queries and Little Else

What can a learner hope for if he uses membership queries? This sounds like a strong enough paradigm, and it also corresponds to several practical situations.

To see where the difficulties are let us consider *lock automata*: these automata recognise just one string of length n. With a two letter alphabet there are 2^n such automata, each of size $n+1$. It is easy to see that an algorithm attempting to distinguish one such automaton has to discard the $2^n - 1$ other automata, and will need $2^n - 1$ queries.

The key result is due to Angluin [5].

Lemma 3.2 *If a class $\mathcal{L}$ contains a non-empty set $L_\cap$ and n sets $L_1, \ldots, L_n$ such that $\forall i, j \in [n]$ $L_i \cap L_j = L_\cap$, any algorithm using membership, weak equivalence and subset queries needs in the worst case to make at least n-1 queries.*

Proof The Oracle will answer each query as follows:

- to MQ(x) with x in $L_\cap$, TRUE, and the learner cannot discard any language;
- to MQ(x) with x in $L_i \backslash L_\cap$, FALSE, and only language L_i can be discarded by the learner;
- to WEQ($L_\cap$), FALSE, and only language $L_\cap$ can be discarded by the learner;
- to WEQ(L_i), FALSE, and only language L_i can be discarded;
- to SSQ($L_\cap$), TRUE and no language can be discarded;
- to SSQ(L_i), FALSE, with counterexample x_i from $L_i \backslash L_\cap$ and only language L_i can be discarded.

The following corollary is due to the lock automata.

Corollary 3.1 *$\mathcal{DFA}_n(\Sigma)$ cannot be identified by a polynomial number of membership, weak equivalence and subset queries.*

3.5.2 Learning with Equivalence Queries

Obviously, due to Lemma 3.2, weak equivalence queries alone are insufficient to hope to learn anything of interest. But what about strong ones?

Pitt [50] noticed that a negative answer to the question of learning from a polynomial amount of strong equivalence queries would also lead to a negative answer to the question of identifying in the limit from informed presentations: an informed presentation of a language $\mathbb{L}(G)$ is just a complete presentation of all the strings in $\Sigma^\star$ with, in each case, a label indicating if the string is in $\mathbb{L}(G)$ or not. A learning algorithm is said to identify the class $\mathcal{G}$ in the limit if there is a polynomial p and an algorithm which, for each target grammar G, after each newly presented pair (x, label(x)) returns a hypothesis H and furthermore always stabilizes itself on a given hypothesis which is equivalent to the target grammar. An implicit prediction error is made when the label of the new example disagrees with the label the hypothesis grammar returns. An algorithm would make a polynomial number of implicit prediction errors (IPE) when the number of IPE is less than $p(G)$.

Lemma 3.3 ([50]) *If a class is identifiable in the limit from an informed presentation making just a polynomial number of implicit prediction errors, it is identifiable from strong equivalence queries with just a polynomial number (in the size of the target) of queries.*

Proof (Supposing this is not true) We can build a presentation containing just the counterexamples (with their labels) received from the different examples and then notice that each counterexample corresponds to an implicit prediction error; an implicit prediction error is made by a learning algorithm using a presentation each time its current hypothesis fails to classify the newly presented example.

We state the following without proof.

Theorem 3.1 ([7]) *$\mathcal{DFA}(\Sigma)$ cannot be identified by a polynomial number of strong equivalence queries.*

The proof is beyond the scope of this paper and uses the concept of approximate fingerprints. It should be noted that the same applies for NFA or context-free grammars.

3.5.2.1 The Halving Algorithm

But information-theoretic arguments such as those used to prove Lemma 3.2 should be handled with care, as, when complexity constraints are not taken into account, it is possible to learn anything by using a traditional dichotomy approach.

Let us first suppose that we have to learn a language which is just a subset of $E = \{w_1, ...w_n\}$. There are therefore 2^n different possible targets.

Let us denote by $\mathcal{H}_k$ the hypothesis space at step k, that is, the set of solutions consistent with the information obtained so far. Initially, at step 0, $\mathcal{H}_0 = 2^E$.

Given set $\mathcal{H}_k$, we say that w is a *positive consensus* string if it belongs to at least half the languages in $\mathcal{H}_k$, and a *negative consensus* string if it belongs to less than half of the languages in $\mathcal{H}_k$.

We denote C_k the set of positive consensus strings at step k.

Now if $\text{EQ}(C_k)$ is made at step $k+1$:

- either the counterexample returned by the Oracle will be a positive consensus string, which, being a counterexample, eliminates all the languages in which it is contained, therefore at least half the languages in $\mathcal{H}_k$,
- or the counterexample returned by the Oracle will be a negative consensus string, which, being a counterexample to C_k is supposed to belong to the target language, and again all the languages in $\mathcal{H}_k$ which reject it are eliminated.

Thus

$$|\mathcal{H}_{k+1}| \leq \frac{|\mathcal{H}_k|}{2}$$

It is easy to see that k steps will be enough to learn.

Now suppose that the class of targets is a bit more general. Still, the targets will be encoded by grammars whose length is going to be reasonable ($<p(n)$), and therefore we have $|\mathcal{H}_0| < 2^{p(n)}$, so the above argument can be adapted and we can devise a dichotomy strategy allowing us to make sure to shatter $\mathcal{H}_0$ in just a polynomial (in n) number of steps. As a consequence $\mathcal{H}_0$ is both more complicated than a powerset, yet does not contain all subsets.

The problem (which makes the halving algorithm impracticable) is that the information theoretic nature of hypothesis C_k makes it impossible, in general, to build. In other words, C_k will not be representable by a grammar of the class we are interested in learning.

It is usually required that equivalence queries are *proper* (i.e., are made inside the solution class) in order to avoid problems of this type.

3.6 Learning DFA from an MAT

We present in a simplified manner algorithm LSTAR invented by Angluin [6]. An alternative presentation using tree like structures can be found in [45].

3.6.1 LSTAR

An observation table is a specific tabular representation of an automaton. An example is given in Table 3.1.

The table is interpreted as follows: by concatenating the name of a row r with the name of a column c we obtain a string rc. This string is in the language we are considering if the corresponding cell $\text{OT}[r][c]$ contains a 1 and does not if it is a 0. If the table complies with certain conditions an automaton can be extracted from the table. The corresponding automaton is depicted in Fig. 3.3. A procedure allowing

Table 3.1 The observation table and the corresponding automaton

	λ	a
λ	0	1
a	1	0
b	1	0
aa	0	1
ab	1	0

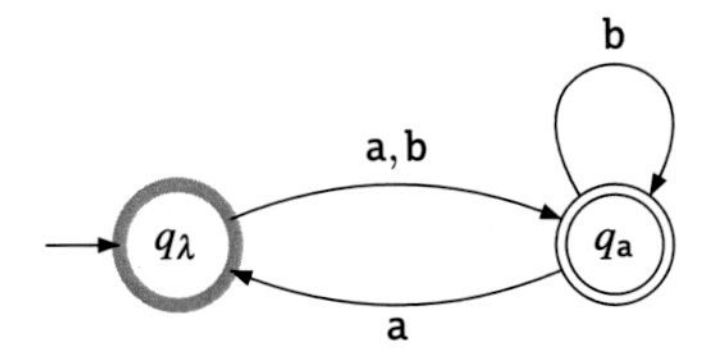

Fig. 3.3 The automaton corresponding to the observation table

to extract a DFA from a table (when possible) is described in [32] (page 272). The idea is to use the names of the rows as states and the columns in order to decide that certain states are equivalent (in the sense of the Nerode equivalence). The rest is simple. Obviously the difficulty resides in the fact that a table can't always be built

Formally, an observation table is a triple ⟨STA, e, OT⟩ with:

- STA=RED ∪ BLUE is a set of strings, denoting labels of states;
- RED $\subset \Sigma^\star$ is a finite set of states;
- BLUE = RED $\cdot\, \Sigma \setminus$ RED is the set of states successors of RED that are not RED;
- EXP $\subset \Sigma^\star$ is the experiment set;
- OT: STA ×EXP → {0, 1, ∗} is a function such that:
- $\text{OT}[u][e] = \begin{cases} 1 \text{ if } ue \in L \\ 0 \text{ if } ue \notin L \\ * \text{ otherwise (not known).} \end{cases}$

Definition 3.7 (*Automaton consistent with a table*) Given an automaton $\mathscr{A}$ and an observation table ⟨STA, EXP, OT⟩, $\mathscr{A}$ is *consistent* with ⟨STA, EXP, OT⟩ when the following holds:

- $\text{OT}[u][e] = 1 \implies ue \in \mathbb{L}(\mathscr{A})$;
- $\text{OT}[u][e] = 0 \implies ue \in \mathbb{L}(\mathscr{A})$.

Building an automaton from a table ⟨STA, EXP, OT⟩ can be done if certain conditions are met:

- the set of strings marking the states in STA must be prefix-closed;
- the set EXP is suffix-closed;
- the table must be complete and therefore have no holes;
- the table must be closed and consistent.

Definition 3.8 (*Holes*) A *hole* in a table ⟨STA, EXP, OT⟩ is a pair (u, e) with $u \in$ STA, $e \in$ EXP such that $\text{OT}[u][e] = *$.

A table is *complete* if it has no holes.

Definition 3.9 (*Closed table*) Two prefixes u and v are *equivalent* if $\mathrm{OT}[u] = \mathrm{OT}[v]$. We will denote this by $u \equiv_{\mathrm{EXP}} v$.
A table ⟨STA, EXP, OT⟩) is *closed* if given any row u of BLUE there is some row v in RED such that $u \equiv_{\mathrm{EXP}} v$.

Checking if the table is closed is straightforward. But what can the algorithm do once it has found that the table is not closed? Let s be the row (of BLUE) that does not appear in RED, we can add s to RED, and $\forall a \in \Sigma$, add sa to BLUE.

By repeating this until the table is closed, we are done. Notice that the number of iterations is bounded by the size of the automaton. This is the goal of procedure LSTAR-CLOSE, not described here.

Definition 3.10 (*Consistent table*) A table is *consistent* if every equivalent pair of prefixes in RED remains equivalent in STA after appending any symbol.
$\mathrm{OT}[s_1] = \mathrm{OT}[s_2] \implies \forall a \in \Sigma, \mathrm{OT}[s_1 a] = \mathrm{OT}[s_2 a]$.

What do we do when we have an inconsistent table? If it is inconsistent, then let $a \in \Sigma$ be the symbol for which $\mathrm{OT}[s_1] = \mathrm{OT}[s_2]$ but $\mathrm{OT}[s_1 a] \neq \mathrm{OT}[s_2 a]$. Let e be the experiment for which the inconsistency has been found ($\mathrm{OT}[s_1 a][e] \neq \mathrm{OT}[s_2 a][e]$). Then by adding experiment ae to the table, rows $\mathrm{OT}[s_1]$ and $\mathrm{OT}[s_2]$ become different. Indeed, $\mathrm{OT}[s_1][ae] \neq \mathrm{OT}[s_2][ae]$. This is the goal of procedure LSTAR-CONSISTENT, not described here.

Once the learner has built a complete, closed and consistent table, it can construct the DFA and this can be used to make an equivalence query!

Obviously, if the Oracle returns a positive answer to the algorithm's equivalence query, it can halt. If she returns a counterexample (u), then the learner should add as RED states all the prefixes of u, and complete the BLUE section accordingly (with all strings pa ($a \in \Sigma$) such that p is a prefix of u but pa is not. In this way at least one new RED line *obviously different* from all the others will have been added. This is the goal of procedure LSTAR-USEEQ, not described here.

Algorithme 1 : LSTAR Learning Algorithm.

Input : –
Output : DFA $\mathscr{A}$
1 LSTAR-INITIALISE;
2 **repeat**
3   **while** *⟨STA, EXP, OT⟩ is not closed or not consistent* **do**
4   **if** *⟨STA, EXP, OT⟩ is not closed* **then**
        ⟨STA, EXP, OT⟩ ←LSTAR-CLOSE(⟨STA, EXP, OT⟩);
5   **if** *⟨STA, EXP, OT⟩ is not consistent* **then**
        ⟨STA, EXP, OT⟩ ←LSTAR-CONSISTENT(⟨STA, EXP, OT⟩)
6   Answer← EQ(⟨STA, EXP, OT⟩);
7   **if** Answer≠ *TRUE* **then** ⟨STA, EXP, OT⟩ ←LSTAR-USEEQ(⟨STA, EXP, OT⟩, Answer)
8 **until** Answer= *TRUE*;
9 **return** LSTAR-BUILDAUTOMATON(⟨STA, EXP, OT⟩)

Theorem 3.2 *Algorithm LSTAR (1) polynomially identifies DFA from membership and equivalence queries.*

In the above algorithm, subroutines LSTAR-INITIALISE, LSTAR-CLOSE, LSTAR-CONSISTENT, LSTAR-USEEQ and LSTAR-BUILDAUTOMATON are used but not defined. These can be found in [32] or reconstructed from the previous definitions and discussion.

Extensions or alternative presentations of LSTAR can be found in [10, 45] and there has been further theoretical work aimed at counting the number of queries really necessary [11, 19], on identifying the power of the equivalence queries [36], or relating the query model to other ones [24, 33]. Several open problems related to learning grammars and automata in this setting have been proposed [31].

3.6.2 Trading Off Equivalence Queries

We can trade-off the equivalence queries used by LSTAR for sampling queries in exchange for a small mistake. In that case, we will also have to be able to sample.

It can be noticed that equivalence queries are used in the algorithm LSTAR in order to check at each moment where the algorithm has come up with a closed and consistent solution whether this solution is the correct one. An attractive alternative is to use sampling in order to have some sort of a statistical equivalence query. The algorithm would work as follows: at each moment an equivalence query is required, draw instead m labelled examples $x_1, \ldots, x_m$ and check if the current hypothesis labels them in the same way as the target does (the true labelling).

If $\forall i \in [m]\ x_i \in \mathbb{L}(G) \iff x_i \in \mathbb{L}(G_T)$ then the error is most likely small... But just how small? We want this error to be at most ε. Let us suppose for contradiction that the true error is more than ε. Then the probability of selecting randomly one example where G and G_T coincide is less than $1 - \varepsilon$ and the probability of selecting randomly m examples where G and G_T coincide (all the time) is now less than $(1 - \varepsilon)^m$.

But now we have the following bound: $(1 - \varepsilon)^m \leq e^{-\varepsilon m}$. So by bounding this value by δ we have $m \geq \frac{1}{\varepsilon}\ln(\frac{1}{\delta})$.

Therefore, by sampling at least $\frac{1}{\varepsilon}\ln(\frac{1}{\delta})$ labelled examples and testing them against the current hypothesis, the algorithm can make a *stochastic* equivalence query.

But the number of equivalence queries that are needed is unknown in practice. And since each time an equivalence query is made, a number of calls to EX(), dependent on the δ parameter, has to be made. In order to make sure the total confidence is at least $1 - \delta$, the size of each sample (or of the calls to EX()) should be at least $m_i = \frac{1}{\varepsilon}\big(\ln(\frac{1}{\delta}) + i\ln 2\big)$. By doing this, one ensures that the (global) confidence parameter δ is at most $\Sigma_{i>0}(1 - \varepsilon)^{m_i}$.

3.6.3 *Adapting to Correction Queries*

It can be shown that correction queries alone cannot compensate for the absence of both membership and equivalence queries: consider a random string $w \in \Sigma^n$ and language $L_{\overline{w}} = \Sigma^n \setminus \{w\}$; it is easy to see that the only way to have useful information about w is by querying w itself. But they can be used instead of membership queries to advantage. In many cases the extra information one obtains allows us to make fewer queries [14, 56].

3.6.4 *Some Implementation Issues*

One difficulty with the implementation of LSTAR comes from maintaining the redundancy in the table. Actually it is not necessary to implement the actual table. A better idea is to manage three association tables and use hash functions:

- A first table MQ contains the result of the membership queries. It can be consulted in near constant time to know if a particular string has been queried or not and, if it has, if it belongs or not to the language.
- A second table PREF contains the different names of rows, and for each row, the status: is it RED or BLUE?
- A third table just contains the different experiments that have been made.

The actual observation table is only simulated by a function $\mathrm{OT}(u, v)$ which will return the value $\mathrm{MQ}[uv]$. The other tables are used to avoid unnecessary searches.

3.7 Extensions

3.7.1 *Post Zulu Extensions*

Competition Zulu [28] was launched in 2009 in order to learn more about active learning. The task proposed to the competitors consisted of learning DFA from membership queries only.

The server [1] could be accessed by any user/learner. The server acts as an Oracle for membership queries. A player can log in and ask for a target DFA. The server then computes how many queries it needs to learn a reasonable machine (reasonable means that it makes less than 30 % classification errors), and invites the player to interact in a learning session in which he can ask up to that number of queries. At the end of the learning process the server gives the learner a set of unlabelled strings (a test set). The labels the learner submits are used to compute his score.

[1] http://labh-curien.univ-st-etienne.fr/zulu.

As a starting point, the baseline algorithm, which is a simple variation of LSTAR, with some sampling done to simulate equivalence queries, is given to the user, who can therefore play with some simple JAVA (i.e., he doesn't have to develop from scratch).

A number of new ideas emerged from Zulu[2]: some authors attempted to search for the most interesting strings to be provided as counter-examples; others attempted to avoid entirely filling the table. A third line of ideas consisted in using the counterexamples differently, by adding extra columns instead of extra rows.

Even if Zulu still leaves a number of routes open for further improvement, it is clear that the ideas proposed allowed a significant gain: with less (queries), achieve more (larger automata, with larger alphabets).

3.7.2 Co-learning

When there are multiple active learners involved, one should consider cooperation questions: one learner learns from the other. The questions themselves can be used as information. This problem was formally proposed in [31].

Consider the situation where two adversaries have to negotiate something. The goal of each is to learn the model of the opponent while giving away as little information as possible. The situation can be modelled as follows:

Let L_1 be the language of player 1 and L_2 be the language of player 2. We suppose here that the languages are regular and can be represented by deterministic finite automata with respectively n_1 and n_2 states. The goal for each is to learn the common language, i.e., language $L_1 \cap L_2$.

The rule is that each player can only query the opponent by asking questions using his own language. This means that when player 1 names string w, $w \in L_1$. In turn, player 2 will answer whether, or not, string w belongs to the language L_2.

Using this setting as a starting point, one should study the possible strategies and outcomes of the different types of players.

This hypothetical game can be illustrated by situations described in control and robotics [25].

3.7.3 Learning Other Types of Grammars

In this survey, we have essentially presented work on learning DFA. But other results are known:

[2]The different papers presented at the workshop can be found at http://users.dsic.upv.es/workshops/icgi2010/.

- Algorithm LSTAR has been adapted to learn probabilistic automata [34, 39], multiplicity automata [17], NFA, tree automata, transducers [58] with, in each case, subtle variations.
- Sakakibara [53] learns context-free grammars from queries; more recently, Clark [26] presents a new learning algorithm in which queries can be contextual queries, in which two substrings can be compared and the Oracle is interrogated as to whether these substrings are interchangeable, i.e., can be used in an identical context.
- Vilar extends queries to translation tasks in [58]; Akram and de la Higuera learn probabilistic transducers [2] using probabilistic queries. In this case, the idea is that these queries (what is the global probability of u as a prefix?) can be later replaced by samples.
- In testing and model checking, queries are used to learn a variety of finite state machines [1, 42, 55].
- Yokomori [62] learns 2-tape automata from both queries and counterexamples, and in [61] non-deterministic finite automata from queries also in polynomial time, but depending on the size of the associated DFA.
- Maler and Pnueli [48] learn Büchi automata from queries over infinite strings.
- In the case of balls, it can be proved [13] that (1) $\mathcal{BALL}(\Sigma)$ is not polynomially identifiable from MQ and EQ, and (2) $\mathcal{BALL}(\Sigma)$ is polynomially identifiable from CQ. The technical proofs will be related to the fact that finding the smallest ball containing a given set of strings is an intractable problem. Note however that if the learner is given one string from a ball, then he can learn using a polynomial number of MQ only.

3.7.4 More About Querying

The relationships between PAC learning, equivalence queries and active learning have been studied in [9, 36]. Balcázar *et al.* have studied the *query complexities* for different combinations of queries [10, 11].

In order to learn probabilistic automata and grammars, the queries should somehow give the learner some indication of the distributions. Between the special sorts of queries for this, extended membership queries were introduced in [17] and extended prefix language queries in [34].

Statistical queries, as introduced by [43], allow us to deal with noise or with imperfect sampling, and are an important tool to prove PAC results for probabilistic automata and grammars.

3.8 Conclusions

Query learning is currently becoming an exciting field in which there are many open challenges. Some are theoretical and may concern the actual paradigm: how does it fit in and compare with other learning paradigms?

Other questions correspond to obtaining better definitions: the way the polynomial bounds are defined is still a matter of discussion.

Obviously new algorithms are needed, either for the existing tasks (and types of queries) or for alternative and new tasks.

Finally, the number of situations where the world can be represented or modelled as a finite state machine is increasing; in many of these situations, some form or another of interaction with the environment is possible. This leads to being able to consider the question of building the model through this interaction, which can be viewed as a query learning problem.

Acknowledgments The questions raised by the helpful and careful reviewer have been of great help when preparing the final version of this chapter.

References

1. F. Aarts, H. Kuppens, J. Tretmans, F. Vaandrager, and S. Verwer. Learning and testing the bounded retransmission protocol. In Heinz et al. [41], pages 4–18.
2. H. I. Akram, C. de la Higuera, and Claudia Eckert. Actively learning probabilistic subsequential transducers. In Heinz et al. [41], pages 19–33.
3. R. Alquézar and A. Sanfeliu. A hybrid connectionist-symbolic approach to regular grammatical inference based on neural learning and hierarchical clustering. In R. C. Carrasco and J. Oncina, editors, *Grammatical Inference and Applications, Proceedings of* ICGI *'94*, number 862 in LNAI, pages 203–211. Springer-Verlag, 1994.
4. D. Angluin. A note on the number of queries needed to identify regular languages. *Information and Control*, 51:76–87, 1981.
5. D. Angluin. Queries and concept learning. *Machine Learning Journal*, 2:319–342, 1987.
6. D. Angluin. Learning regular sets from queries and counterexamples. *Information and Control*, 39:337–350, 1988.
7. D. Angluin. Negative results for equivalence queries. *Machine Learning Journal*, 5:121–150, 1990.
8. D. Angluin. Queries revisited. *Theoretical Computer Science*, 313(2):175–194, 2004.
9. D. Angluin and M. Kharitonov. When won't membership queries help? In *Proceedings of 24th* ACM *Symposium on Theory of Computing*, pages 444–454, New York, 1991. ACM Press.
10. J. L. Balcázar, J. Diaz, R. Gavaldà, and O. Watanabe. An optimal parallel algorithm for learning DFA. In *Proceedings of the 7th* COLT, pages 208–217, New York, 1994. ACM Press.
11. J. L. Balcázar, J. Diaz, R. Gavaldà, and O. Watanabe. The query complexity of learning DFA. *New Generation Computing*, 12:337–358, 1994.
12. L. Beccera-Bonache, C. Bibire, and A. Horia Dediu. Learning DFA from corrections. In Henning Fernau, editor, *Proceedings of the Workshop on Theoretical Aspects of Grammar Induction (*TAGI*)*, WSI-2005-14, pages 1–11. Technical Report, University of Tübingen, 2005.
13. L. Becerra-Bonache, C. de la Higuera, J. C. Janodet, and F. Tantini. Learning balls of strings from edit corrections. *Journal of Machine Learning Research*, 9:1841–1870, 2008.

14. L. Becerra-Bonache, A. Horia Dediu, and C. Tîrnauca. Learning DFA from correction and equivalence queries. In Sakakibara et al. [54], pages 281–292.
15. L. Becerra-Bonache and T. Yokomori. Learning mild context-sensitiveness: Toward understanding children's language learning. In Paliouras and Sakakibara [49], pages 53–64.
16. T. Berg, O. Grinchtein, B. Jonsson, M. Leucker, H. Raffelt, and B. Steffen. On the correspondence between conformance testing and regular inference. In *Proceedings of Fundamental Approaches to Software Engineering, 8th International Conference, FASE 2005*, volume 3442 of LNCS, pages 175–189. Springer-Verlag, 2005.
17. F. Bergadano and S. Varricchio. Learning behaviors of automata from multiplicity and equivalence queries. SIAM *Journal of Computing*, 25(6):1268–1280, 1996.
18. L. Bréhélin, O. Gascuel, and G. Caraux. Hidden Markov models with patterns to learn boolean vector sequences and application to the built-in self-test for integrated circuits. *Pattern Analysis and Machine Intelligence*, 23(9):997–1008, 2001.
19. N. H. Bshouty, R. Cleve, R. Gavaldà, S. Kannan, and C. Tamon. Oracles and queries that are sufficient for exact learning. *Journal of Computer and System Sciences*, 52:421–433, 1996.
20. J. Carme, R. Gilleron, A. Lemay, and J. Niehren. Interactive learning of node selecting tree transducer. In IJCAI *Workshop on Grammatical Inference*, 2005.
21. J. Carme, R. Gilleron, A. Lemay, and J. Niehren. Interactive learning of node selecting tree transducer. *Machine Learning Journal*, 66(1):33–67, 2007.
22. D. Carmel and S. Markovitch. Model-based learning of interaction strategies in multi-agent systems. *Journal of Experimental and Theoretical Artificial Intelligence*, 10(3):309–332, 1998.
23. D. Carmel and S. Markovitch. Exploration strategies for model-based learning in multiagent systems. *Autonomous Agents and Multi-agent Systems*, 2(2):141–172, 1999.
24. J. Castro and D. Guijarro. PACS, simple-PAC and query learning. *Information Processing Letters*, 73(1–2):11–16, 2000.
25. J. Chandlee, J. Fu, K. Karydis, Cesar Koirala, J. Heinz, and H. G. Tanner. Integrating grammatical inference into robotic planning. In Heinz et al. [41], pages 69–83.
26. A. Clark. Distributional learning of some context-free languages with a minimally adequate teacher. In J. Sempere and P. García, editors, *Grammatical Inference: Theoretical Results and Applications, Proceedings of* ICGI *'10*, volume 6339 of LNCS, pages 24–37. Springer-Verlag, 2010.
27. A. Clark, F. Coste, and L. Miclet, editors. *Grammatical Inference: Algorithms and Applications, Proceedings of* ICGI *'08*, volume 5278 of LNCS. Springer-Verlag, 2008.
28. D. Combe, C. de la Higuera, and J.-C. Janodet. Zulu: An interactive learning competition. In *Proceedings of* FSMNLP *'09*, volume 6062 of LNCS, pages 139–146. Springer-Verlag, 2009.
29. C. de la Higuera. Characteristic sets for polynomial grammatical inference. *Machine Learning Journal*, 27:125–138, 1997.
30. C. de la Higuera. Data complexity issues in grammatical inference. In M. Basu and T. Kam Ho, editors, *Data Complexity in Pattern Recognition*, pages 153–172. Springer-Verlag, 2006.
31. C. de la Higuera. Ten open problems in grammatical inference. In Sakakibara et al. [54], pages 32–44.
32. C. de la Higuera. *Grammatical inference: learning automata and grammars*. Cambridge University Press, 2010.
33. C. de la Higuera, J.-C. Janodet, and F. Tantini. Learning languages from bounded resources: the case of the DFA and the balls of strings. In Clark et al. [54], pages 43–56.
34. C. de la Higuera and J. Oncina. Learning probabilistic finite automata. In Paliouras and Sakakibara [49], pages 175–186.
35. T. Dean, K. Basye, L. Kaelbling, E. Kokkevis, O. Maron, D. Angluin, and S. Engelson. Inferring finite automata with stochastic output functions and an application to map learning. In W. Swartout, editor, *Proceedings of the 10th National Conference on Artificial Intelligence*, pages 208–214, San Jose, CA, 1992. MIT Press.
36. R. Gavaldà. On the power of equivalence queries. In *Proceedings of the 1st European Conference on Computational Learning Theory*, volume 53 of *The Institute of Mathematics and its Applications Conference Series*, pages 193–203. Oxford University Press, 1993.

37. C. L. Giles, S. Lawrence, and A.C. Tsoi. Noisy time series prediction using recurrent neural networks and grammatical inference. *Machine Learning*, 44(1):161–183, 2001.
38. E. M. Gold. Language identification in the limit. *Information and Control*, 10(5):447–474, 1967.
39. O. Guttman, S. V. N. Vishwanathan, and R. C. Williamson. Learnability of probabilistic automata via oracles. In S. Jain, H.-U. Simon, and E. Tomita, editors, *Proceedings of* ALT *2005*, volume 3734 of LNCS, pages 171–182. Springer-Verlag, 2005.
40. A. Hagerer, H. Hungar, O. Niese, and B. Steffen. Model generation by moderated regular extrapolation. In R. Kutsche and H. Weber, editors, *Proceedings of the 5th International Conference on Fundamental Approaches to Software Engineering (*FASE *'02)*, volume 2306 of LNCS, pages 80–95, Heidelberg, Germany, 2002. Springer-Verlag.
41. J. Heinz, C. de la Higuera, and T. Oates, editors. *Grammatical Inference: Theoretical Results and Applications, 11th International Conference, ICGI 2012, University of Maryland, College Park, United States. Proceedings*, volume 21. JMLR.org, 2012.
42. F. Howar, B. Steffen, and M. Merten. From ZULU to RERS—lessons learned in the zulu challenge. In *4th International Symposium on Leveraging Applications, ISoLA 2010*, volume 6415 of LNCS, pages 687–704. Springer-Verlag, 2010.
43. M. Kearns. Efficient noise-tolerant learning from statistical queries. In *Proceedings of the Twenty-Fifth Annual* ACM *Symposium on Theory of Computing*, pages 392–401, 1993.
44. M. Kearns and L. Valiant. Cryptographic limitations on learning boolean formulae and finite automata. In *21st* ACM *Symposium on Theory of Computing*, pages 433–444, 1989.
45. M. J. Kearns and U. Vazirani. *An Introduction to Computational Learning Theory*. MIT Press, 1994.
46. E. B. Kinber. On learning regular expressions and patterns via membership and correction queries. In Clark et al. [27], pages 125–138.
47. V. I. Levenshtein. Binary codes capable of correcting deletions, insertions, and reversals. *Doklady Akademii Nauk SSSR*, 163(4):845–848, 1965.
48. O. Maler and A. Pnueli. On the learnability of infinitary regular sets. In *Proceedings of* COLT, pages 128–136, San Mateo, 1991. Morgan–Kaufmann.
49. G. Paliouras and Y. Sakakibara, editors. *Grammatical Inference: Algorithms and Applications, Proceedings of* ICGI *'04*, volume 3264 of LNAI. Springer-Verlag, 2004.
50. L. Pitt. Inductive inference, DFA's, and computational complexity. In *Analogical and Inductive Inference*, number 397 in LNAI, pages 18–44. Springer-Verlag, 1989.
51. H. Raffelt and B. Steffen. LearnLib: A library for automata learning and experimentation. In *Proceedings of* FASE *2006*, volume 3922 of LNCS, pages 377–380. Springer-Verlag, 2006.
52. R. L. Rivest and R. E. Schapire. Inference of finite automata using homing sequences. *Information and Computation*, 103:299–347, 1993.
53. Y. Sakakibara. Inferring parsers of context-free languages from structural examples. Technical Report 81, Fujitsu Limited, International Institute for Advanced Study of Social Information Science, Numazu, Japan, 1987.
54. Y. Sakakibara, S. Kobayashi, K. Sato, T. Nishino, and E. Tomita, editors. *Grammatical Inference: Algorithms and Applications, Proceedings of* ICGI *'06*, volume 4201 of LNAI. Springer-Verlag, 2006.
55. B. Steffen, F. Howar, and M. Merten. Introduction to active automata learning from a practical perspective. In *SFM 2011. Advanced Lectures*, volume 6659 of LNCS, pages 256–296. Springer-Verlag, 2011.
56. C. Tirnauca. A note on the relationship between different types of correction queries. In Clark et al. [27], pages 213–223.
57. L. G. Valiant. A theory of the learnable. *Communications of the Association for Computing Machinery*, 27(11):1134–1142, 1984.
58. J. M. Vilar. Query learning of subsequential transducers. In L. Miclet and C. de la Higuera, editors, *Proceedings of* ICGI *'96*, number 1147 in LNAI, pages 72–83. Springer-Verlag, 1996.
59. R. Wagner and M. Fisher. The string-to-string correction problem. *Journal of the* ACM, 21:168–178, 1974.

60. M. Warmuth. Towards representation independence in PAC-learning. In K. P. Jantke, editor, *Proceedings of* AII *'89*, volume 397 of LNAI, pages 78–103. Springer-Verlag, 1989.
61. T. Yokomori. Learning non-deterministic finite automata from queries and counterexamples. *Machine Intelligence*, 13:169–189, 1994.
62. T. Yokomori. Learning two-tape automata from queries and counterexamples. *Mathematical Systems Theory*, pages 259–270, 1996.

Chapter 4
On the Inference of Finite State Automata from Positive and Negative Data

Damián López and Pedro García

Abstract The works by Thrakhtenbrot–Barzdin and Gold can be considered to be the first works on the identification of Finite Automata from given data. The main drawback of their results is that they may obtain hypotheses that may be inconsistent with the provided data. This drawback was solved by the *RPNI* and Lang algorithms. Aside from these works, other works have introduced more efficient algorithms with respect to the training data. The direct consequence of this improvement has lead to algorithms that have lower error rates. Recently, some works have tackled the identification of NFAs instead of using the traditional DFA model. In this line of research, the inference of Residual Finite State Automata (RFSA) provides a canonical non-deterministic model. Other works consider the inference of teams of NFAs to be a method that is suitable to solve the grammatical inference of finite automata. We review the main approaches that solve the inference of finite automata by using positive and negative data from the target language. In this review, we will describe the above-mentioned formalisms and induction techniques.

4.1 Introduction

The problem of language identification from positive and negative information was proposed in the 1970s. This task is related to the search for a minimal DFA that is compatible with a given positive and negative set of data. Automata synthesis from data is, in fact, a classic problem of Automata Theory. Nowadays this problem is incorporated within the Grammatical Inference (GI) topics.

One common approach to GI is based on the merging of states of an initial representation that strictly recognizes the training set. In this approach, two states are merged when there is no evidence that the associated language is different, and,

D. López (✉) · P. García
Departamento de Sistemas Informáticos y Computación,
Universidad Politécnica de Valencia, Valencia, Spain
e-mail: dlopez@dsic.upv.es

P. García
e-mail: pgarcia@dsic.upv.es

J. Heinz and J.M. Sempere (eds.), *Topics in Grammatical Inference*,
DOI 10.1007/978-3-662-48395-4_4

therefore, one of them can be considered to be redundant. The above-mentioned evidence comes from the existence or absence of a positive or negative sample in the training set. This means that mergible states in a given situation might not be merged with additional extra information. Such *inconvenient* merges usually imply a larger size of the automata output and a lower recognition rate.

The work of Gold in the 1960s and 1970s led to several relevant contributions in the field of regular GI. First, he proved that the problem of finding a minimal deterministic finite automaton (DFA) that is consistent with some input positive (D_+) and negative (D_-) data is NP-hard [32]. He also proposed the *identification in the limit* model [31], which is a valuable tool for proving the correction of new GI algorithms. Furthermore, Gold proposed a GI algorithm [32] that outputs the minimal deterministic automaton for the language in polynomial time when a representative enough set of data is provided. This algorithm is of great interest because several subsequent algorithms can be viewed as modifications of it.

In the same decades, Trakhtenbrot and Barzdin [44] proposed a DFA synthesis algorithm that considers all the strings of the language whose length is bounded by a given integer. Even though this work does not present a grammatical inference algorithm it is known that the Gold and the Trakhtenbrot–Barzdin algorithms work in a similar way, despite the fact that they use very different representations of the same information [24].

In the 1980s, some new negative results confirmed the theoretical difficulty of the regular language GI problem. Angluin proved that, for a given set of data, the problem of finding the minimal DFA is NP-hard even for target automata of two states. She also proved that the training set must be complete and that whenever a few samples are removed from the training set, the problem also becomes NP-hard [4]. Another result by Angluin is the proof that the task is also hard when the algorithm is allowed to question an oracle about the membership to the target language of certain samples [6] or to question an oracle about the equivalence of the current hypothesis and the target language [8].

Since those negative theoretical results led to obvious pessimism, in order to design new GI algorithms for applied tasks, several approaches were considered. Some works focused on the inference of stochastic DFA (e.g. [9]). Other papers studied active learning, that is, the possibility to question an oracle whenever it is considered necessary [6–8]. There were also works devoted to the characterization of language classes learnable from positive information [5, 25, 30, 48]. Other works studied the inference of transducers [42]. In our review, we do not consider these methods and techniques, which were reviewed recently by de la Higuera [17], instead we focus on those methods that output (deterministic or non-deterministic) automata using positive and negative information from the target language.

In 1992, two works appeared independently of each other. Oncina and García proposed the *RPNI* algorithm [41] and Lang proposed a modification of the Trakhten–Barzdin algorithm known as the Traxbar algorithm [38]. In *RPNI*, the output depends on the *quality* of the training set, and it has been proved that whenever this training set is *characteristic*, the algorithm converges to the minimal DFA for the target language [41].

In the quest to avoid incorrect mergings as much as possible, de la Higuera et al. [18] proposed a modification of the canonical order in order to favor those pairs of states that have high evidence of equivalence. Therefore, they proposed attaching a score to each pair of states, and subsequently merging the pairs of states with the highest score. There are multiple criteria for attaching this score, but, generally speaking, high (low) score values are expected whenever the training set contains great (little) evidence that the languages of both states are the same. Preliminary experimental results were encouraging but not conclusive.

Price revisited this same idea and proposed the EDSM (Evidence Driven State Merging) algorithm in the context of the Abbadingo One contest [37]. In this competition, which was organized by Lang in 1998, several GI algorithms tried to solve a set of regular inference problems. There were two winners: the EDSM algorithm and a parallel algorithm based on non-deterministic search. The fact that the winners were based on a strategy of merging states encouraged the community to again take up the design of identification algorithms. In fact, taking into account the contest results, Lang proposed the Blue-Fringe algorithm [11, 37]. Several other works were also based on the EDSM algorithm [1, 11]. In [29], García et al. revisited these works and the inference of automata teams was proposed.

Eventually, the GI community also studied the inference of non-deterministic finite automata (NFA). In 1994, Yokomori published a work in this field proposing a method that is based on counterexamples and questions to an oracle [16, 47]. Coste and Fredouille also studied this issue [13–15]; they proposed their NFA inference method taking into account results on the inference of unambiguous finite automata.

In this field of NFA inference, Denis et al. [19, 20, 22] have focused their work on the inference of a subset of NFA called RFSA (residual finite state automata). The interest in this subclass of NFA comes from the fact that there is a canonical RFSA for each regular language.

Currently, even though few research groups are working on the inference of NFA, there is some research on the minimalization of NFAs that are attempting to obtain a canonical form [43]; there are works that present algorithms that converge to RFSAs [2]; algorithms that infer UFAs (Unambiguous Finite Automata) [1, 14, 15]; other works that extend the *RPNI* strategy but consider non-deterministic merging [2]; and other works that study the inference of NFAs using subautomata juxtaposition [46]. In [28], a new general method named *OIL* (Order Independent Learning) is proposed. This method constructs the maximal automaton for the training set (instead of the prefix tree acceptor) and then considers any order in the merging of the states.

This chapter is organized as follows. First, in Sect. 4.2 we summarize the basic notation and definitions used. In Sect. 4.3, we review the inference methods that output deterministic automata (beginning with the seminal results of Gold and Trakhtenbrot–Barzdin, and including more recent works that consider heuristics or teams of automata). Section 4.4 is devoted to reviewing those methods that output non-deterministic automata. Section 4.5 experimentially compares the quality of the reviewed methods taking into account the results obtained when a dataset of languages proposed in the literature is used. Some conclusions and suggestions for continuing this study are presented at the end of the chapter.

4.2 Definitions and Notation

Definitions that are not contained in this section can be found in [34]. Definitions and previous works concerning RFSA can be found in [19, 20, 22]. For the reader who is interested in the Universal Automaton, we suggest [39, 43].

Let Σ be a finite alphabet and let Σ^* be the free monoid generated by Σ with concatenation as the internal operation and λ as the neutral element. A *language* L over Σ is a subset of Σ^*. The elements of L are called *words* or *strings*. Given $x \in \Sigma^*$, if $x = uv$ with $u, v \in \Sigma^*$, then u (resp. v) is called the *prefix* (resp. *suffix*) of x. Let us denote the set of prefixes (suffixes) of x by $Pr(x)$ (resp. $Suf(x)$), and the natural extension of the prefixes (suffixes) operation to a language L by $Pr(L)$ (resp. $Suf(L)$). Let us also recall here the definition of the *canonical order* over Σ^* as being the order that first classifies the shorter strings and considers the alphabetic order for those strings of the same length.

A *finite automaton* (*NFA*) is a 5-tuple $A = (Q, \Sigma, \delta, I, F)$, where Q is a finite set of states, Σ is an alphabet, $I \subseteq Q$ is the set of initial states, $F \subseteq Q$ is the set of final states and $\delta : Q \times \Sigma \to 2^Q$ is the transition function, which will also be seen as $\delta \subseteq Q \times \Sigma \times Q$. Given an automaton A and a state $q \in Q$, we denote the right language of q in A as R_q^A, that is, the language accepted by the automaton $A = (Q, \Sigma, \delta, \{q\}, F)$. The language accepted by the automaton A will be denoted as $L(A)$ and can be defined as $L(A) = \bigcup_{q \in I} R_q^A$. Two automata are *equivalent* if they recognize the same language. Given any two states p, q of the automaton, let $\prec$ be the relation defined as $p \prec q$ if and only if $R_p^A \subseteq R_q^A$.

Given any two automata $A_1 = (Q_1, \Sigma_1, \delta_1, I_1, F_1)$ and $A = (Q_2, \Sigma_2, \delta_2, I_2, F_2)$, the *disjoint union of automata* is defined as the automaton $A_1 \uplus A_2 = (Q_1 \cup Q_2, \Sigma_1 \cup \Sigma_2, \delta_1 \cup \delta_2, I_1 \cup I_2, F_1 \cup F_2)$.

Let $S \subset \Sigma^*$ be finite. The *maximal automaton* for S is the NFA $MA(S) = (Q, A, \delta, I, F)$ where: $Q = \bigcup_{x \in S}\{(u, v) \in \Sigma^* \times \Sigma^* \ : \ uv = x\}$, $I = \{(\lambda, x) \ : \ x \in S\}$, $F = \{(x, \lambda) \ : \ x \in S\}$, and where $\delta((u, av), a) = (ua, v)$ for $(u, av) \in Q$. Note that, when so defined, $L(MA(S)) = S$.

A *deterministic* finite automaton (*DFA*) is an automaton such that $card(I) = 1$ and, for every state q and every symbol a, the number of transitions $\delta(q, a)$ is at most one.

Given a language L over Σ, the (left) quotient of L by a string u is defined as the language $u^{-1}L = \{v \in \Sigma^* \ : \ uv \in L\}$. Let us also define the set $S_L = \{u^{-1}L : u \in \Sigma^*\}$ and the set $U_L = \{u_1^{-1}L \cap \cdots \cap u_k^{-1}L \ : \ k \geq 0, \ u_1, \ldots, u_k \in \Sigma^*\}$. Note that whenever the language L is regular, the sets S_L and U_L are finite.

The minimal DFA for a language L is $A = (S_L, \Sigma, \delta, \{q_0\}, F)$, where: $q_0 = \lambda^{-1}L = L$; $F = \{u^{-1}L \ : \ u \in L\}$; and $\delta(u^{-1}L, a) = (ua)^{-1}L$ for any state in S_L and any symbol a of the alphabet. The *universal automaton* [12] for a language L is $\mathscr{U}_L = (U_L, \Sigma, \delta, I, F)$, where $I = \{p \in U_L \ : \ p \subseteq L\}$, $F = \{p \in U_L \ : \ \lambda \in p\}$ and the set of transitions is defined as $\delta(u^{-1}L, a) = \{p \in U_L \ : \ p \subseteq (ua)^{-1}L\}$. We note here that, given a regular language L, for every automaton A that recognizes L, there is a morphism ψ such that $\psi(A)$ is a subautomaton of the universal automaton for L [39, 43].

A Moore machine is a 6-tuple $M = (Q, \Sigma, \Delta, \delta, q_0, \Phi)$, where Σ (resp. Δ) is the input (resp. output) alphabet, δ is a partial function that maps $Q \times \Sigma$ in Q, and Φ is a function that maps Q in Δ called *output function*. The behavior of M is given by the partial function $t_M : \Sigma^* \rightarrow \Delta$ defined as $t_M(x) = \Phi(\delta(q_0, x))$, for every $x \in \Sigma^*$ such that $\delta(q_0, x)$ is defined.

Given two disjoint finite sets of words D_+ and D_-, we define the (D_+, D_-)*-prefix Moore machine* ($PTMM(D_+, D_-)$) as the Moore machine having $\Delta = \{0, 1, ?\}$, $Q = \text{Pr}(D_+ \cup D_-)$, $q_0 = \lambda$, and $\delta(u, a) = ua$ if $u, ua \in Q$ and $a \in \Sigma$. For every state u, the value of the output function associated to u is 1, 0 or ? (undefined) depending on whether u belongs to D_+, to D_-, or to $Q - (D_+ \cup D_-)$.

A Moore machine $M = (Q, A, \{0, 1, ?\}, \delta, q_0, \Phi)$ is *consistent* with (D_+, D_-) if $\forall x \in D_+$ we have $\Phi(x) = 1$ and $\forall x \in D_-$ we have $\Phi(x) = 0$. Note that a *DFA* $A = (Q, \Sigma, \delta, q_0, F)$ can be simulated by a Moore machine $M = (Q, \Sigma, \{0, 1\}, \delta, q_0, \Phi)$, where $\Phi(q) = 1$ if $q \in F$ and $\Phi(q) = 0$ otherwise. Thus, the language defined by M is $L(M) = \{x \in \Sigma^* : \Phi(\delta(q_0, x)) = 1\}$.

A Mealy machine is a 6-tuple $M = (Q, \Sigma, \Delta, \delta, q_0, \Phi)$, where all the elements are defined as they are in the Moore machine except Φ. In this machine, the function Φ maps $Q \times \Sigma$ in Δ. The behavior of M is given by the partial function $t_M : \Sigma^* \rightarrow \Delta$ that, given $x = a_1, a_2, \ldots, a_n \in \Sigma^*$, is defined as $t_M(x) = \Phi(q_0, a_1)\Phi(q_1, a_2) \ldots \Phi(q_{n-1}, a_n)$, where $q_0, q_1, \ldots, q_n$ is such that $\delta(q_{i-1}, a_i) = q_i$ for $1 < i \leq n$.

4.3 Inference of Deterministic Automata

In this section we review the most important contributions to the inference of DFAs. First, we present the seminal works by Gold [32] and Trakhtenbrot–Barzdin [44]. We then analyze the *RPNI* algorithm [41] and the algorithm proposed by Lang in [38]. Both of these are extensions of the works by Gold and Trakhtenbrot–Barzdin. This section also reviews the works that guide the merging of states by heuristics, based on the general scheme of the *RPNI* and Lang algorithms.

4.3.1 Non-merging Algorithms: Trakhtenbrot–Barzdin and Gold

The algorithm we describe below can be ascribed to Trakhtenbrot and Barzdin [44] and also to Gold [32]. The algorithm proposed by Trakhtenbrot and Barzdin was published in a book on automata theory and in a context that is different from the GI context. Trakhtenbrot and Barzdin tackled the automata synthesis task using a uniform and complete set of strings from the language. This set is represented using a finite tree that is complete up to a certain string length, and it contains the information of membership to the language for each string in the set. This set of data concerning positive and negative information is then represented using the *PTMM* that considers the sets D_+ and D_- (which is the input of the algorithm).

In an independent way (and in the context of the GI field), Gold proposes an algorithm that converges to the minimal DFA of a regular language using complete presentation samples. Gold uses an evidence table to represent the available data, and his algorithm outputs a Mealy machine. The variety of formalisms that are used may hide the equivalence of the two algorithms from the reader. Here we use a prepresentation that is close to the Trakhtenbrot–Barzdin representation. Algorithm 4.3.1 shows this GI method, which hereafter we will call *TBG*.

Algorithm 4.3.1 Trakhtenbrot–Barzdin and Gold algorithm

1: **Input:** Two disjoint finite sets (D_+, D_-)
2: **Output:** A consistent Moore Machine
3: **Method**
4: $M_0 = PTMM(D_+, D_-) = (Q_0, \Sigma, \{0, 1, ?\}, \delta, q_0, \Phi_0)$;
5: $R = \{\lambda\}$;
6: **while** $\exists s' \in R\Sigma - R \ : \ \forall s \in R, \ od(s, s', M_0) = True$ **do**
7: choose s';
8: $R = R \cup \{s'\}$;
9: **end while**
10: $Q = R$;
11: $q_0 = \lambda$;
12: **for** $s \in R$ **do**
13: $\Phi(s) = \Phi_0(s)$;
14: **for** $a \in \Sigma$ **do**
15: **if** $sa \in R$ **then** $\delta(s, a) = sa$
16: **else** $\delta(s, a) =$ any $s' \in R$ such that $od(sa, s', M_0) = False$
17: **end if**
18: **end for**
19: **end for**
20: $M = (Q, \Sigma, \{0, 1, ?\}, \delta, q_0, \Phi)$;
21: **if** M is consistent with (D_+, D_-) **then**
22: Return(M)
23: **else**
24: Return(M_0);
25: **end if**
26: **End Method.**

The *TBG* algorithm calls a function to determine whether or not two states are *obviously distinguishable*. Taking M_0 as the initial representation of the training set and two states, s and s', this function is defined as:

$$od(s, s', M_0) = True \Leftrightarrow \exists x \in \Sigma^* \ : \ \begin{cases} \Phi(\delta(s, x)), \Phi(\delta(s', x)) \in \{0, 1\} \\ \Phi(\delta(s, x)) \neq \Phi(\delta(s', x)) \end{cases}$$

It is worth noting here that this algorithm does not guarantee the consistency of the output with respect to the input data. The result of the inference process is first checked for consistency (line 21) and when the obtained machine is not consistent the $PTMM(D_+, D_-)$ is output (line 24). Also note that the algorithm is not deterministic. In line 7, the algorithm chooses any state in $R\Sigma - R$ that is obviously different from the states in R. The usual way to deterministically implement this command is to choose s' as the first state in canonical order. In the same way, line 16 is usually implemented to choose the first indistinguishable state in canonical order to sa as the state reached by s using symbol a. Example 4.1 illustrates the behaviour of Algorithm 4.3.1.

Example 4.1 Let us take the finite sample $D_+ = \{\lambda, 00, 10, 11, 010\}$ and $D_- = \{0, 1, 001\}$. Algorithm 4.3.1 first constructs the Moore machine $PTMM(D_+, D_-)$ (shown in Fig. 4.1) where the positive, negative and undefined states are represented with double circles, single circles, and dashed circles respectively. Note that the numbering of the states is consistent with the canonical order of the strings that reach each state. As mentioned above, we will follow this canonical order in order to select the states to be analyzed.

Initially $R = \{1\}$ and $R\Sigma - R = \{2, 3\}$. The empty string allows us to say that $od(1, 2, M_0) = True$. Therefore, $R = \{1, 2\}$ and $R\Sigma - R = \{3, 4, 5\}$. The string λ also shows that $od(1, 3, M_0) = True$, but $od(2, 3, M_0) = False$. This is followed by an analysis of states 1 and 4, which concludes that $od(1, 4, M_0) = False$. State 5 is then considered and it is detected that $od(1, 5, M_0) = True$ (due to the string λ), but that $od(2, 5, M_0) = False$. The output of the algorithm is shown in Fig. 4.2. Note that the output automaton is non-consistent with the input data.

Note that a slight modification of the training data is enough for the algorithm to obtain consistency. Let us consider a new training set where $D_+ = \{\lambda, 00, 10, 11, 010\}$ and $D_- = \{0, 1, 01, 001\}$. Note that the difference lies in string 01, which is now in the negative sample.

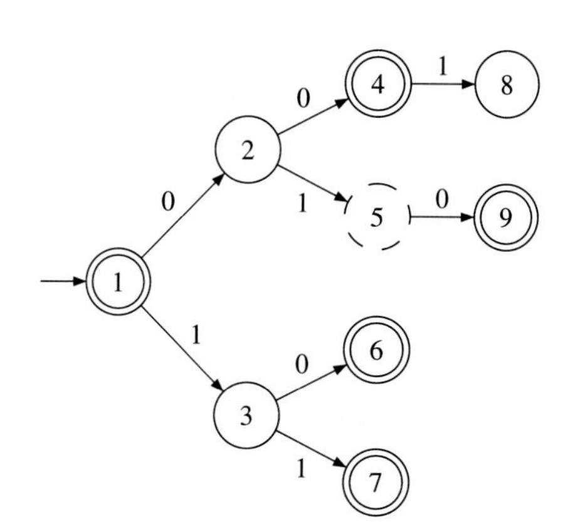

Fig. 4.1 PTMM example when the sets $D_+ = \{\lambda, 00, 10, 11, 010\}$ and $D_- = \{0, 1, 001\}$ are considered

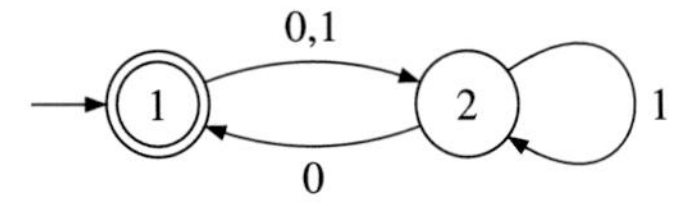

Fig. 4.2 The DFA obtained by the *TBG* algorithm using $D_+ = \{\lambda, 00, 10, 11, 010\}$ and $D_- = \{0, 1, 001\}$. Note that string 11 should be accepted and it is not

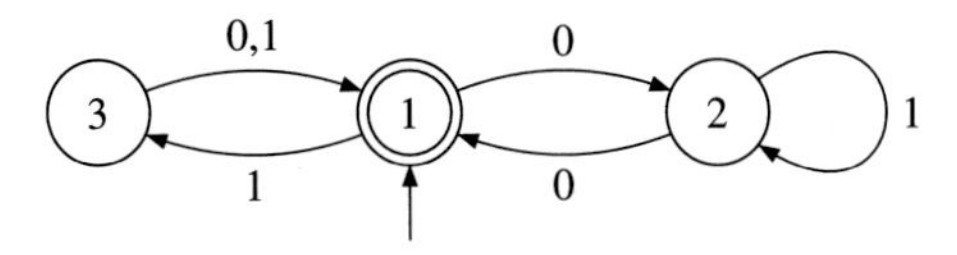

Fig. 4.3 The DFA obtained by the *TBG* algorithm using $D_+ = \{\lambda, 00, 10, 11, 010\}$ and $D_- = \{0, 1, 01, 001\}$ as the training set

In this case, when the finite sample $D_+ = \{\lambda, 00, 10, 11, 010\}$ and $D_- = \{0, 1, 01, 001\}$ is taken into account, Algorithm 4.3.1 first constructs the Moore machine $PTMM(D_+, D_-)$, which is essentially the same as in Fig. 4.1 but where state 5 is a negative state. Initially $R = \{1\}$ and $R\Sigma - R = \{2, 3\}$. The empty string allows us to say that $od(1, 2, M_0) = True$. Therefore, $R = \{1, 2\}$ and $R\Sigma - R = \{3, 4, 5\}$. The string λ also leads to detecting that $od(1, 3, M_0) = True$ and string 1 leads to detecting that $od(2, 3, M_0) = True$. Thus, $R = \{1, 2, 3\}$ and $R\Sigma - R = \{4, 5, 6, 7\}$. This is followed by the analysis of states 1 and 4, and $od(1, 4, M_0) = False$. State 5 is then considered and it is detected that $od(1, 5, M_0) = True$ (due to λ), but $od(2, 5, M_0) = False$. Finally, states 6 and 7 are indistinguishable to state 1, that is, $od(1, 6, M_0) = False$ and $od(1, 7, M_0) = False$. The output of the algorithm is shown in Fig. 4.3. Note that the output automaton is consistent with the input data.

The reason for the inconsistency of the *TBG* algorithm is due to the fact that it carries out all the comparison among states on M_0. The initial automaton (the *PTMM* of the input data) is not modified as a result of these comparisons. In other words, the algorithm does not merge states while it checks consistency. Therefore, for any state whose output is not determined by the training set, it is possible to relate different outputs.

4.3.2 *Merging Algorithms:* RPNI *and Lang*

As already mentioned, the main drawback of the *TBG* algorithm is that consistence with respect to the input data is not guaranteed. In these cases, the algorithm gives up any generalization and outputs a strict representation of the input data.

In the early 1990s, two algorithms that were proposed independently but were essentially identical solved this problem. These algorithms are: the *RPNI* algorithm (Regular Positive and Negative Inference) [41] and the *Traxbar* algorithm [38]. The main difference of these methods with respect to Algorithm 4.3.1 consists in the

Algorithm 4.3.2 Function for the deterministic merge of states.

1: **Input:** A Moore Machine $M = (Q_M, \Sigma, \{0, 1, ?\}, \delta_M, q_0, \Phi_M)$
2: **Input:** Two states p and q
3: **Output:** A Moore Machine with p and q deterministically merged (if possible)
4: **Output:** False if the merge of states is not possible
5: **Method**
6: **if** $\{\Phi_M(p), \Phi_M(q)\} = \{0, 1\}$ **then** Return(*False*);
   end if
7: $M' = M$;
8: **if** $\Phi_{M'}(p) =?$ **then** $\Phi_{M'}(p) = \Phi_M(q)$;
   end if
9: Substitute with p any reference to q in M'
10: **for all** $a \in \Sigma$ **do**
11:   **if** both $\delta_M(p, a)$ and $\delta_M(q, a)$ are defined **then**
12:     $M' = detmerge(M', \delta_M(p, a), \delta_M(q, a)$
13:     **if** $M' = False$ **then**
14:       Return(*False*);
15:     **end if**
16:   **end if**
17: **end for**
18: Return(M')
19: **End Method.**

merging of those indistinguishable states that are detected. Once one of these merges has been carried out, the algorithm considers this new hypothesis and rejects the previous one. As mentioned above, the *TBG* algorithm always compares each pair of states taking into account the initial *PTMM*.

The *RPNI* algorithm uses a function to merge states while assuring that the automaton retains determinism. This operation is carried out by the *detmerge* function that recursively triggers another merge of states whenever the merge of two states implies non-determinism. Note that a given merge may not be possible because it would imply the merge of a positive state and a negative state. In that case, the function returns *False*.

Algorithms 4.3.3 and 4.3.2 show this method, which can be summarized as follows: Let the states of $PTMM(D_+, D_-)$ be ordered canonically; the deterministic merge of each state (based on this order) with all the previous states is checked. In order to do so, the outputs associated to these states must be compatible (the outputs cannot be simultaneously in $\{0, 1\}$ and also to be different); whenever the outputs are compatible, the states can be merged. Nevertheless, the merge can lead to non-determinism and the algorithm tries new merges to handle this situation. These new merges can lead to some incompatibility, in which case the algorithm returns to the previous situation that triggered the process; the algorithm ends when there is no state to be considered.

The first classification of GI algorithms that is possible to make takes into account whether or not the algorithms modify the initial *PTMM* by merging states each time two *compatible* states are found. In order to compare the behaviour of the two approaches, Example 4.2 depicts a run of the *RPNI* algorithm.

Algorithm 4.3.3 *RPNI* algorithm

1: **Input:** Two disjoint finite sets (D_+, D_-)
2: **Output:** A consistent Moore Machine
3: **Method**
4: $M = PTMM(D_+, D_-)$;
5: //$\{u_0, u_1, ..., u_r\}$ states of M in canonical order, $u_0 = \lambda$//;
6: $R = \{u_0\}$;
7: $B = R\Sigma - R$;
8: **while** B not empty **do**
9: $q =$ canonical order first state in B;
10: $B = B - \{q\}$
11: *merged* = *False*;
12: **for all** p in R traversed in canonical order **do**
13: **if** $detmerge(M, p, q) \neq False$ **then**
14: *merged* = *True*;
15: $M = detmerge(M, p, q)$;
16: BreakFor
17: **end if**
18: **end for**
19: **if** *not merged* **then**
20: $R = Append(R, q)$;
21: **end if**
22: $B = R\Sigma - R$;
23: **end while**
24: Return(M)
25: **End Method.**

Example 4.2 Let us consider the following sets of samples:

$$D_+ = \{a, aba, abba, abbba\}$$
$$D_- = \{\lambda, b, aa, ab, ba, bb, aaa, abb, baa, bba\}$$

A run of the *RPNI* algorithm first constructs the *PTMM* for D_+ and D_-. Figure 4.4 shows the resulting *PTMM*.

When the first merging $detmerge(M, 1, 2)$ is tried, it fails due to the fact that $\Phi(1) = 0$ and $\Phi(2) = 1$. The next merge to try is $detmerge(M, 1, 3)$, which also fails. Note that both $detmerge(M, 1, 4)$ and $detmerge(M, 2, 4)$ are not possible, but $detmerge(M, 3, 4)$ is possible. This merge triggers the merge of state 6 and state 8. The next state to consider is state 5, which cannot be deterministically merged and is therefore promoted to the set R. State 6 is then considered and deterministically merged with state 3, which triggers the merge of state 3 and state 11. The machine obtained at this point is shown in Fig. 4.5.

The consideration of state 7 leads to the merge of states 3 and 7, which also triggers the merge of states 3 and 12. Now, the states in the set B are 9 and 10. The first state to consider is state 9 which can be deterministically merged with state 2. State 10 can be deterministically merged with state 5. This merge triggers the following merges:

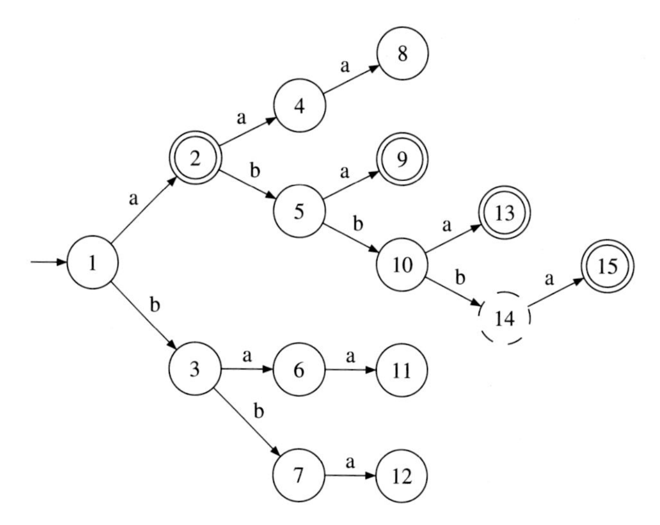

Fig. 4.4 The *PTMM* for $D_+ = \{a, aba, abba, abbba\}$ and $D_- = \{\lambda, b, aa, ab, ba, bb, aaa, abb, baa, bba\}$

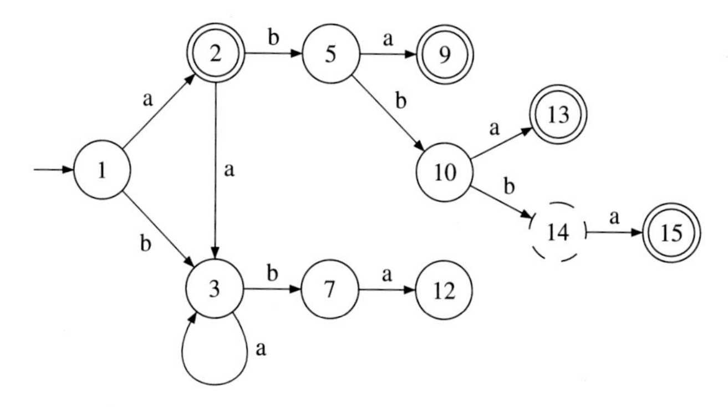

Fig. 4.5 The DFA obtained when the pairs of states ⟨3, 4⟩, ⟨3, 6⟩ of the initial *PTMM* (Fig. 4.4) are deterministically merged

states 2 and 13; states 5 and 14; and, finally, states 2 and 15. These merges end the process because set *B* is now empty. Thus the output of the algorithm is the machine shown in Fig. 4.6.

4.3.3 *Algorithms Guided by Heuristics:* EDSM *and* Blue-Fringe

In a run, *RPNI* merges those equivalent states (i.e., the states for which there is no sample that contradicts this equivalence).

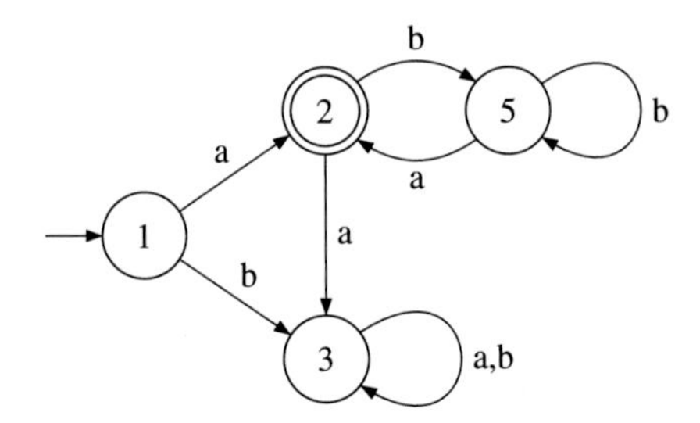

Fig. 4.6 The output of *RPNI* algorithm with input $D_+ = \{a, aba, abba, abbba\}$ and $D_- = \{\lambda, b, aa, ab, ba, bb, aaa, abb, baa, bba\}$

When the training set is not sufficiently representative of the language, it may imply that some *inconvenient* merges could be done. Those merges produce a negative effect on the inference process, leading to lower recognition rates.

In order to avoid as many of these merges as possible, de la Higuera et al. [18] proposed modifying the canonical order used by *RPNI* to favor the merge of states for which there exists great evidence of equivalence. This evidence is measured by a score to be computed for each pair of states. It is expected that the higher the score value, the greater the evidence of equivalence.

The initial results were not conclusive. Nevertheless, this approach was revisited by Price [37] who proposed the *EDSM* (Evidence-Driven State Merging) algorithm. This algorithm had very good behaviour in the Abbadingo contest and was one of the two winners. The *EDSM* strategy is summarized in Algorithm 4.3.4. In this algorithm, we denote by $\Phi(p) \approx \Phi(q)$ that states p and q are compatible, where states p and q are compatible if and only if $\Phi(p) = 1$ (resp. $\Phi(p) = 0$) implies that $\Phi(q) \neq 0$ (resp. $\Phi(q) \neq 1$).

As stated in line 6, the algorithm iterates while mergible states in the current automaton, let's say $A = (Q, \Sigma, \delta, q_0, F)$, are found. Each pair of mergible states p and q is evaluated by the function *FindScore* that, briefly speaking, takes into account the number of coincidences of states with defined output in the automata $A = (Q, \Sigma, \delta, p, F)$ and $A = (Q, \Sigma, \delta, q, F)$. Once all the mergible pairs of states are evaluated, the algorithm merges the pair of states with the highest score, that is, the pair of states with greatest evidence of equivalence. The algorithm ends when all mergible states have been considered.

The main drawback of the *EDSM* strategy is due to the cost of evaluating each pair of mergible states for each merge carried out. The first modification that was proposed to avoid this was to consider only those pairs of states at a given distance W from the initial state. This version is known as *W-EDSM* [11]. A better strategy for selecting the pairs of states to merge is known as *red-blue* [11, 37], which led to the *Blue-Fringe* method which is described in Algorithm 4.3.5. This algorithm is considered to be the *state of the art* with respect to the inference of DFA by merging of states.

Algorithm 4.3.5 uses the *PTMM* of the training sample. First, the *Blue-Fringe* algorithm considers the *red* set that contains only states of the hypothesis. Thus, the algorithm initializes the *red* set using the initial state of the machine. The *blue* set is

Algorithm 4.3.4 *EDSM* algorithm.

1: **Input:** Two disjoint finite sets (D_+, D_-)
2: **Output:** A consistent Moore Machine
3: **Method**
4: $A = PTMM(D_+, D_-)$;
5: $ok = True$;
6: **while** *ok* **do**
7: $score = \{\}$;
8: **for** $(p, q) \in Q \times Q$ **do**
9: **if** $\Phi(p) \approx \Phi(q) \wedge p \neq q$ **then**
10: $score = score \cup \{((p, q), FindScore(M, p, q))\}$;
11: **end if**
12: **end for**
13: **if** $score = \{\}$ **then**
14: $ok = False$;
15: **else**
16: $(p, q) = MaximumScorePair(score)$;
17: $A = detmerge(M, p, q)$;
18: **end if**
19: **end while**
20: Return(A);
21: **End Method.**

obtained by taking into account the *red* set, which contains those non-*red* states of the hypothesis that are reachable from any state in the *red* set. The algorithm ends when the *blue* set is empty.

In each iteration, the algorithm searches for a *blue* state that is non-mergible with any *red* state. The first of such states detected is promoted to the *red* set and the *blue* set is recalculated. Note that any state in the *blue* set that is not mergible should belong to the hypothesis in order to maintain consistence with the supplied data.

If this is not the case, then all the *blue* states are mergible with (at least) one *red* state. At this stage, the algorithm merges the pair of states with the greatest evidence of equivalence. The quantification of the evidence of equivalence takes into account the coincidences of the output values of the states in the prefix machine and also takes into account the depth of the *red* state (minimal number of transitions to reach the red state from the initial state). When the algorithm ends, it returns the resulting Moore machine.

It is worth noting here that the *RPNI* algorithm can be considered to be a *red-blue* method. In fact, note that if canonical order is considered (which is the usual order considered in the *Blue-Fringe* implementations) and the score computation is not carried out in the the *Blue-Fringe* algorithm, then the algorithms do not differ from each other.

The next example shows the different experimental behaviours of the *Blue-Fringe* and *RPNI* algorithms. Note that the guided merging leads to a more efficient use of the available data.

Algorithm 4.3.5 *Blue-Fringe* algorithm

```
1: Input: Two disjoint finite sets (D+, D−)
2: Output: A consistent Moore Machine
3: Method:
4: M = PTMM(D+, D−) = (Q, Σ, {0, 1, ?}, δ, q0, Φ);
5: red = {λ}
6: score = ∅
7: blue = {q ∈ Q : q = δ(p, a), p ∈ red ∧ a ∈ Σ} − red;
8: while blue ≠ ∅ do
9:   for q ∈ blue do
10:    merged = False;
11:    for p ∈ red do
12:      if (p, q) have a score in score then
13:        merged = True
14:      else
15:        if p and q are mergible ∧ p ≠ q then
16:          score = score ∪ {((p, q), 100 ∗ FindScore(M, p, q) + 99 − depth(p))};
17:          merged = True
18:        end if
19:      end if
20:    end for
21:    if not merged then
22:      red = red ∪ {q}
23:      BreakFor
24:    end if
25:  end for
26:  if merged then
27:    (p, q) = MaximumScorePair(score);
28:    M = detmerge(M, p, q);
29:    score = ∅
30:  end if
31:  blue = {q ∈ Q|q = δ(p, a) for p ∈ red ∧ a ∈ Σ} − red
32: end while
33: Return(M);
34: End Method.
```

Example 4.3 Let us consider the following training sets:

$$D_+ = \{b, bb, aab, aba, bbb\}$$
$$D_- = \{bba\}$$

Figure 4.7 shows $PTMM(D_+, D_-)$. If the *RPNI* algorithm is run with this input, the first merge, $detmerge(M, 1, 2)$, is successful: the merging of the pair of states $(1, 2)$ triggers the merging of the pairs $(1, 4)$, $(3, 5)$, $(3, 7)$. The output of the *RPNI* algorithm is shown in Fig. 4.8.

When the *Blue-Fringe* algorithm is run, initially $red = \{1\}$ and $blue = \{2, 3\}$. The score of the pair $(1, 2)$ is 0; therefore, an evidence of $0 \cdot 100 + 99 - 0 = 99$ is assigned to the pair of states $(1, 2)$. The pair of states $(1, 3)$ lead to the comparison

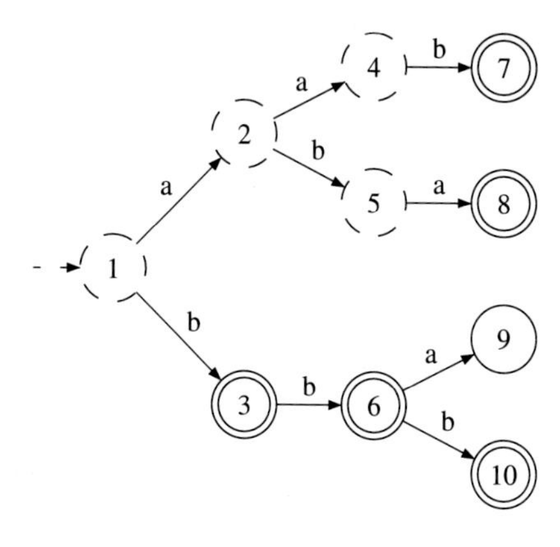

Fig. 4.7 The *PTMM* for $D_{+} = \{b, bb, aab, aba, bbb\}$ and $D_{-} = \{bba\}$

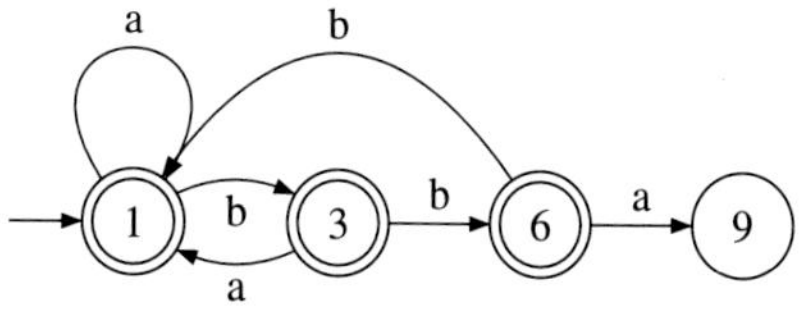

Fig. 4.8 The *RPNI* output when $D_{+} = \{b, bb, aab, aba, bbb\}$ and $D_{-} = \{bba\}$ are considered

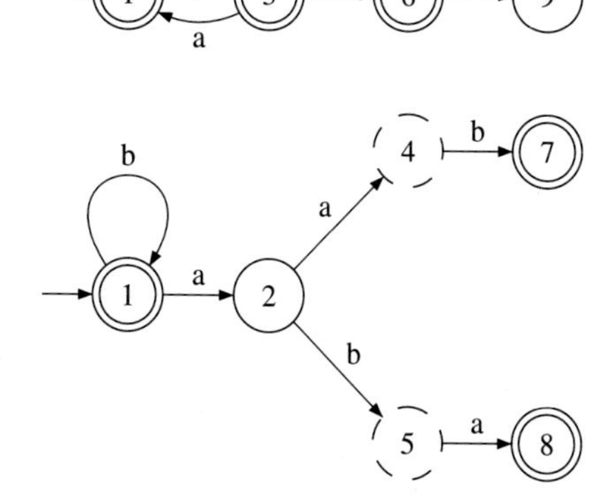

Fig. 4.9 The DFA obtained by the deterministic merging of states 1 and 3 of the machine in Fig. 4.7

of the output values of the pairs (3, 6) and (6, 10), with two coincidences ($\Phi(3) = \Phi(6) = 1$ and $\Phi(6) = \Phi(10) = 1$); therefore, an evidence of $2 \cdot 100 + 99 - 0 = 299$ is assigned. The hypothesis that is obtained by the deterministic merging of states 1 and 3 is shown in Fig. 4.9.

From *red* $= \{1\}$ and *blue* $= \{2\}$, it follows that state 2 cannot be merged with any *red* state, and, therefore, state 2 is promoted to the *red* set. Thus, *red* $= \{1, 2\}$ and *blue* $= \{4, 5\}$. The evidence computed for the pairs of states (1, 4), (2, 4), and (2, 5) are 199, 98, and 98, respectively (note that state 1 and 5 are not mergible). This leads to the merge of states 1 and 4. The new *red* and *blue* sets are *red* $= \{1, 2\}$ and *blue* $= \{5\}$. State 5 can only be merged with state 2, which gives the final automaton shown in Fig. 4.10.

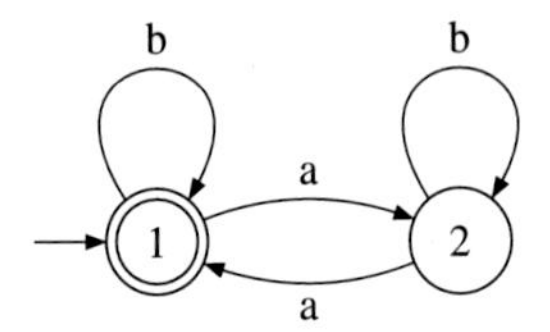

Fig. 4.10 The *Blue-Fringe* output when $D_+ = \{b, bb, aab, aba, bbb\}$ and $D_- = \{bba\}$ are considered

4.3.4 Inference of Teams of Automata

The last method we review in this section is the inference of teams of automata proposed by García et al. in [29]. This work takes into account that there is usually not enough information available to infer the automaton that accepts the target language (the characteristic sample is not available). The approach considers that different orders in the state merging process are able to capture some features of the target language while neglecting others. Therefore, it should be possible to improve the classification rates by considering a set (team) of automata and a vote scheme that weights the result obtained by each individual automaton in the set.

This approach is based on the proof that, under certain conditions, a red-blue algorithm that consistently but arbitrarily merges states in the prefix tree acceptor of the sample converges.

Similar to any red-blue algorithm, the approach described in [29] considers the merging of states but arbitrarily chooses a blue state and tries to merge it with a red one. Only when all merges are proved to not be possible is the state promoted to the red set. This scheme improves the computational efficiency with respect to other methods that give priority to promotion (e.g. the *Blue-Fringe* method).

It is worth noting here that the time complexity of the method is ($\mathcal{O}(k \times n^2)$, where k stands for the number of automata in the team and n stands for the size of the training set), which is better than the best algorithm so far.

Once the team of automata is obtained, several methods can be used to carry out the recognition of test strings. In [29], the authors consider a *fair vote* scheme together with several weighted vote schemes. The experimentation shows that the method obtains better results when the bigger automata (those bigger than the average size in the inferred team) are discarded and the vote of each remaining automaton is inversely proportional to its size.

The authors claim that the use of automata teams increases the probability of good results since the inclusion of the characteristic sample in the input data is not usually assured. This is corroborated by the experimental results shown in [29]. These results are summarized in Sect. 4.5. The authors compare their approach with current state-of-the-art algorithms. It can be seen that their method outperforms previous research approaches.

4.3.5 Identification in the Limit

Despite the differences between the algorithms reviewed hitherto, all three algorithms follow a red-blue scheme. These algorithms identify the class of regular languages in the limit. The respective proofs follow the same approach, that is, for every algorithm and any regular language, there exists a characteristic set (or sample) of positive and negative data. Whenever any of the three algorithms is run with such a characteristic set as input, the algorithm outputs the minimal DFA for the language. Furthermore, any training set that contains the characteristic set also obtains the same output.

Recently, García et al. studied the convergence of data-driven, red-blue algorithms [23]. In their work, the authors propose a general scheme to implement red-blue algorithms. In this general algorithm, any order can be used to traverse the states of the automata. Therefore, it is possible to implement as many red-blue algorithms as orders among the states can be defined. Please, note that *Blue-Fringe* is just an instance of this family of algorithms.

In their work, García et al. prove the existence of a characteristic sample for any red-blue inference algorithm (whether or not it is data-driven) that uses an *a* priori fixed order to traverse the states, thus proving the convergence of any such algorithms. This result is enunciated in Theorem 4.1.

Theorem 4.1 (García et al. [23]) *Any DFA GI algorithm, such that the promotion of states is independent from the input set, has a polynomial characteristic set, no matter the order followed to carry out the merge of states.*

In order to compute the characteristic set for any given language L, the authors present a method that considers a *minimal set of test states* of the minimal DFA $A = (Q, \Sigma, \delta, q_0, F)$ for L. This set of test states consists of any set of strings such that, if for every $q \in Q$, there exists only one string in the set such that $\delta(q_0, x) = q$. Example 4.4 illustrates this concept.

Example 4.4 Let us consider the automaton in Fig. 4.11. Note that it is possible to obtain several sets of test states by taking into account different orders to traverse the states. For instance, according to the alphabetic order, the set $S = \{\lambda, a, aa\}$ would be a minimal set of test states, and, according the canonical order, $S' = \{\lambda, a, b\}$ would be the minimal test set. Note that, for each state q, each one of the minimal test sets contains the first string that reaches q based on the chosen order (alphabetic or canonic).

Briefly speaking, for any language L, the proof of Theorem 4.1 takes into account a prefix-closed minimal set of test states obtained according to a defined order. The authors prove that Algorithm 4.3.6 outputs two sets $D_+(S)$ and $D_-(S)$ that are a characteristic sample for a red-blue algorithm using this defined order to identify the language L. Example 4.5 depicts how the algorithm works.

Example 4.5 Let us again consider the automaton in Fig. 4.11 and the prefix-closed minimal sets of test states $S = \{\lambda, a, aa\}$. The following table summarizes the process

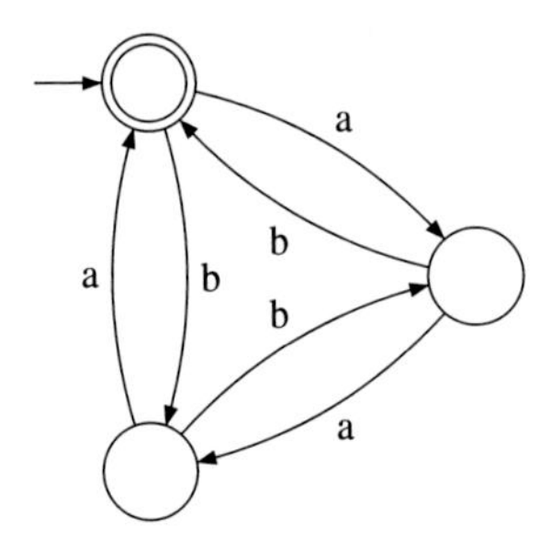

Fig. 4.11 DFA example

Algorithm 4.3.6 Algorithm to obtain the characteristic set for a language L.

Require: The minimal DFA A for L
Ensure: The polynomial characteristic set for L
1: **Method**
2: Let S be the minimal set of test states for A
3: $E = \{\lambda\}$
4: Let $S' = S\Sigma - S$
5: Let T be a matrix indexed by the strings $u \in S \cup S'$ and $e \in E$ that stores the membership of the string ue to the language
6: **while** there exist two undistinguished $u, v \in S$ and a symbol $a \in \Sigma$ such that $T[ua, e] \neq T[va, e]$ for some $e \in E$ **do**
7: $E = E \cup \{ae\}$
8: **end while**
9: **for** each row u and column v in T **do**
10: **if** $T[u, v] = 1$ **then**
11: Add uv to D_+
12: **else**
13: Add uv to D_-
14: **end if**
15: **end for**
16: Return(D_+, D_-);
17: **End Method.**

---

of obtaining the characteristic set. For the sake of clarity, we represent the elements in S and those in $S\Sigma - S$ separately. Initially the only column available is the one with label λ. The 1 and 0 entries in the table represent whether or not the strings obtained by concatenation of the strings that label the row and column belong to the language L.

	λ	b
λ	1	0
a	0	1
aa	0	0
b	0	0
ab	1	0
aaa	1	0
aab	0	1

Note that taking into account just the column labelled λ, the undistinguished elements in S are $\{a, aa\}$. It is possible to distinguish the first element using the suffix b. Once the table is filled in, all the elements in S are distinguished; therefore, the characteristic sample for the language is $D_+(S) = \{\lambda, aaa, aabb, ab\}$ and $D_-(S) = \{a, aa, aaab, aab, abb, b, bb\}$. Figure 4.12 shows the corresponding *PTMM* and the numbering of the states.

When the algorithm is run using the canonical order, then the prefix-closed minimal set of test states to be used is $S = \{\lambda, a, b\}$ and the sets output by the algorithm are $D_+(S) = \{\lambda, ab, ba, bbb\}$ and $D_-(S) = \{a, b, aa, bb, aab, abb, bab\}$. Figure 4.13 shows the corresponding *PTMM* and the order to traverse the states.

Obviously, it is not possible to ensure the inclusion of a characteristic sample in the training set. In any case, in this context, a characteristic sample depends directly on the promotion order of the states that a red-blue algorithm uses. This implies that

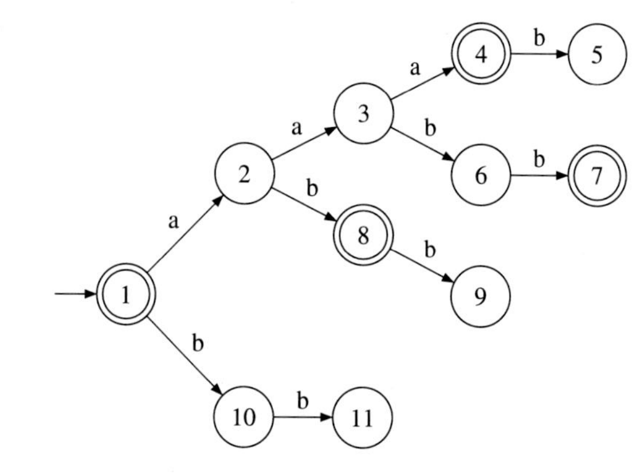

Fig. 4.12 The *PTMM* for $D_+(S) = \{\lambda, aaa, aabb, ab\}$, $D_-(S) = \{a, aa, aaab, aab, abb, b, bb\}$ and an alphabetic numbering of the states

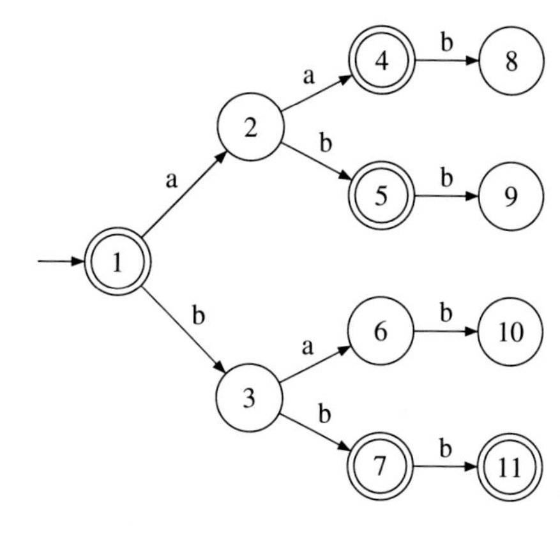

Fig. 4.13 The *PTMM* for $D_+(S) = \{\lambda, ab, ba, bbb\}$, $D_-(S) = \{a, b, aa, bb, aab, abb, bab\}$ and a canonical numbering of the states

the same training set may or may not be representative depending on the promotion order implemented.

Thus, the consideration of several arbitrary orders (and therefore a team of automata) is convenient and motivates the approach presented in [29]; here we note that the experimental results confirm the authors' hypothesis.

Let us finally note the similarity between the inference of teams of automata and the *Blue-Fringe* algorithm. In essence, *Blue-Fringe* looks for evidence of the best order to merge the states. The team approach uses different orders in order to obtain a team of automata and classifies any test string by combining those inferred automata.

4.4 Inference of Non-deterministic Automata

Recently, a new line of work has proposed the inference of non-deterministic automata (NFA) instead of the usual deterministic model. In 1994, Yokomori published a paper that proposes a method that considers queries to an oracle and counterexamples [16, 47]. Coste and Fredouille also studied the inference of NFAs [13–15] proposing a method to infer unambiguous finite automata.

Later, Denis et al. [19, 21] focused their work on the inference of a subclass of NFA, the *Residual Finite State Automata* (RFSA). The consideration of this subclass of NFAs is interesting because it has been proved that there is a unique canonical RFSA for each regular language. In [19], Denis et al. propose an algorithm (known as the *DeLeTe2* algorithm) that converges to a RFSA. Unfortunately, the hypothesis that the algorithm outputs is not always consistent with the supplied training data. To solve this drawback, the authors propose the *DeLeTe2* program [21], which always returns a consistent machine. In the same paper, the authors presented an experimental comparison with respect to other well-known methods such as *RPNI* and *Blue-Fringe*. The experimentation considers data from languages represented by NFAs and regular expressions, and the results obtained show that *DeLeTe2* performs better than the *RPNI* and *Blue-Fringe* methods.

Among the recent work on NFA inference, it is worth citing some works in the field. In [10, 33, 35, 36, 40, 43] results are presented on the minimality and reduction of automata. Other papers study methods that converge to a RFSA [3]. In [1, 14, 15], the authors propose methods to infer *Unambiguous Finite Automata* (UFA). There are also works that propose extensions of the *RPNI* scheme that consider language inclusion relations as a way to increase the training data [26]. Finally, in [46], the authors study the inference of NFAs by juxtaposition of subautomata.

Other works that are related to this topic are [27, 28]. These papers study the influence of the merging order on the convergence of the algorithms. It is proved that whenever some more general conditions are fulfilled, convergence is assured, regardless of the order in which the non-deterministic merges have been done.

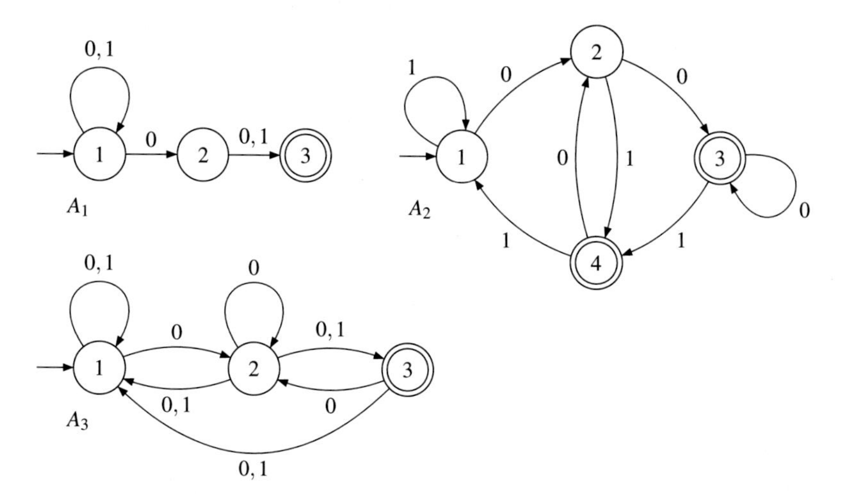

Fig. 4.14 Example from [20]. A_1 is an automaton that recognizes $L = \Sigma^*0\Sigma$, but it is neither a DFA nor a RFSA (note that $\nexists u \in \Sigma^*$ such that $R_3^{A_1} = u^{-1}L$). The automaton A_2 is a DFA that recognizes L (it is also a RFSA by definition). A_3 is the canonical RFSA for L

4.4.1 Inference of Canonical NFAs

Given a language L, a *residual finite state automaton* (RFSA) [20] for the language is a finite automaton A that accepts the language L and, for any state q, there exists a word u such that $R_q^A = u^{-1}L$. The *saturated* RFSA of a minimal DFA A is defined as the subautomaton of the universal automaton $\mathcal{U}_L$ that is induced by the set of states of the automaton A.

In other words, a RFSA A is a non-deterministic automaton such that all its states define a residual language of $L(A)$. Note that every DFA fulfills the conditions for being RFSA. Figure 4.14 shows three automata for the same regular language $L = \Sigma^*0\Sigma$. Note that there is a NFA, a DFA, and the canonical RFSA, all of which identify L.

Two relationships that are defined over the set of states of an automaton link the RFSA with GI. Let $D = (D_+, D_-)$ be a set of samples and let u, v be two strings in $Pr(D_+)$. First, the relation $\prec$ is redefined as $u \prec v$ if there is no string w such that $uw \in D_+$ and $vw \in D_-$. Note that this is important because, in a grammatical inference process, right languages of the states are unknown. Second, if $u \prec v$ and $v \prec u$, the function *obviously different* is defined as $od(u_1, u_2, PTMM(D_+, D_-)) =$ *False* (non-obviously different using Gold's terms). In the sequel, we will denote this condition as $u \simeq v$.

It is well known that every regular language has a finite set of residual languages (see Nerode's theorem in any language theory text, for instance [34]). It is also known that every regular language has a DFA that recognizes the language and that any given DFA is also a RFSA. Therefore, any regular language can be represented using a

RFSA [20]. Among the RFSA that identify a regular language, the smallest one, the *canonical RFSA*, is of special interest. Formally speaking, for a given language $L \subseteq \Sigma^*$, the canonical RFSA of L is the automaton $A = (Q, \Sigma, \delta, I, F)$ where:

- $Q = \{u^{-1}L \,:\, u^{-1}L \text{ is prime}, u \in \Sigma^*\}$, where prime quotients are such that they cannot be obtained by the union of other quotients.
- $\delta(u^{-1}L, a) = \{p \in Q \,:\, p \subseteq (ua)^{-1}L\}$
- $I = \{p \in Q \,:\, p \subseteq L\}$
- $F = \{p \in Q \,:\lambda \in p\}$

The canonical RFSA for a given language L is unique, and it is the smallest RFSA that recognizes L (taking into account the number of states). If $R_q^A = u^{-1}L$, then the string u is said to be characteristic of state q. The size of the canonical RFSA is upper bounded by the size of the minimal DFA for the language and lower bounded by the minimal equivalent NFAs [20].

Under these conditions, several situations may occur: the canonical RFSA may be exponentially smaller than the minimal equivalent DFA; both automata may be of the same size and there may be an exponentially smaller NFA than both the minimal DFA and the canonical RFSA; there may be a minimal NFA with the same number of states but with few transitions. Taking into account the length of the characteristic strings, it is possible for the smallest characteristic string of a given state to have an exponential length with respect to the size (number of states) of the canonical RFSA. This may occur when the characteristic strings are computed taking into account the minimal DFA and also where this automaton has an exponential size with respect to the canonical RFSA.

All these situations show that any inference algorithm that converges to the canonical RFSA does not ensure the output to be minimal. Despite this, it is possible to improve the results whenever the target language L is characterized by the fact that the DFA for L is made up of many composite residual languages. It remains to be studied how common such languages are. The practical use of RFSAs depends on the results of that study.

Denis et al. studied the inference of RFSAs in [22]. In their work, the authors propose a method to infer residual automata from positive and negative data. The authors claim this method converges to the saturated subautomata of the minimal DFA. The main drawback of the method is that, in some circumstances, the method does not guarantee consistency. The implementation of the method provides a solution to this drawback, but the implementation does not completely correspond to the algorithm. Section 4.4.1.1 reviews the initial method, which will hereafter be referred to as the *DeLeTe2* algorithm. Section 4.4.1.2 reviews the implementation of the method, which will be referred to as the *DeLeTe2* procedure.

4.4.1.1 The *DeLeTe2* Algorithm

The *DeLeTe2* procedure is shown in Algorithm 4.4.1. We have rewritten the algorithm in [22] in order to relate it with the other algorithms mentioned above. We point out that (leaving aside implementation details) Algorithm 4.4.1 and the method described in [22] are equivalent.

As the authors state in [22], this automata can be obtained by saturating the minimal DFA A and reducing all the states that are greater than p, where p is the greatest prime state in A.

The algorithm looks for inclusion relations among residual languages and considers them using a saturation operator. As commented above, this method is of special interest when the target automaton has many non-prime residual languages. In this situation, many inclusion relationships may be found, and this leads to a smaller hypothesis. Whenever the target automaton has many prime residual languages, the size of the hypothesis and the size of the minimal DFA is expected to be the same [22].

Algorithm 4.4.1 *DeLeTe2* algorithm.

1: **Input:** Two finite sets of data $D_+ \cup D_-$ over Σ
2: **Output:** A finite automaton
3: **Method:**
4: Let *Pref* be the set of prefixes of $D_+ \cup D_-$
5: $R = \{\lambda\}$
6: $B = (R\Sigma - R) \cap Pref$
7: $Q = R$
8: $I = \delta = F = \emptyset$
9: $A = (Q, \Sigma, I, \lambda, F)$
10: **while** $B \neq \emptyset$ **do**
11: $\quad I = \{q \in Q \ : \ q \prec \lambda\}$
12: $\quad F = Q \cap D_+$
13: $\quad$ **for** $p \in R$ **do**
14: $\quad\quad \delta = \{(p, a, r) \ : \ r \in R \ \wedge \ r \prec pa\}$
15: $\quad$ **end for**
16: $\quad$ **if** A is not consistent with respect to (D_+, D_-) **then**
17: $\quad\quad$ Let $C = \{q \in B \ : \ \nexists p \in R, \ p \simeq q\}$
18: $\quad\quad$ **if** $C = \emptyset$ **then**
19: $\quad\quad\quad$ *Return*(A)
20: $\quad\quad$ **end if**
21: $\quad\quad$ choose $q \in C$
22: $\quad\quad R = R \cup \{q\}$
23: $\quad\quad B = (R\Sigma - R) \cap Pref$
24: $\quad$ **else**
25: $\quad\quad B = \emptyset$
26: $\quad$ **end if**
27: **end while**
28: *Return*(A)
29: **End Method:**

It is worth noting that the *TBG* method (Algorithm 4.3.1) is closely related to the *DeLeTe2* algorithm. Also note that when the *TBG* algorithm is going to add a transition, say for a state s and a symbol a, the algorithm selects one of the possible states which is not distinguishable from sa (line 16 in Algorithm 4.3.1). When the *DeLeTe2* algorithm builds the set of transitions, it adds a transition (p, a, r) for each

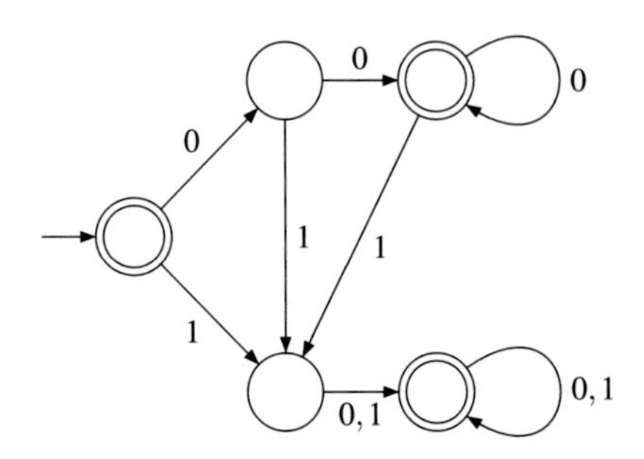

Fig. 4.15 The minimal DFA for $L = \lambda + 00^+ + 0^*1(0+1)^+$

state $r \prec pa$. Thus, if line 16 of the *TBG* algorithm is modified to work the way *DeLeTe2* works, then the *TBG* algorithm would output the minimal saturated DFA for the language (when a characteristic sample is input).

The reason why the *DeLeTe2* algorithm can output smaller automata than the minimal DFA is that this algorithm builds a hypothesis whenever a new state is added. The algorithm ends when one of these hypotheses is consistent with the data. Nevertheless, an important drawback of the *DeLeTe2* algorithm is that, in some cases, it outputs non-consistent automata. The following example illustrates this.

Example 4.6 Let us consider the language $L = \lambda + 00^+ + 0^*1(0+1)^+$. The minimal DFA for L is shown in Fig. 4.15.

When the sets $D_+ = \{\lambda, 00, 10, 11, 010\}$ and $D_- = \{0, 1, 01, 001\}$ are provided, the *RPNI* algorithm outputs the automaton shown in Fig. 4.3. Note that the automaton does not accept the string 10111 that belongs to L. Although not explicitly considered in the *DeLeTe2* algorithm, in order to illustrate the run of the algorithm we will take into account the *PTMM* of the sample shown Fig. 4.1 (if the state 5 is considered to be negative). Thus, the *DeLeTe2* algorithm first considers the state λ. The algorithm sets $Q = \lambda$ and also adds this state to the set of initial states (it fulfills that $\lambda \prec \lambda$) and to the final states (λ is in D_+). The algorithm takes into account the relationships among the state λ and the set of *blue* states $\{0, 1\}$. Since no relationship is detected, no transitions are added. Then, the first hypothesis is the automaton $A = (\{\lambda\}, \{0, 1\}, \{\}, \{\lambda\}, \{\lambda\})$. Since this hypothesis is not consistent with the data, a new state is added.

When the first state that is obviously distinguishable from the those in Q (according a canonical order) is considered, the state 0 is taken into account. The relationships that are to be analyzed consider the sets of states $\{\lambda, 0\}$ and $\{1, 00, 01\}$. Thus, it is detected that $0 \prec 1$, $\lambda \simeq 00$, $0 \prec 00$ and $0 \simeq 01$. The second hypothesis is shown in Fig. 4.16.

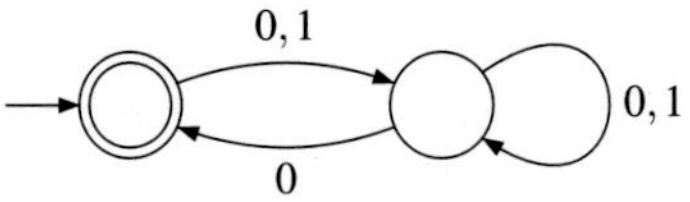

0, 1
0, 1
0

Fig. 4.16 The second hypothesis of the *DeLeTe2* algorithm. Note that the automaton does not accept 001

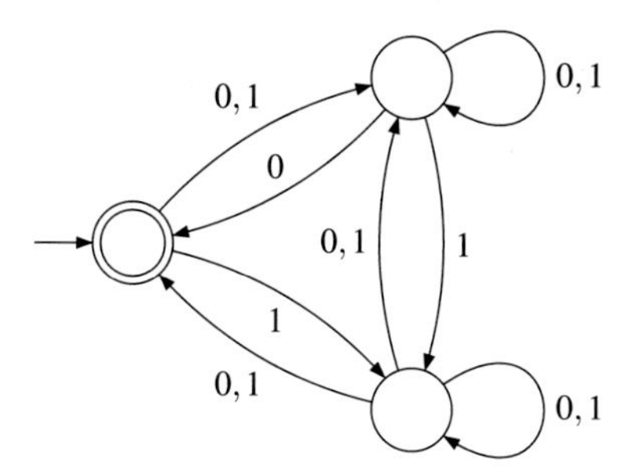

Fig. 4.17 The RFSA output by the *DeLeTe2* algorithm for $D_+ = \{\lambda, 00, 10, 11, 010\}$ and $D_- = \{0, 1, 01, 001\}$

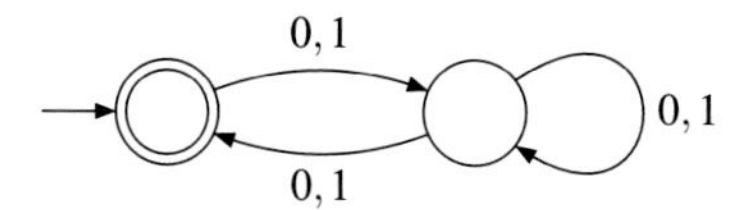

Fig. 4.18 The automaton output by the *DeLeTe2* algorithm using. $D_+ = \{\lambda, 00, 10, 11, 010\}$ and $D_- = \{0, 1, 001\}$. Note that the automaton is not consistent with the data

This hypothesis is also non-consistent with respect to the data, and the next added state is 1. The relationships among the sets of states $\{\lambda, 0, 1\}$ and $\{00, 01, 10, 11\}$ are to be taken into account. Thus, it is detected that $1 \simeq 01$, $\lambda \simeq 10$, $0 \prec 10$, $1 \prec 10$, $\lambda \simeq 11$, $0 \prec 11$ and $1 \prec 11$. The third hypothesis, which is shown in Fig. 4.17, is consistent with the data and therefore is output. Please note that this automaton accepts L and that it is smaller than the minimal DFA.

Let us consider the same example but where the string 01 does not belong to D_-. Figure 4.18 shows the automaton output in this case. This automaton is not consistent with the data. It accepts the string 001 which is in D_-.

Convergence of the Inference of RFSAs To prove the convergence of *DeLeTe2*, it is possible to use an argument that is similar to the one used to prove the *RPNI* convergence. Therefore, it is possible to build a characteristic sample that ensures accessibility to all of the states of the target automaton and that contains information of the possible relations of inclusion.

In [22], the authors prove that the number of strings in the characteristic sample is polynomial with respect to the size of the minimal DFA for the language. Note that, for a given language L, the minimal DFA for L can be exponentially bigger than the canonical RFSA for L.

4.4.1.2 An Improved Implementation

As shown above, the *DeLeTe2* algorithm does not guarantee consistence. In order to solve this drawback, the authors propose an improved implementation of the algorithm, which is usually referred to as the *DeLeTe2* procedure. The authors do not study the theoretical properties of the implementation, and experimentation cannot guarantee the convergence of the *DeLeTe2* procedure to the saturated subautomata of the minimal DFA.

The *DeLeTe2* procedure also represents the hypothesis obtained by RFSAs. Its source code is available on the home web page of the authors. To our knowledge, there is only one work that has studied the behaviour of this implementation [2]. In that work, the author takes into account the source code and detects that both (the *DeLeTe2* algorithm and the *DeLeTe2* procedure) behave differently. The main features of the *DeLeTe2* procedure are summarized below:

- It is a regular language GI algorithm based on state merging.
- The algorithm considers the *PTMM* of the training set and outputs a RFSA. The algorithm takes advantage of several properties of the residual languages associated to each state. Specifically, the transitivity of the inclusion relationship is very useful.
- The hypotheses that the algorithm obtains are always consistent with respect to the supplied data.

The *DeLeTe2* procedure strategy considers each pair of states of the *PTMM*, in canonical order, in order to determine the relationship of the states, that is, the *inclusion*, *no inclusion*, or *unknown* relation of the residual languagest that are associated to the states of the pair.

When an *unknown* relationship is going to be considered, it is temporarily set to *inclusion* and propagated. If this assumption generates inconsistency, the initial assumption is discarded, along with any other relationship found by the triggered one. We note that this kind of assumption may imply several consequences, such as the definition of unknown outputs of the *PTMM* or the discovery of new *inclusion* or *no inclusion* relationships (among others). The time complexity of the *DeLeTe2* procedure is bounded by $\mathcal{O}(n^4)$.

4.4.2 Inference by Juxtaposition of Automata

One advantage of using non-deterministic models to infer regular languages is that this model is more concise than the deterministic one. It is worth noting that, in this sense, the use of RFSA does not ensure that we obtain good (small) representations of the target languages. This is because, for a given language, the size of the minimal RFSA can be exponentially bigger than the size of a minimal NFA for the same language. As an example, consider the following family of languages indexed by n:

$$\{0^i : i \text{ has a divisor greater than 1 and lower than n}\}$$

In this family, the number of states of the NFAs grows polynomially with respect to n, while the number of states of the RFSAs grows exponentially with respect to n.

Algorithm 4.4.2 Word Associated Subautomata Regular Inference (*WASRI*) general scheme.

1: **Input:** Two finite sets $D_+ = \{x_1, x_2, \ldots, x_n\}$ and D_-
2: **Output:** A finite automaton consistent with the input sets
3: **Method:**
4: $A = (Q, \Sigma, \delta, I, F)$ where $Q = \delta = I = F = \emptyset$
5: **for** $i = 1$ to n **do**
6:   **if** $x_i \notin L(A)$ **then**
7:     **for** at least $j = 1$ **do**
8:       Infer an automaton A_i^j consistent with $D_+ = \{x_i\}$ and D_-
9:       $A = A \uplus A_i^j$
10:       //let us recall that $\uplus$ stands for disjoint union of the automata//
11:     **end for**
12:   **end if**
13: **end for**
14: **for all** component (automaton) $A_i \in A$ **do**
15:   **if** Component A_i is not necessary to accept D_+ **then**
16:     Remove the component A_i from A
17:   **end if**
18: **end for**
19: Return A
20: **End Method.**

In [46], the authors propose a family of algorithms that infer the regular language class in the limit. The general scheme of this family is shown in Algorithm 4.4.2. Briefly speaking, for each string of the positive sample u, the method obtains at least one irreducible consistent automaton (i.e., an automaton such that the merging of any two states in it makes the resulting automaton accept negative strings). In order to infer such an automaton, the method takes into account the automaton that only recognizes the positive string u and merges states while it is possible. The automaton to be output is obtained by disjoint union of the inferred automata.

This method is quite flexible because the input parameters of the method are the number of automata to infer for each word as well as the order of state merging. The authors prove that convergence is ensured even if the these parameters change.

The method also includes an option to filter the collection of automata obtained (the loop in line 14 is just one possibility to do so). For example, if several automata for each word have been obtained (using different merging order criteria), it is possible to consider a criterion to select one (or some) of them (for instance, the smallest in size).

Algorithm 4.4.2 implements this filter by the deletion of some of the automata in the output collection (of course, as long as the resulting automaton still accepts D_+).

The complexity of the *WASRI* algorithm is $\mathcal{O}(kn^2|D_-|)$, where k is an integer, n is the length of the longest word of D_+, and $|D_-|$ is the sum of the lengths of the negative words of the sample. It is worth mentioning that the time complexity depends on the length of each word and not on the sum of the lengths of the input (i.e., the size of the prefix tree acceptor of the sample). This fact makes *WASRI* a very fast algorithm.

An interesting example to illustrate the behaviour of *WASRI* is the language of strings in 0^* whose length is a multiple of either 2, 3, or 5. The minimal DFA for L has the same number of states as the canonical RFSA (30 states). Taking into account the sets $D_+ = \{0^2, 0^3, 0^5\}$ and $D_- = \{0, 0^{11}\}$, Algorithm 4.4.2 returns an automaton with 10 states that identifies the language. We also note that *RPNI* and *DeLeTe2* (running with the same input) return automata that are far from convergence.

Convergence of the *WASRI* Algorithm To prove the convergence of any *WASRI* implementation, for any given target language L, the authors consider the universal automaton for the language $\mathcal{U}_L$. This automaton allows a *universal sample* D_+ to be computed. It is proved that, from $MA(D_+)$ every irreducible automaton in $\overline{L}$ that can be obtained accepts the target language. The authors also show that this theoretical condition can be replaced by the construction of a (finite) set of negative samples D_-, which contains at least one string in $merge(p, q, MA(D_+)) - L$ for any pair of states p and q in $MA(D_+)$.

In other words, when D_- is taken into account, for any string in L, it is possible to obtain an irreducible automaton that accepts a sublanguage of the target language. This process can be iterated as many times as necessary using the strings in D_+. Note that convergence is ensured because any automaton that recognizes a sublanguage of L can be projected into the universal automaton for L and there exists a finite set of subautomata of the universal automaton.

4.4.3 Order-Independent Merging Inference

In [28], the authors propose a state-merging algorithm that, given a *universal sample* as input, converges to a nondeterministic finite automaton that recognizes the target language independently of the order in which the states are merged.

The definition of universal sample is closely related to the notion of *automata irreducibility*, where an automaton A is said to be irreducible if the automaton is such that the merge of any pair of states leads to an automaton that accepts a super-language of the target language. The authors extend this notion and define an automaton A as being *irreducible in a regular language* L if and only if $L(A) \subseteq L$ and the merge of any pair of states leads to an automaton that accepts a super-language of L.

Taking this into account, for every regular language L, a universal sample for L is a finite set $D_+ \subseteq L$ with the property that any partition over the maximal automaton for D_+ that produces an irreducible automaton in L produces an automaton that identifies L.

It is worth noting that, while the irreducibility of an automaton in a regular target language L needs $\overline{L}$ to be known, it is possible to prove that it is enough to achieve irreducibility using only a finite set of negative samples.

In [28], the authors use some theoretical tools (especially the universal automaton for the language) that help to prove the convergence of the method. These tools also help to clarify and simplify ideas about the convergence of other inference algorithms, including those that may be proposed in the future. Even though different orders of merging states may lead to different hypotheses (automata), note that, when convergence is achieved, the language accepted by those automata will be the target language.

In the same work, the authors use the theoretical results to propose the *Order-Independent Learning* (*OIL*) family of algorithms. This scheme is described in Algorithm 4.4.3. The authors prove that this algorithm identifies the family of regular languages in the limit.

When a set of blocks of positive and negative samples for the target language L is given to the algorithm, an automaton that recognizes L in the limit is obtained. Note that a block may contain just a single word, so the algorithm is presented here in a very general way.

Briefly, the method first builds the maximal automaton for D_+^1 and merges the states in a random order until the algorithm obtains an irreducible automaton in D_-^1. Then the algorithm performs the following steps for every new block:

1. If the existing automaton is consistent with the new block, nothing has to be done.
2. If it is consistent with the new set of negative samples, the algorithm considers only the positive words that were not accepted by the previous hypothesis. Then the algorithm builds the maximal automaton for the new set of positive words, adds the new negative words to D_-, and finds a partition of the states of the automaton until an irreducible automaton in D_- is obtained.
3. Otherwise, the algorithm runs a recursive call taking the following into account in each step: the corresponding set of positive samples and the whole set of negative samples. This part of the algorithm (line 19) overcomes the fact that even though we may have a universal sample, the negative samples may not lead to consistency.

Algorithm 4.4.3 Order-Independent Learning (*OIL*) general scheme.

1: **Input:** A sequence of finite sets $\langle (D_+^1, D_-^1), (D_+^2, D_-^2), \ldots, (D_+^n, D_-^n) \rangle$
2: **Output:** A finite automaton consistent with the input sets
3: **Method:**
4: $A = MA(D_+^1)$
5: $D_- = D_-^1$
6: Find a partition π of the states of A irreducible in $\overline{D_-}$
7: $A = A/\pi$
8: **for** $i = 2$ to n **do**
9: $D_- = D_- \cup D_-^i$
10: **if** A is consistent with (D_+^i, D_-^i) **then** Continue **end if**
11: **if** A is consistent with D_-^i **then**
12: $S_+ = D_+^i - L(A)$
13: $A' = (Q', \Sigma, \delta', I', F') = MA(S_+)$
14: $A = (Q \uplus Q, \Sigma, \delta \uplus \delta', I \uplus I', F \uplus F')$
15: //where $\uplus$ stands for disjoint union //
16: Find a partition π of the states of A irreducible in $\overline{D_-}$
17: $A = A/\pi$
18: **else**
19: $A = OIL(D_+^1, D_-), (D_+^2, D_-), \ldots, (D_+^i, D_-))$
20: **end if**
21: **end for**
22: Return A
23: **End Method.**

Note that Algorithm 4.4.3 is presented in a very general way and there are many possible ways to obtain a partition of the set of states (lines 6 and 16 of the algorithm).

One possible implementation of the algorithm could consider a random ordering of the states in $MA(D_+^1)$. Thus, the algorithm would analyze the merging of the states in the chosen order in order to obtain an irreducible automaton in D_-.

At every step, the algorithm considers only those words of the new block (e.g. D_+^i) that are not accepted by the hypothesis at that moment. The algorithm then proceeds to analyze the merging of the states (according to a random ordering) of the maximal automaton obtained from the remaining words.

Whenever the current hypothesis is not consistent with the new block of negative samples, the previous hypothesis is discarded. A recursive call takes into account the whole set of negative samples seen up to that point. The following example illustrates this implementation.

Example 4.7 Let $L = a^* + b^*$, and let the input sample be divided into the following blocks:

$$D_+^1 = \{a, bb, aa\} \quad D_-^1 = \{ab, bba\}$$
$$D_+^2 = \{b, aaa\} \quad D_-^2 = \{aaab, aab\}$$
$$D_+^3 = \{\lambda, bbb, aaaa\} \quad D_-^3 = \{abb, ba\}$$

The *OIL* algorithm starts considering the first block (D_+^1, D_-^1). The maximal automaton $MA(D_+)$ for this block is shown in Fig. 4.19, in which the order among

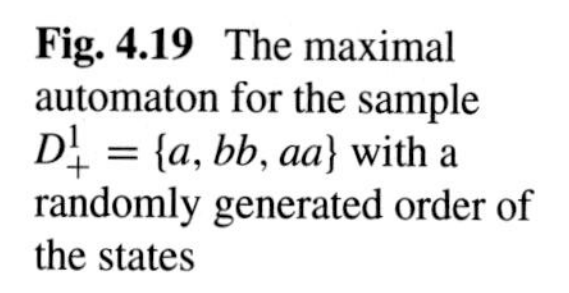

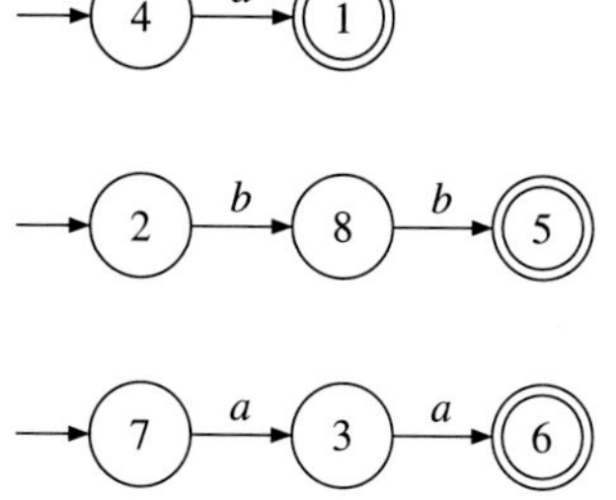

Fig. 4.19 The maximal automaton for the sample $D_+^1 = \{a, bb, aa\}$ with a randomly generated order of the states

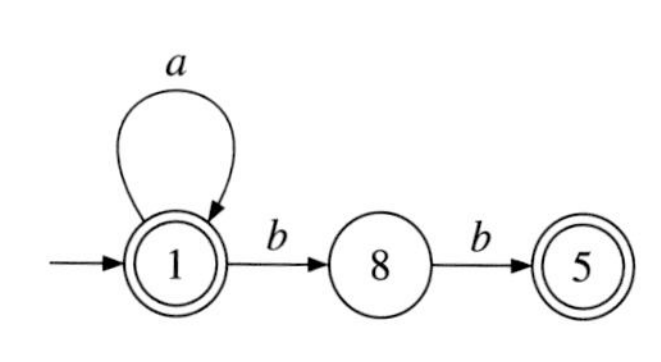

Fig. 4.20 The first hypothesis obtained by *OIL*

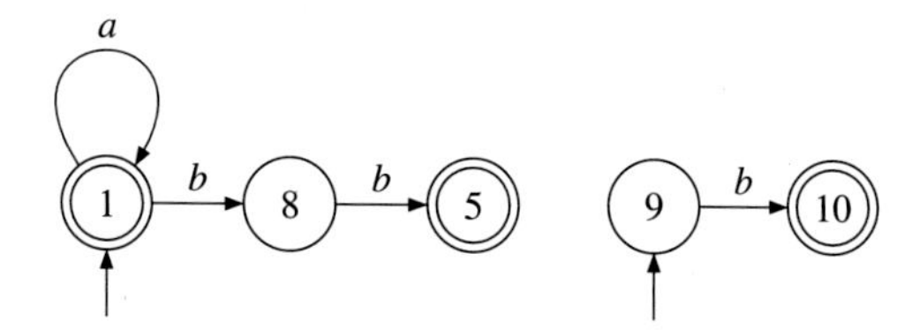

Fig. 4.21 The automaton that considers the relevant information in the second block of samples

the states has been randomly generated. The algorithm follows this order and makes all the possible merges. Thus, states 1 and 2 are (non-deterministically) merged as well as states 1 and 3, states 1 and 4, states 1 and 6, and states 1 and 7.

Note that, in this process, state 1 cannot be merged with state 5 because it would lead to accepting the word $bba \in D_-^1$. The pair of states 1 and 8 cannot be merged for a similar reason. The algorithm obtains the automaton shown in Fig. 4.20, which is irreducible with respect to D_-^1.

Once the first hypothesis has been obtained, the algorithm starts processing the second block. First, the algorithm checks for consistency with the current hypothesis. Thus, the automaton shown in Fig. 4.20 is consistent with the negative samples $D_-^2 = \{aaab, aab\}$. Second, the algorithm selects those strings in D_+^2 that are not accepted by the current hypothesis. With the remaining positive strings, a new maximal automaton is built and incorporated into the hypothesis. Note that the ordering of the states has been randomly generated, starting with $N + 1$, where N is the number of states of the maximal automaton for the previous block ($MA(D_+^1)$). The resulting automaton is shown in Fig. 4.21.

Now, the algorithm, which is controlled by the set of negative samples (i.e. $D_- = D_-^1 \cup D_-^2 = \{ab, bb, aab, aaab\}$), tries to merge the states in this order. The possible merges at this point are state 1 with state 10, and state 5 with state 9. Note that the merging of state 1 with state 9 would cause the word *ab* to be accepted. The merging of the previous states produces the second hypothesis depicted in Fig. 4.22.

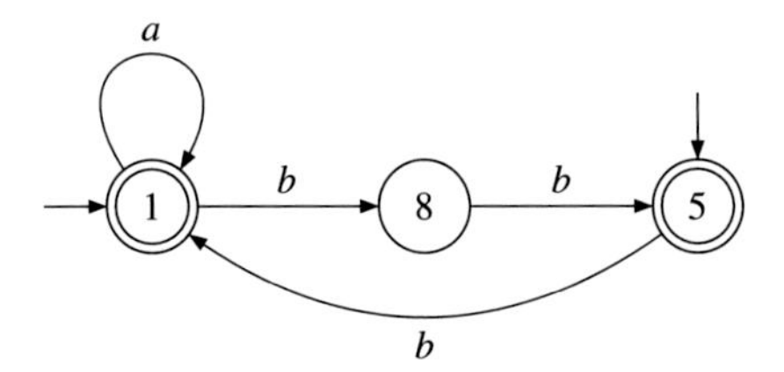

Fig. 4.22 The second hypothesis obtained by *OIL*

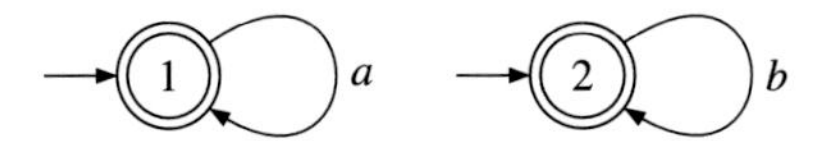

Fig. 4.23 The final automaton output by *OIL*

The algorithm starts the processing of the third block of samples. The algorithm checks the third block of the sample for consistency with the current hypothesis and detects that there are words in the negative sample that are accepted by the hypothesis (for instance *abb*). Therefore, the algorithm processes the positive samples D_+^1, D_+^2, and D_+^3 again. This process is controlled by the whole set of negative samples, that is, $D_- = \{ab, abb, aab, aaab, abb, ba\}$.

If we consider the previous enumeration of the states in the maximal automaton, in this run states 1 and 2 cannot be merged since the resulting automaton would accept *abb*. States 1 and 5 cannot be merged either since *bba* would be accepted. States 1 and 3, states 1 and 4, states 1 and 7, states 2 and 5, and states 2 and 8 are merged. Figure 4.23 shows the output of the algorithm that recognizes the target language.

Convergence of the *OIL* Algorithm Despite the non-deterministic nature of any implementation of *OIL*, the convergence of this algorithm is proved in [28]. The argument is quite similar to the one used to prove the convergence of *WASRI*. Thus, for any given target language L, the authors consider the universal sample D_+ of L.

As described above, the *OIL* algorithm considers a pair of sets of positive D_+^i and negative samples D_-^i in each iteration. Let us denote the sets of positive and negative information available at a given iteration i with $D_+^{\leq i}$ and $D_-^{\leq i}$, respectively.

First, note that the universal sample for the target language L will eventually be available after n iterations, that is $D_+ \subseteq D_+^{\leq n}$. Second, as stated in Sect. 4.4.2, there exists a finite subset of negative samples D_- that avoids any undesired merging of states over the maximal automaton $MA(D_+)$. This set will also be available after $m \geq n$ iterations ($D_- \subseteq D_-^{\leq m}$).

Note that *OIL* can output an automaton that recognizes L whenever enough negative samples are provided. If convergence has not yet been achieved before the m iteration, then the current hypothesis will accept words that are not in L, and, therefore, the hypothesis will not be consistent with $D_-^{\leq m}$. In this case, the algorithm runs a recursive call taking into account $D_{-}^{\leq m}$ and $D_{-}^{\leq m}$, which will return a correct automaton.

4.5 (Experimental) Comparison of Different Approaches

In order to compare the behaviour of their results, in [22], Denis et al. built a corpus of data. This corpus has been subsequently used by several other authors, and, therefore, it is quite useful for comparing a variety of GI algorithms. This corpus contains data for 120 languages represented by NFAs and 120 represented by regular expressions. Each set of 120 languages was processed and distributed into four sets of 30 languages. A training set of 50 strings was generated for each language in the first set, and training sets of 100, 150 and 200 strings were generated for each language in the second, third, and fourth sets. The maximum length of the strings was set to 30. A test set of 1000 strings was also generated for every language in the corpus. Each set was guaranteed to contain, at least, 20 % of positive and negative strings.

Coste and Fredouille [14] extend this dataset to consider languages represented by DFAs and Unambiguous Finite Automata (UFAs). Their dataset is built following the distribution described above. Therefore, this new dataset contains information for 240 target languages (120 represented by DFAs and 120 represented by UFAs). Each one of the test sets for each language also contains 1000 strings.

In order to test the evolution behaviour of the algorithms with respect to the size data available, the training sets should be incremental. Note that neither of these corpora are incremental, and, furthermore, each set may contain several occurrences of the same string. In order to solve this problem, in [2], the author takes into account the distinct languages represented in these corpora and generates new training and test data for each language. The new corpus considers 102 languages represented by regular expressions, 120 languages represented by NFAs, and 119 languages represented by DFAs. For each of these languages, a training set of 500 strings was generated, labelled by the target language and incrementally distributed into five sets of 100, 200, 300, 400, and 500 strings. In addition, a disjoint set of 1000 strings was also generated to be used as the test set. The length of all the strings varied from 0 to 18. This corpus is the one that is used in the experimental comparison below.

In order to evaluate the *learning rate*, we first show the results of a naive baseline algorithm, the *Majority* algorithm. This algorithm considers the number of positive and negative strings in the training set. Thus, if there are more positive than negative strings, the algorithm classifies the whole test set as positive (otherwise, it accordingly classifies the whole test set as negative). Table 4.1 shows the results.

The first comparison we show includes the *RPNI*, *Blue-Fringe* and *DeLeTe2* algorithms (data from [2]). Table 4.2 shows the results. Taking into account the subset of regular expressions, all three algorithms obtained very good recognition rates. Nevertheless, *DeLeTe2* obtained the best results even with the smaller training sets. *DeLeTe2* was also the best algorithm when the NFA subset was considered. Note

Table 4.1 The results of the (naive) baseline algorithm using the experimental dataset

Tr. set ident.	*Majority* Rec. rate (%)
er_100	66.33
er_200	66.35
er_300	66.46
er_400	66.27
er_500	66.16
nfa_100	66.94
nfa_200	67.00
nfa_300	66.97
nfa_400	67.04
nfa_500	67.03
dfa_100	72.13
dfa_200	72.13
dfa_300	72.13
dfa_400	72.13
dfa_500	72.14

Table 4.2 The experimental results and average size of the inferred automata for the *Blue-Fringe*, *RPNI*, and *DeLeTe2* algorithms

	Blue-Fringe		*RPNI*		*DeLeTe2*	
	Rec. rate (%)	av. $\lvert A\rvert$	Rec. rate (%)	av. $\lvert A\rvert$	Rec. rate (%)	av. $\lvert A\rvert$
er_100	87.94	10	83.35	12.39	91.65	30.23
er_200	94.81	9.97	93.91	11.52	96.96	24.48
er_300	96.46	11.05	96,32	11.17	97.80	31.41
er_400	97.74	10.43	97.45	10.98	98.49	27.40
er_500	98.54	10.47	98.11	10.95	98.75	29.85
nfa_100	68.15	18.83	66.50	20.31	73.95	98.80
nfa_200	72.08	28.80	69.27	32.35	77.79	220.93
nfa_300	74.55	36.45	72.90	40.86	80.86	322.13
nfa_400	77.53	42.58	74.59	49.75	82.66	421.30
nfa_500	80.88	47.54	76.75	55.91	84.29	512.55
dfa_100	69.12	18.59	66.29	20.27	62.94	156.89
dfa_200	77.18	25.83	71.01	31.11	64.88	432.88
dfa_300	88.53	25.10	80.61	33.33	66.37	706.64
dfa_400	94.42	21.36	87.39	31.90	69.07	903.32
dfa_500	97.88	18.75	91.67	29.61	72.41	1027.42

that *RPNI* is the algorithm that obtained the lowest recognition rates with this subset. Finally, we considered the DFA subset. With this subset, the algorithm with the best behaviour up to that point becames the worst one and *Blue-Fringe* became the best one.

When the recognition rates obtained with the baseline algorithm are compared, note that the recognition rates of the three algorithms using the subset of regular expressions beat the rates obtained by the baseline algorithm. When the NFA subset is taken into account, on the one hand, both the *RPNI* and the *Blue-Fringe* algorithms had poor learning rates using the smaller training sets of the corpus. On the other hand, *DeLeTe2* notably improved the results of the baseline. Finally, the smaller training sets of the DFA subset were apparently not big enough for any of the algorithms, and all three obtained even lower recognition rates than the baseline. Despite this, *RPNI* and *Blue-Fringe* quickly improved their results and obtained close to 100 % results with the bigger training sets. Nevertheless, *DeLeTe2* was only able to reach the baseline rate with the bigger set.

In Table 4.3, we also show the comparative results of an instance of *WASRI*, *Blue-Fringe*, and *DeLeTe2*. This data is from [45] and considers the regular expression and the NFA subsets of the corpus.

Before commenting on the results, in order to properly compare the behaviour of the algorithms, the instance of *WASRI* used in the experiments should be described in detail. In this experimentation, two automata were inferred for each string in the training set. The first inference process considered the states of the maximal automaton for the string in canonical order. The second inference process considered the states in inverse canonical order. The smaller automaton was chosen. It should be noted that *WASRI* was the algorithm that best behaved with small training sets. An overall comparison shows that *WASRI* behaved better than *Blue-Fringe*, and also behaved in a way that is quite similar to *DeLeTe2* with respect to the recognition rates

Table 4.3 The experimental results of an instance of *WASRI* compared with the *Blue-Fringe* and the *DeLeTe2* algorithms

	Blue-Fringe	*WASRI*		*DeLeTe2*
	Rec. rate (%)	Rec. rate (%)	av. $\|A\|$	Rec. rate (%)
er_100	87.94	93.4	25.3	91.65
er_200	94.81	95.7	33.4	96.96
er_300	96.46	96.2	47.0	97.80
er_400	97.74	97.1	51.3	98.49
er_500	98.54	97.6	48.6	98.75
nfa_100	68.15	74.7	82.9	73.95
nfa_200	72.08	77.1	170.4	77.79
nfa_300	74.55	79.7	267.2	80.86
nfa_400	77.53	81.1	350.1	82.66
nfa_500	80.88	83.5	431.2	84.29

Table 4.4 The experimental results of *OIL* compared with the *Blue-Fringe* and *DeLeTe2* algorithms

	Blue-Fringe Rec. rate (%)	*OIL*			*DeLeTe2* Rec. rate
		Smallest $\|A\|$		Fair vote	
		Rec. rate (%)	av. $\|A\|$	Rec. rate (%)	
er_100	87.94	93.79	8.27	93.32	91.65
er_200	94.81	97.83	7.80	97.27	96.96
er_300	96.46	98.77	7.68	98.68	97.80
er_400	97.74	99.20	7.55	99.10	98.49
er_500	98.54	99.66	6.82	99.53	98.75
nfa_100	68.15	75.00	21.46	76.42	73.95
nfa_200	72.08	78.05	35.23	79.94	77.79
nfa_300	74.55	81.27	45.81	82.94	80.86
nfa_400	77.53	83.87	52.40	85.58	82.66
nfa_500	80.88	85.64	58.81	87.06	84.29
dfa_100	69.12	60.17	28.01	60.34	62.94
dfa_200	77.18	63.05	49.63	63.54	64.88
dfa_300	88.53	66.01	65.17	67.41	66.37
dfa_400	94.42	69.12	78.66	70.53	69.07
dfa_500	97.88	72.29	88.30	73.66	72.41

achieved. With regard to the size of the automata inferred, note that this instance of *WASRI* obtained smaller automata than *DeLeTe2*, but bigger than *RPNI* and *Blue-Fringe*.

In [27], the authors carried out a comparative experimentation of *OIL* and other algorithms. Due to the non-deterministic nature of *OIL* five runs of the algorithm were carried out in their work for each language in the corpus. The results of two different approaches are given: the first approach considers the classification rate of the smallest automaton obtained; the second approach carries out the classification using a poll where the weight of the vote of all the automata is the same. Table 4.4 summarizes these results.

Using the regular expression or NFA subsets of the corpus, both approaches behaved in a similar way and improved previous results obtained by *RPNI*, *Blue-Fringe*, *DeLeTe2*, or *WASRI*. Nevertheless, the performance of *OIL* with the DFA subset was quite similar to the performance of *DeLeTe2* and far from other algorithm rates.

We recall the results obtained by the inference of a team of automata. In [29], the authors proposed their method and used the same dataset described above to compare the experimental behaviour of the approach with *Blue-Fringe* and *DeLeTe2*. Several sizes of the team were considered, and the classification was carried out taking into account a weighted vote scheme (inverse to the square of the sizes of the automata). An option to select those automata that were smaller than the average size of the team

Table 4.5 The results obtained by the inference of teams of automata method

Set	Team Inference (81 FA)			Blue-Fringe	DeLeTe
	No select.	Sel. FA size smaller average			
	cl. rate	cl. rate	#*FA*		
er_100	95.19	95.55	50.00	87.94	91.65
er_200	98.32	98.45	58.24	94.81	96.96
er_300	98.78	98.87	60.58	96.46	97.80
er_400	99.24	99.32	63.45	97.74	98.49
er_500	99.42	99.49	65.91	98.54	98.75
nfa_100	77.24	77.45	40.51	68.15	73.95
nfa_200	80.96	81.25	41.28	72.08	77.79
nfa_300	83.46	83.77	41.26	74.55	80.86
nfa_400	85.14	85.50	41.47	77.53	82.66
nfa_500	86.71	86.98	42.25	80.88	84.29
dfa_100	76.68	76.83	40.48	69.12	62.94
dfa_200	83.13	84.04	36.00	77.18	64.88
dfa_300	90.04	91.59	36.00	88.53	66.37
dfa_400	95.24	96.48	36.31	94.42	69.07
dfa_500	98.16	98.68	37.69	97.88	72.41

was also studied. Table 4.5 shows the results obtained using a team of 81 automata. This table shows the size of the resulting team when the bigger automata are discarded (column #*FA*). Note that regardless of whether or not the selection of automata was carried out, the approach improved the classification rates obtained by both *Blue-Fringe* and *DeLeTe2*.

As Table 4.5 shows, comparing the average size of the automata inferred and leaving aside the inference of teams of automata where the size depends on the size of the team, the smallest automata were obtained by the *Blue-Fringe* algorithm followed by the *RPNI* and *OIL* algorithms (which obtained the smallest automata for the regular expression subset). The methods that obtained the largest hypotheses were *WASRI* and *DeLeTe2*.

4.6 Conclusions

The inference of finite deterministic automata has been thoroughly studied since it was proposed in the 1970s. As a result of this work, many methods have been proposed that provide solutions to the problem with good time and space complexity.

The consideration of different orderings in the process of state merging has been proved to be relevant, so these can be applied to practical tasks. Nevertheless, no order has been proved to be the best, and the final selection depends on the nature and characteristics of the practical task. Recently, it has been proved that, no matter what order is considered, convergence is not an issue when some minimum conditions are fulfilled.

Non-deterministic automata can be exponentially more concise than deterministic automata, and this can be of particular interest in some practical contexts. This is the main reason why the inference of non-deterministic finite automata arouses interest nowadays. Some works have proposed algorithms that consider different approaches.

The main drawback for the inference of non-deterministic automata is the lack of a canonical representative for any regular language. The definition and inference of residual automata is an attempt to obtain such a canonical representation. Other approaches are focused on obtaining a set of several automata, with the aim of achieving a team that is able to identify the target language so that each individual automata concisely recognizes a fragment of that language. Another drawback of these methods is that they have worse time complexity than those that infer deterministic automata.

In our opinion, the inference of non-deterministic models deserves to be studied further, and the study of time-efficient NFA inference or the inference of *small* NFAs may provide interesting results.

References

1. J. Abela, F. Coste, and S. Spina. Mutually compatible and incompatible merges for the search of the smallest consistent DFA. *LNAI*, 3264:28–39. 7th International Colloquium, ICGI-04. Springer. 2004.
2. G. I. Álvarez. *Estudio de la Mezcla de Estados Determinista y No Determinista en el Diseño de Algoritmos para Inferencia Gramatical de Lenguajes Regulares*. PhD Thesis in Computer Science, Departamento de Sistemas Informáticos y Computación. Universidad Politécnica de Valencia, 2007. in Spanish.
3. G. I. Álvarez, J. Ruiz, A. Cano, and P. García. Non-deterministic regular positive negative inference NRPNI. In J.F. Díaz, C. Rueda, and A.A. Buss, editors, *Proc. of the XXXI Conferencia Latinoamericana de Informática (CLEI2005)*, pages 239–249, 2005. ISBN:958-670-422-X.
4. D. Angluin. On the complexity of minimum inference of regular sets. *Information and Control*, 39:337–350, 1978.
5. D. Angluin. Inductive inference of formal languages from positive data. *Information and Control*, 45:117–135, 1980.
6. D. Angluin. Learning regular sets from queries and counterexamples. *Information and Computation*, 75:87–106, 1987.
7. D. Angluin. Queries and Concept Learning. *Machine Learning*, 2(4):319–342, 1988.
8. D. Angluin. Negative Results for Equivalence Queries. *Machine Learning*, 5(2):121–150, 1990.
9. R. C. Carrasco and J. Oncina. Learning stochastic regular grammars by means of a state merging method. *Lecture Notes in Artificial Intelligence*, 862:139–152, 1994.
10. J. M. Champarnaud and F. Coulon. NFA reduction algorithms by means of regular inequalities. *Theoretical Computer Science*, 327(3):241–253, 2004. Erratum in TCS 347:437-440 (2005).
11. O. Cichello and S. C. Kremer. Inducing grammars from sparse data sets: A survey of algorithms and results. *Journal of Machine Learning Research*, 4:603–632, 2003.
12. J. H. Conway. *Regular Algebras and Finite Machines*. Chapman & Hall, London, 1974.
13. F. Coste and D. Fredouille. Efficient ambiguity detection in C-NFA. a step towards the inference of non-deterministic automata. *LNAI*, 1891:25–38. 5th International Colloquium, ICGI-00. Springer. 2000.

14. F. Coste and D. Fredouille. Unambiguous automata inference by means of state-merging methods. *LNCS*, 2837:60–71, 2003. 14th European Conference ECML-03.
15. F. Coste and D. Fredouille. What is the search space for the inference of non-deterministic, unambiguous and deterministic automata? Technical Report 4907, INRIA, 2003. Internal Report Project Symbiose.
16. C. de la Higuera. A bibliographical study of grammatical inference. *Pattern Recognition*, 38:1332–1348, 2005.
17. C. de la Higuera. *Grammatical Inference. Learning Automata and Grammars*. Cambridge University Press, 2010.
18. C. de la Higuera, J. Oncina, and E. Vidal. Identification of DFA: data-dependent vs data-independent algorithms. *LNAI*, 1147:313–325. 3rd International Colloquium, ICGI-96. Springer. 1996
19. F. Denis, A. Lemay, and A. Terlutte. Learning regular languages using non-deterministic finite automata. *LNAI*, 1891:39–50. 5th International Colloquium, ICGI-00. Springer. 2000
20. F. Denis, A. Lemay, and A. Terlutte. Residual finite state automata. *Fundamenta Informaticae*, 51(4):339–368, 2002.
21. F. Denis, A. Lemay, and A. Terlutte. Learning regular languages using non-deterministic finite automata. *LNCS*, 1891:39–50, 2004.
22. F. Denis, A. Lemay, and A. Terlutte. Learning regular languages using RFSA. *Theoretical Computer Science*, 313(2):267–294, 2004.
23. P. García, D. López, and M. Vázquez de Parga. Polynomial characteristic sets for DFA identification. *Theoretical Computer Science*, 448:41–46, 2012.
24. P. García, A. Cano, and J. Ruiz. A comparative study of two algorithms for automata identification. *LNAI*, 1891:115–126. 5th International Colloquium, ICGI-00. 2000.
25. P. García and J. Ruiz. Learning in varieties of the form $V * LI$ from positive data. *Theoretical Computer Science*, 362(1–3):100–114, 2006.
26. P. García, J. Ruiz, A. Cano, and G. I. Álvarez. Is learning RFSAs better than learning DFAs? *LNCS*, 3845:343–344, 2005. 10th International Conference on Implementation and Application of Automata (CIAA-05).
27. P. García, M. Vázquez de Parga, G. I. Álvarez, and J. Ruiz. Learning regular languages using nondeterministic finite automata. *LNCS*, 5148:92–102, 13th International Conference on Implementation of Automata (CIAA'08). Springer. 2008.
28. P. García, M. Vázquez de Parga, G. I. Álvarez, and J. Ruiz. Universal automata and NFA learning. *Theoretical Computer Science*, 407(1-3):192–202, 2008.
29. P. García, M. Vázquez de Parga, D. López, and J. Ruiz. Learning automata teams. *LNAI*, 6339:52–65. 10th International Colloquium, ICGI-10. Springer. 2010.
30. P. García and E. Vidal. Inference of k-testable Languages in the Strict Sense and application to Syntactic Pattern Recognition. *IEEE Trans. Pattern Analysis and Machine Intelligence*, pages 920–925, 1990.
31. E. M. Gold. Language identification in the limit. *Information and Control*, 10:447–474, 1967.
32. E. M. Gold. Complexity of automaton identification from given data. *Information and Control*, 37:302–320, 1978.
33. G. Gramlich and G. Schnitger. Minimizing NFA's and regular expressions. *Journal of Computer and System Sciences*, 73(6):908–923, 2007.
34. J. E. Hopcroft and J. D. Ullman. *Introduction to Automata Theory, Languages and Computation*. Addison-Wesley, 1979.
35. L. Illie and S. Yu. Reducing NFAs by invariant equivalences. *Theoretical Computer Science*, 306(1-3):373–390, 2003.
36. T. Kameda and P. Weiner. On the state minimization of nondeterministic finite automata. *IEEE Trans. on Computers*, 19(7):617–627, 1970.
37. K. J. Lang, B. A. Pearlmutter, and R. A. Price. Results of the Abbadingo One DFA learning competition and a new evidence-driven state merging algorithm. *LNAI*, 1433:1–12. 4th International Colloquium, ICGI-98. Springer. 1998.

38. K.J. Lang. Random DFA's can be approximately learned from sparse uniform examples. In *Proceedings of the fifth annual workshop on Computational learning theory*, pages 45–52, 1992.
39. S. Lombardy and J. Sakarovitch. The universal automaton. In Jörg Flum, Erich Grädel, and Thomas Wilke, editors, *Logic and Automata*, volume 2 of *Texts in Logic and Games*, pages 457–504. Amsterdam University Press, 2008.
40. O. Matz and A. Potthoff. Computing small nondeterministic finite automata. In *Tools and Algorithms for the Construction and Analysis of Systems - TACAS '95*, volume NS-95-2 of *BRICS Notes Series*, pages 74–88, 1995.
41. J. Oncina and P. García. Inferring regular languages in polynomial updated time. *Pattern recognition and image analysis*, volume 1, pages 49–61. World Scientific, 1992.
42. J. Oncina, P. Garcia, and E. Vidal. Learning subsequential transducers for pattern recognition interpretation tasks. *IEEE Transactions on Pattern Analysis and Machine Intelligence*, 15(5):448–458, 1993.
43. L. Polák. Minimalizations of NFA using the universal automaton. *Int. J. Found. Comput. Sci.*, 16(5):999–1010, 2005.
44. B.A. Trakhtenbrot and Ya. M. Barzdin. *Finite automata. Behavior and Synthesis*. North-Holland, 1973.
45. M. Vázquez de Parga. *Autómatas finitos: Irreducibilidad e inferencia*. PhD Thesis in Computer Science, Departamento de Sistemas Informáticos y Computación. Universidad Politécnica de Valencia, 2008. in Spanish.
46. M. Vázquez de Parga, P. García, and J. Ruiz. A family of algorithms for non-deterministic regular language inference. *LNCS*, 4094:265–274, 2008. 12th International Conference on Implementation of Automata (CIAA'06).
47. T. Yokomori. Learning non-deterministic finite automata from queries and counterexamples. *Machine Learning*, 13:169–189, 1994.
48. T. Yokomori. Polynomial-time identification of very simple grammars from positive data. *Theoretical Computer Science*, 298:179–206, 2003.

Chapter 5
Learning Probability Distributions Generated by Finite-State Machines

Jorge Castro and Ricard Gavaldà

Abstract We review methods for inference of probability distributions generated by probabilistic automata and related models for sequence generation. We focus on methods that can be proved to learn in the inference in the limit and PAC formal models. The methods we review are state merging and state splitting methods for probabilistic deterministic automata and the recently developed spectral method for nondeterministic probabilistic automata. In both cases, we derive them from a high-level algorithm described in terms of the Hankel matrix of the distribution to be learned, given as an oracle, and then describe how to adapt that algorithm to account for the error introduced by a finite sample.

5.1 Introduction

Finite state machines in their many variants are accepted as one of the most useful and used modeling formalisms for sequential processes. One of the reasons is their versatility: They may be deterministic, nondeterministic, or probabilistic, they may have observable or hidden states, and they may be acceptors, transducers, or generators. Additionally, many algorithmic problems (determinization, minimization, equivalence, set-theoretic or linear-algebraic operations, etc.) are often computationally feasible for these models.

Learning from samples or observations of their behavior is one of the most important associated problems, both theoretically and practically, since good methods for the task offer a competitive alternative to expensive modeling by experts. It has been intensely studied in various communities, and particularly in the grammatical inference one. Here we concentrate on learning probabilistic finite automata that generate probabilistic distributions over strings, where more precisely the task is to come up

J. Castro (✉) · R. Gavaldà
LARCA Research Group, Universitat Politècnica de Catalunya-BarcelonaTech, Barcelona, Spain
e-mail: castro@lsi.upc.edu

R. Gavaldà
e-mail: gavalda@lsi.upc.edu

J. Heinz and J.M. Sempere (eds.), *Topics in Grammatical Inference*,
DOI 10.1007/978-3-662-48395-4_5

with a device generating a similar distribution. We focus on two formal models of learning (the identification in the limit paradigm and the PAC learning models) rather than on heuristics, practical issues, and applications.

The main goal of the chapter is to survey known results and to connect the research on merging/splitting methods, mainly by the grammatical inference community, with the recently proposed methods collectively known as "spectral methods". The latter have emerged from several communities that present them in very different lights, for example as instances of the method of moments, of principal component analysis methods, of tensor-based subspace learning, etc. Our goal is to present it as emerging as a relatively natural extension of the work on automata induction that starts with Angluin's $L^\star$ method [3] for DFA and continues with the work by Beimel et al. [11] on learning multiplicity automata from queries. The Hankel matrix representation of functions and the generalization from probabilistic automata to weighted automata are essential ideas here.

The chapter is organized as follows. Section 5.2 presents the preliminaries on languages, probability, finite-state machines, and learning models.

Section 5.3 surveys the existing results on identification-in-the-limit and PAC learning. We discuss the evidence pointing to the hardness of PAC learning probabilistic automata when the only measures of complexity of the target machine are the number of states and alphabet size. We then indicate how PAC learning becomes feasible if other measures of complexity are taken into account.

Section 5.4 introduces the two main notions on which we base the rest of our exposition: the Hankel matrix of a function from strings to real values, and weighted automata, which generalize DFA and probabilistic automata. We present three results that link automata size and properties of the Hankel matrix: the well-known Myhill–Nerode theorem for DFA; a theorem originally due to Schützenberger and rediscovered several times linking weighted automata size and rank (in the linear algebraic sense) of the Hankel matrix; and a similar characterization of the size of deterministic weighted automata in terms of the Hankel matrix, which we have not seen stated so far—although it may be known.

Section 5.5 presents a high-level algorithm for learning probabilistic deterministic finite automata using the Hankel matrix which distills the reasoning behind many of the state merging/splitting methods described in the literature. We then present (variants of) the ALERGIA method [13, 14] and of the method by Clark and Thollard [16] building on this formulation; additionally, we describe them using a recently introduced notion of statistical query learning for distributions, which we believe makes for a clearer presentation.

Section 5.6, in analogy with the previous one, presents a high-level algorithm for learning weighted automata using the Hankel matrix. We then derive a formulation of the spectral method as a way of dealing with the effect of finite size samples in that high-level algorithm. We also discuss a few optimizations and extensions in recent woks in Sect. 5.6.3.

Finally, Sect. 5.7 mentions a few open questions for further research. In an Appendix we describe the well-known Baum–Welch heuristics for learning HMM. Even though it does not fit into the formal models we discuss, the comparison with the other methods presented is interesting.

Since the literature in this topic is large, we have surely omitted many relevant references, either involuntarily or because they were out of our focus (rigorous results in formal models of learning). For example, recent works using Bayesian approaches to learning finite automata have not been covered because, while promising, they seem still far from providing formal guarantees. We have also omitted most references to probability smoothing, a most essential ingredient in any implementation of such methods; information on smoothing for automata inference can be found in [16, 23, 36, 42].

The reader is referred to the surveys by Vidal et al. [45, 46], Dupont et al. [24], and the book by de la Higuera [18] for background on the models and results by the grammatical inference community. A good source of information about spectral learning of automata is the thesis of Balle [17] and the paper [9]. The 2012 and 2013 editions of the NIPS conference hosted workshops dedicated to spectral learning, including but more general than automata learning.

5.2 Preliminaries

We denote by $\Sigma^\star$ the set of all strings over a finite alphabet Σ. Elements of $\Sigma^\star$ will be called strings or words. Given $x, y \in \Sigma^\star$ we will write xy to denote the concatenation of both strings. We use λ to denote the empty string which satisfies $\lambda x = x\lambda = x$ for all $x \in \Sigma^\star$. The length of $x \in \Sigma^\star$ is denoted by $|x|$. The empty string is the only string with $|\lambda| = 0$. We denote by Σ^t the set of strings of length t. A *prefix* of a string $x \in \Sigma^\star$ is a string u such that there exists another string v such that $x = uv$. String v is a *suffix* of x. Hence, for example, $u\Sigma^\star$ is the set of all strings having u as a prefix. A subset X of $\Sigma^\star$ is prefix-free whenever for all $x \in X$, if y is a prefix of x and $y \in X$ then $y = x$.

Several measures of divergence between probability distributions are considered. Let D^1 and D^2 be distributions over $\Sigma^\star$. The *Kullback–Leibler (KL) divergence* or *relative entropy* is defined as

$$\mathrm{KL}(D^1 \| D^2) = \sum_{x \in \Sigma^\star} D^1(x) \log \frac{D^1(x)}{D^2(x)},$$

where the logarithm is taken to base 2 and by definition $\log(0/0) = 0$.

The total variation distance is $\mathrm{L}_1(D^1, D^2) = \sum_{x \in \Sigma^\star} |D^1(x) - D^2(x)|$. The supremum distance is $\mathrm{L}_\infty(D^1, D^2) = \max_{x \in \Sigma^\star} |D^1(x) - D^2(x)|$. While KL is neither symmetric nor satisfies the triangle inequality, measures L_1 and L_∞ are true distances. We recall Pinsker's inequality, $\mathrm{L}_1 \leq \sqrt{2\mathrm{KL}}$, bounding the total variation distance in terms of the relative entropy. Thus, as L_1 obviously upperbounds L_∞, the relative entropy is, up to a factor, the most sensitive divergence measure among the ones considered here to distribution perturbations, and convergence criteria based on the KL value are most demanding.

Frequently, machine descriptions are provided in terms of vectors and matrices of real numbers and computations are defined by matrix products. We use square brackets to denote a specific component of a vector or matrix. For instance, component j of vector α is $\alpha[j]$. Row x of a matrix T is denoted by $T[x,:]$ and column y is $T[:,y]$. Vectors are always assumed to be columns. If α is a vector, a row vector α^T is obtained by transposing α.

5.2.1 Learning Distributions in the PAC Framework

We introduce the PAC model for learning distributions, an adaptation of Valiant's PAC model for concept (function) learning [44]. Let $\mathscr{D}$ be a class of distributions over some fixed set X. Assume $\mathscr{D}$ is equipped with some measure of *complexity* assigning a positive number $|D|$ to any $D \in \mathscr{D}$. We say that an algorithm A *PAC learns* a class of distributions $\mathscr{D}$ using $S(\cdot)$ examples and time $T(\cdot)$ if, for all $0 < \varepsilon, \delta < 1$ and $D \in \mathscr{D}$, with probability at least $1 - \delta$, the algorithm reads $S(1/\varepsilon, 1/\delta, |D|)$ examples drawn i.i.d. from D and after $T(1/\varepsilon, 1/\delta, |D|)$ steps outputs a hypothesis $\widehat{D}$ such that $L_1(D, \widehat{D}) \leq \varepsilon$. The probability is over the sample used by A and any internal randomization. As usual, PAC learners are considered efficient if the functions $S(\cdot)$ and $T(\cdot)$ are polynomial in all of their parameters.

Sometimes we will consider the relative entropy instead of the variation distance as a measure of divergence, which creates a different learning problem. Which specific measure we are considering in each PAC result will be clear from the context. Although the majority of PAC statements in the chapter are provided for L_1 all of them can be also shown for the KL divergence measure, at the cost of longer proofs. As the main proof ideas are the same for both measures, we have chosen mainly L_1 versions for simplicity.

Concerning the measure of complexity of distributions, standard practice is to fix a formalism for representing distributions, such as finite-state machines, and then consider the smallest, in some sense, representation of a given distribution in that formalism (usually exactly, but possibly approximately). In the case of finite-state machines, a seemingly reasonable measure of "size" is the number of states times the number of alphabet symbols, as that roughly measures the size of the transition table, hence of the automaton description. A first objection is that transition tables contain numbers, so this notion does not match well the more customary notion of bit-length complexity. One could refine the measure by assuming rational probabilities (which have reasonable bit-length complexity measures) or by truncating the probabilities to a number of digits that does not distort the probability distribution by more than about ε. We will see however that numbers of states and symbols alone do not fully determine the learning complexity of probabilistic finite-state machines, and that the bit-length of the probabilities seems irrelevant. This will motivate the (non-standard) introduction of further complexity parameters, some defined in terms of the finite-state machine, and some in terms of the distribution itself.

5.2.2 Identification in the Limit Paradigm

An alternative framework for learning distributions is the identification in the limit paradigm, originally introduced by Gold [27] for the setting of language learning. Later, the model was adapted in [19] to the learning distributions scenario. Basically, the model demands that with probability 1, given an infinite sample from the target, the learning algorithm with input the first m examples of the sample exactly identifies the target distribution when m is large enough. We consider here a slightly weaker definition, mainly because we consider state machines defined on real numbers instead of rational ones as in [19].

As before, let $\mathscr{D}$ be a class of distributions. We say that an algorithm A *identifies in the limit* a class of distributions $\mathscr{D}$ if for all $0 < \varepsilon < 1$ and $D \in \mathscr{D}$, given an infinite sample $x_1, x_2, \ldots$ of D

1. with probability 1, there exists $m_0(\varepsilon)$ such that for all $m \geq m_0(\varepsilon)$ algorithm A with input $x_1, \ldots, x_m$ outputs a hypothesis $\widehat{D}$ such that $\mathrm{L}_1(D, \widehat{D}) \leq \varepsilon$.
2. A runs in polynomial time in its input size.

It is easy to check that any PAC learning algorithm also achieves identification in the limit: Assume that identification in the limit does not occur. We have that with probability $\delta > 0$ there are arbitrarily large values m such that A with input $x_1, \ldots x_m$ outputs a hypothesis $\widehat{D}$ such that $\mathrm{L}_1(D, \widehat{D}) > \varepsilon$. This implies that PAC learning does not hold.

5.2.3 Probabilistic Automata

A *probabilistic finite automaton* (PFA) of size n is a tuple $\langle Q, \Sigma, \tau, \alpha_0, \alpha_\infty \rangle$ where Q is a set of n states, Σ is a finite alphabet, $\tau : Q \times \Sigma \times Q \to [0, 1]$ is the transition probability function and α_0 and α_∞ are respectively $[0, 1]^n$ vectors of initial and final probabilities. We require that $\sum_{i \in Q} \alpha_0[i] = 1$ and, for every state i, $\alpha_\infty[i] + \sum_{a \in \Sigma, j \in Q} \tau(i, a, j) = 1$. States i such that $\alpha_0[i] > 0$ ($\alpha_\infty[i] > 0$) are initial (final) sates. To each PFA D corresponds an underlying non-deterministic finite automaton (NFA) called the *support* of D, defined as $\langle Q, \Sigma, \delta, Q_I, Q_F \rangle$ where Q_I and Q_F are respectively the set of initial and final states, and transition function δ is defined as $\delta(i, a) = \{j | \tau(i, a, j) > 0\}$.

The transition probability function of a PFA D can be extended to strings in $\Sigma^\star$. Given $x \in \Sigma^\star$, we define $\tau(i, xa, j) = \sum_{k \in Q} \tau(i, x, k)\tau(k, a, j)$ and $\tau(i, \lambda, j) = 1$ if $i = j$ and 0 otherwise. A probability mass can be assigned to words in Σ defining:

$$D(x) = \sum_{i \in Q_I, j \in Q_F} \alpha_0[i]\tau(i, x, j)\alpha_\infty[j].$$

Defining transition matrices T_a for each $a \in \Sigma$ as $T_a[i, j] = \tau(i, a, j)$ and provided $x = x_1 \ldots x_m$ where $x_t \in \Sigma$ for $t = 1 \ldots m$, a more convenient expression in terms of matrix products for the probability mass of x is

$$D(x) = \alpha_0^T T_{x_1} \cdots T_{x_m} \alpha_\infty$$

which we write in short as $\alpha_0^T T_x \alpha_\infty$. Note that transition matrices are square $|Q| \times |Q|$ and define the transition function. So, an alternative tuple description for a PFA D in terms of transition matrices is $\langle \Sigma, \{T_a\}_{a \in \Sigma}, \alpha_0, \alpha_\infty \rangle$.

The probability mass induced by D defines a semi-distribution in $\Sigma^\star$: it satisfies $0 \leq \sum_{x \in \Sigma^\star} D(x) \leq 1$. Whenever from every state i there is non-zero probability of reaching a final state, D defines a true probability distribution, i.e. $\sum_{x \in \Sigma^\star} D(x) = 1$. In the rest of the chapter we will consider only PFAs defining true probability functions. More generally, every state i of a PFA D defines a probability distribution $D_i(x) = \gamma_i^T T_x \alpha_\infty$ where γ_i is the i-indicator vector $\gamma_i[j] = 1$ if $i = j$ and 0 otherwise. We have $D(x) = \sum_{i \in Q} \alpha_0[i] D_i[x]$.

5.2.4 Probabilistic Deterministic Automata and Distinguishability

A *probabilistic deterministic finite automaton* (PDFA for short) is a PFA whose support is a deterministic finite automaton (DFA). We note that for a PDFA we can assume without loss of generality—in short, "without loss of generality" (w.l.o.g.)—that the initial probability vector α_0^T is (1, 0, 0,..., 0) and each row of each transition matrix T_a has at most one non-zero component.

It is known [24] that PFA cannot exactly compute all probability distributions over $\Sigma^\star$, and that there are PFA computing distributions that cannot be exactly computed by any PDFA. That is, PFA are strictly more expressive than PDFA.

The following parameter will be useful to measure the complexity of learning a particular PDFA. It appears in [16] as defined here, although a similar idea is implicit in [36].

Definition 5.1 We say that distributions D^1 and D^2 are *μ-distinguishable* if $\mu \leq L_\infty(D^1, D^2)$. A PDFA D is μ-distinguishable when for each pair of states i and j their corresponding distributions D_i and D_j are μ-distinguishable. The *distinguishability* of a PDFA is defined as the supremum over all μ for which the PDFA is μ-distinguishable.

The distinguishability parameter can sometimes be exponentially small in the number of states. There exists reasonable evidence suggesting that polynomially learnability of PDFA in the number of states alone may not be achievable [30, 41]. However, PAC results have been obtained [16] when the inverse of distinguishability of the target is also considered, as we will see. We will also discuss parameters that play similar roles in PFA learning.

5.2.5 Hidden Markov Models

Hidden Markov models (HMM) are representations of choice for Markovian stochastic processes whose latent states are non-observable but their effects are. Visible effects are the observations arising from the process.

There are several formal definitions for HMM in the literature, but all of them are equivalent up to a polynomially transformation [24]. Often, observations in HMM are associated to states rather than transitions but for simplicity, we choose a definition closest to that of PFA. A HMM is a tuple $\langle Q, \Sigma, \tau, \alpha_0 \rangle$ where the four parameters have the same meaning as in the PFA definition. For any state i, it always holds $\sum_{a \in \Sigma, j \in Q} \tau(i, a, j) = 1$. Thus, one can see a HMM as a PFA having no stopping probabilities, i.e. as an infinite duration process. Transition probability matrices $\{T_a\}_{a \in \Sigma}$ are defined as in the PFA case, i.e. $T_a[i, j] = \tau(i, a, j)$.

A HMM defines, for each integer $t \geq 0$, a probability distribution on Σ^t. The probability mass assigned to string $x = x_1 \ldots x_t$ is $\alpha_0^T T_{x_1} \ldots T_{x_t} \mathbf{1}$ where $\mathbf{1}$ is the all-ones vector. This is the probability that the machine generates an infinite string of observations whose length-t prefix is x.

5.2.6 Weighted Automata

Weighted automata (WA) are the most general class of finite state machines we consider. They encompass all the models we have introduced before. A weighted automaton T over Σ with n states is a tuple $\langle \Sigma, \{T_a\}_{a \in \Sigma}, \alpha_0, \alpha_\infty \rangle$ where $T_a \in \mathbb{R}^{n \times n}$ and α_0 and α_∞ are vectors in $\mathbb{R}^n$. To each weighted automaton T corresponds a real-valued function defined on $\Sigma^\star$. Given $x = x_1 \ldots x_m \in \Sigma^\star$:

$$f_T(x) = \alpha_0^T T_{x_1} \cdots T_{x_m} \alpha_\infty = \alpha_0^T T_x \alpha_\infty.$$

We note that WA can be defined over semirings other than the real numbers, but whether WA are learnable (in any particular learning model) strongly depends on the semiring. For example, the high-level algorithms for real-valued WA that we present in Sects. 5.5.1 and 5.6.1 work in fact on any field, including finite fields. On the other hand, WA over the boolean semiring $(\vee, \wedge, 0, 1)$ are at least as expressive as DNF formulas, whose learnability in the PAC [44] or query [3] concept learning models are major open problems in computational learning theory.

A weighted automaton is deterministic (in short, a DWA) when rows of its transition matrices have at most one nonzero value and its initial vector α_0^T is $(1, 0, \ldots, 0)$.

5.3 A Panoramic View of Known Results

Around 1970, Baum and Welch described a practical heuristic for learning Hidden Markov Models generating infinite sequences; their properties were studied in [10], where it was shown to perform hill-climbing with respect to maximum likelihood. As such, it cannot be shown to learn in either of the two models we have presented (identification in the limit and PAC) but, even today, it is possibly the most used method in practice. Although it is described in many sources, we state it in our formalism in the Appendix for completeness and for comparison to the other methods we review.

Rudich [37] was, to our knowledge, the first to prove that Hidden Markov Models generating infinite sequences are identifiable in the limit. His method seems far from the PAC criterion in the sense that not only are there no bounds on the error on a finite sample, but even processing each observation involves cycling over all exponentially many possible state structures, hence it is very inefficient.

A decade later, Carrasco and Oncina [13, 14] described ALERGIA, a practical method for learning PDFA generating distributions on $\Sigma^\star$. We will describe a slight variant of ALERGIA, which we name the Red-Blue algorithm, in Sect. 5.5.2.1. ALERGIA has been highly influential in the grammatical inference community. Some of the (many) works that build on the ideas of ALERGIA to learn PDFA are [20, 31, 43]. All these algorithms are efficient in the sense that they work in time polynomial in the size of the sample, although no bounds are given on the number of samples required to reach convergence up to some ε. Apparently independently, [39] proposes the Causal State Splitting Reconstruction algorithm (CSSR) for inferring the structure of *deterministic* Hidden Markov Models. Although the underlying ideas are similar, their work differs in that the input consists of a single biinfinite string, not a sample of finite strings.

Still in the paradigm of identification in the limit, we mention the work by Denis and Esposito [21] who learn Residual Automata, a class strictly more powerful than PDFA but less than general PFA.

If we move to the PAC paradigm, the first result to mention on learning PFA is that of Abe and Warmuth [1], who show that they are PAC learnable in polynomial space. The method essentially iterates over all (exponentially many) state structures with a hypothesized number of states n, fitting probabilities in each according to the sample, and producing the one which assigns maximum likelihood to the sample. It can be shown that, for a sample of size polynomial in n, $|\Sigma|$, $1/\varepsilon$, and $\log(1/\delta)$, this hypothesis satisfies the PAC criterion of learning. The difficulty of learning PFA is thus computation, not information.

The same paper [1] shows that it is NP-complete to PAC learn PFA when the size of the alphabet Σ is unbounded, which in effect implies that all methods will require time exponential in $|\Sigma|$. Kearns et al. [30] showed that the problem may be hard even for 2-letter alphabets. They show that learning PDFA generalizes the *noisy parity learning problem,* which has received a fair amount of attention and for which all algorithms known so far take exponential time. Furthermore, only a small set of simple probabilities are required in the reduction (say, $\{i/8 \mid i = 0 \dots 8\}$). Terwijn [41] showed that indeed the problems of learning HMM and acyclic PDFA are hard under plausible cryptographic assumptions.

The previous results paint a rather discouraging panorama with respect to PAC learning of PFA and even PDFA. But note that, critically, these negative results all imply the hardness when the complexity of the target distribution (hence, the polynomiality of the PAC model) is defined to be the number of states times the alphabet size for the smallest machine generating it. A line of research starting in [36] aims at giving positive results by using other measures of complexity, in particular taking other parameters of the distribution that may not even be directly observable from a generating machine. Alternatively, one can view this line of research as designing sensible algorithms for learning PDFA/PFA, then analyzing their running time in search of the relevant features of the target distribution, instead of deciding, a priori, what the bounds on the running time should look like.

To be precise, Ron et al. [36] gave an algorithm for learning acyclic PDFA that can be shown to PAC learn with respect to the KL-divergence in time and sample size polynomial in the inverse of the distinguishability of the target machine, besides the usual parameters.

Later, Clark and Thollard [16] extended the result to cyclic automata; they introduce an additional dependence on the expected length of the strings in the distribution, L. We will describe the Clark–Thollard algorithm in detail in Sect. 5.5.2.2, under the name Safe-Candidate algorithm.

Extensions and variants of the Clark–Thollard algorithm include the following. Palmer and Goldberg [35] show that the dependence on L can be removed if learning only w.r.t. the L_1 distance is required. In another direction, Guttman et al. [28] show that PAC learning is still possible in terms of L_2-distinguishability, which is more restrictive that the L_∞-distinguishability we use here. The variations presented in [15, 26], while retaining the PAC guarantees, aim at being more efficient in practice. In [5] the algorithm is extended to machines whose transitions have random durations, determined by associated probability distributions that must be learned too. In [7], the algorithm is transported to the so-called *data stream* paradigm, where data (strings) arrive in sequence and the algorithm is required to use sublinear memory and low time per item.

In substantial breakthroughs, Mossel and Roch [33], Denis et al. [22], Hsu et al. [29], and Bailly et al. [4] gave algorithms having formal proofs of PAC learning the full class of PFA. The sample size and running times of the algorithms depend polynomially in the inverse of some quantity of a spectral flavor associated to the Hankel matrix of the target distribution. This is, for example, the determinant in [33] and the inverse of its nth singular value in [29]. Denis et al. [22] do not explicitly state a PAC learning result, but in our opinion they refrain from doing so only because they lack a proper name for the feature of the distribution that determines their algorithms' running times.

5.4 The Hankel Matrix Approach

In this section we review the notion of the Hankel matrix and weighted automata, which will be the main tools for describing generic methods for learning PDFA and PFA.

The *Hankel matrix* [25] H_f of a function $f : \Sigma^\star \to \mathbb{R}$ is a matrix representation of f in $\mathbb{R}^{\Sigma^\star \times \Sigma^\star}$ defined as:

$$H_f[x, y] = f(xy), \forall x, y \in \Sigma^\star.$$

Despite the fact that the Hankel matrix H_f is a very redundant representation—value $f(z)$ is represented $|z| + 1$ times—the computability of f by finite state machines is determined by algebraic properties of H_f.

Assume function f is a language, i.e. $f : \Sigma^\star \to \{0, 1\}$. For each $x \in \Sigma$ we consider the subset $x^{-1}f = \{y \in \Sigma^\star | f(xy) = 1\}$. The well-known Myhill–Nerode theorem claims that f is a regular language if and only if there are only a finite number of different sets $x^{-1}f$ when x varies on $\Sigma^\star$. The number of different sets $x^{-1}f$ also determines the minimum DFA size required for computing f. Thus, translating this theorem in terms of the Hankel matrix, we have the following.

Theorem 5.1 *Function $f : \Sigma^\star \to \{0, 1\}$ is regular if and only if the Hankel matrix H_f has a finite number n of different rows. Moreover, the minimum size of a DFA computing f is n.*

A similar characterization can be shown for functions on $\Sigma^\star$. Given a function $f : \Sigma^\star \to \mathbb{R}$, an equivalence relation on $\Sigma^\star$ can be defined as follows. Words x and y are related—written $x \sim_f y$—iff their corresponding rows in the Hankel matrix H_f are the same up to an nonzero scalar factor, i.e. $H_f[x, :] = c_{x,y} H_f[y, :]$ for some scalar $c_{x,y} \neq 0$. Let $\Sigma^\star/\sim_f$ be the quotient set. When matrix H_f has an identically zero row, this set has a class—the zero class—representing all zero rows. We propose the following theorem that characterizes the computability of f by deterministic weighted automata in terms of the cardinality of the quotient set.

Theorem 5.2 *Let $f : \Sigma^\star \to \mathbb{R}$ be any function. There is a DWA computing f if and only if the cardinality of $\Sigma^\star/\sim_f$ is finite. Moreover, the minimum size of a DWA computing f is the number n of nonzero classes in $\Sigma^\star/\sim_f$.*

Proof We shorten $\sim_f$ to $\sim$ in this proof. We prove first the if part. Assume that words x and y are $\sim$-related and let $c_{x,y}$ be a scalar such that $H_f[x, :] = c_{x,y} H_f[y, :]$. It is easy to see that

1. the value of $c_{x,y}$ is uniquely determined except when x and y are in the zero class, in which case we define $c_{x,y}$ as 0, and
2. for any strings x_1, x_2, x_3, u, v, if $x_1u \sim x_2$ and $x_2v \sim x_3$, then $x_1uv \sim x_3$ and $c_{x_1uv,x_3} = c_{x_1u,x_2} c_{x_2v,x_3}$.

The last property uses the redundancy of the Hankel matrix, namely, for every z,

$$\begin{aligned} H_f[x_1uv, z] &= f(x_1u \cdot vz) = c_{x_1u,x_2} f(x_2 \cdot vz) = c_{x_1u,x_2} f(x_2v \cdot z) \\ &= c_{x_1u,x_2} c_{x_2v,x_3} f(x_3 \cdot z) = c_{x_1u,x_2} c_{x_2v,x_3} H_f[x_3, z] \end{aligned}$$

from where, by definition of $\sim$, it is $x_1uv \sim x_3$ with scalar $c_{x_1uv,x_3} = c_{x_1u,x_2} c_{x_2v,x_3}$.

Let $x^1, \ldots x^n$ be representatives of the n nonzero classes of $\Sigma^\star/\sim$. W.l.o.g. we assume $n \geq 1$—otherwise f is the null function—and $x^1 = \lambda$. We define for each $a \in \Sigma$ the transition matrix $T_a \in \mathbb{R}^{n\times n}$ as $T_a[i, j] = c_{x^i a, x^j}$ if $x^i a \sim x^j$ and 0 otherwise. It is immediate to see that every row of T_a has at most one nonzero value.

Let γ_i be the i-indicator vector, with $\gamma_i[j] = 1$ if $i = j$ and 0 otherwise. We show by induction on $|w|$ that for any word w we have

$$\gamma_i^T T_w = \begin{cases} \mathbf{0} & \text{if } x^i w \text{ is in the zero class} \\ c_{x^i w, x^j}\, \gamma_j^T & \text{for the } j \text{ such that } x^i w \sim x^j \text{ otherwise.} \end{cases}$$

The equality is obvious for $|w| = 0$, since T_w is the identity. Let $w = au$ for some alphabet symbol a. We consider two cases. In the first one we assume $x^i a$ belongs to a nonzero class represented by x^k. Observe that by definition of T_a we have $\gamma_i^T T_a = c_{x^i a, x^k} \gamma_k^T$. Assuming $x^k u$ belongs to a nonzero class represented by x^j, we have

$$\gamma_i^T T_{au} = \gamma_i^T T_a T_u = c_{x^i a, x^k} \gamma_k^T T_u = c_{x^i a, x^k}\, c_{x^k u, x^j} \gamma_j^T .$$

where the last equality follows from the induction hypothesis. By property (2) above, $x^i au \sim x^j$ and $c_{x^i au, x^j} = c_{x^i a, x^k}\, c_{x^k u, x^j}$ as required. If $x^k u$ is in the zero class, by transitivity $x^i au$ also belongs to the zero class. By the induction hypothesis

$$\gamma_i^T T_{au} = \gamma_i^T T_a T_u = c_{x^i a, x^k} \gamma_k^T T_u = c_{x^i a, x^k}\, \mathbf{0} = \mathbf{0}$$

as required. This concludes the inductive claim in the first case. For the second one, note that $\gamma_i^T T_a$ is the zero vector and the claim is obvious.

Consider the DWA T with matrices T_a defined above, initial vector $\alpha_0 = \gamma_1$ and $\alpha_\infty^T = (f(x^1), \ldots, f(x^n))$. Then for any string w, if $w \sim x^k$,

$$T(w) = \alpha_0^T\, T_w\, \alpha_\infty = \gamma_1^T\, T_w\, \alpha_\infty = c_{\lambda w, x^k}\, \gamma_k^T\, \alpha_\infty = c_{w, x^k}\, f(x^k),$$

and then we have

$$f(w) = H_f[w, \lambda] = c_{w,x^k} H_f[x^k, \lambda] = c_{w,x^k} f(x^k) = T(w).$$

We consider now the only-if part. Let $\langle \{T_a\}_{a\in\Sigma}, \alpha_0, \alpha_\infty \rangle$ be a DWA of size n computing a function f. We say that a matrix is deterministic when all rows have at most one nonzero value. We note that the product of deterministic matrices is also deterministic. So, for any string x, matrix T_x is deterministic and $\alpha_0^T\, T_x$ must be either $c_x \gamma_j^T$ for some nonzero scalar c_x and integer $j \in \{1 \dots n\}$ or the zero vector. Let h be an integer function on $\Sigma^\star$ such that $h(x) = j$ when $\alpha_0^T T_x = c_x \gamma_j^T$ for some nonzero scalar c_x and $h(x) = 0$ when $\alpha_0 T_x$ is the zero vector. Note that the cardinality of the range of h is at most $n + 1$. We show that if $h(x) = h(y)$ then rows x and y of the Hankel matrix H_f are the same up to a nonzero factor. The result is obvious when $h(x) = h(y) = 0$. Assume $h(x) = h(y) = j > 0$. For any string z,

$$\begin{aligned} H_f[x, z] &= \alpha_0^T T_x T_z \alpha_\infty = c_x \gamma_j^T T_z \alpha_\infty = (c_x/c_y) c_y \gamma_j^T T_z \alpha_\infty \\ &= (c_x/c_y) \alpha_0^T T_y T_z \alpha_\infty = (c_x/c_y) H_f[y, z]. \end{aligned}$$

Thus, up to nonzero scalar factors, the Hankel matrix of f has at most n nonzero rows. □

As an example, consider $\Sigma = \{a\}$ and function f defined on $\Sigma^\star$ having values $f(\lambda) = 0$ and $f(a^k) = 1/2^k$ when $k \geq 1$. It is easy to check that, up to nonzero scalar factors, the Hankel matrix H_f has only two different rows, $(0, 1/2, 1/4, \dots)$ and $(1/2, 1/4, 1/8, \dots)$, corresponding, respectively, to words λ and a. Following the proof of Theorem 5.2 the DWA defined by $\alpha_0^T = (1, 0)$, $\alpha_\infty^T = (0, 1/2)$ and matrix T_a having rows (0, 1) and (0, 1/2) computes function f.

Finally, the full class of weighted automata can also be characterized by an algebraic parameter, in this case the number of linearly independent rows—i.e. the rank—of the Hankel matrix. The characterization of weighted automata in terms of the rank has been shown by several authors [11, 12, 25, 38]. We follow the exposition in [11].

Theorem 5.3 *Let $f : \Sigma^\star \to \mathbb{R}$ be a function with Hankel matrix H_f. Function f can be computed by a weighted automaton if and only if the rank n of H_f is finite. Moreover, the minimum size of a WA computing f is n.*

Proof Only-if. Let $\langle \{T_a\}_{a \in \Sigma}, \alpha_0, \alpha_\infty \rangle$ be a weighted automaton of size n computing f. We define backward and forward matrices $B \in \mathbb{R}^{\infty \times n}$ and $F \in \mathbb{R}^{n \times \infty}$ with rows, respectively columns, indexed by words in $\Sigma^\star$ as:

$$\begin{aligned} B[x, :] &= \alpha_0^T T_x, \\ F[:, y] &= T_y \alpha_\infty. \end{aligned}$$

From the fact that $H_f[x, y] = \alpha_0 T_x T_y \alpha_\infty = B[x, :] F[: y]$ we conclude that $H_f = BF$ and therefore $\text{rank}(H_f) \leq \text{rank}(F) \leq n$.

For the if part let $x^1, \dots, x^n$ be words in $\Sigma^\star$ indexing n linearly independent rows of H_f. W.l.o.g. we can assume $x^1 = \lambda$, as otherwise f is the null function. We consider the weighted automaton of size n defined by vectors $\alpha_0^T = (1, 0, \dots, 0)$, $\alpha_\infty^T = (f(x^1), \dots, f(x^n))$ and transition matrices $T_a \in \mathbb{R}^{n \times n}$ for $a \in \Sigma$ with values $T_a[i, j] = a_j^i$ satisfying:

$$H_f[x^i a, :] = \sum_j a_j^i H_f[x^j, :].$$

These values exist because by hypothesis, H_f is a rank n matrix and strings x^j for $j = 1, \ldots, n$ are indexes of n linear independent rows.

We show that $f(x^i v) = \gamma_i^T T_v \alpha_\infty$ for $i = 1, \ldots, n$, where γ_i is the i-indicator vector. Once the proof is completed, the if part of the theorem follows from considering the equality for $i = 1$. We induct on the length of v. When $|v| = 0$ the equality is immediate. Assume $v = au$ for some $a \in \Sigma$. We have:

$$f(x^i v) = f(x^i au) = H_f[x^i a, u] = \sum_j a_j^i H_f[x^j, u] = \sum_j a_j^i f(x^j u).$$

By induction hypothesis, $f(x^j u) = \gamma_j^T T_u \alpha_\infty$. Thus,

$$f(x^i v) = \sum_j a_j^i (\gamma_j^T T_u \alpha_\infty) = (\sum_j a_j^i \gamma_j^T) T_u \alpha_\infty = (\gamma_i^T T_a) T_u \alpha_\infty = \gamma_i^T T_v \alpha_\infty.$$

□

We remark that Theorems 5.2 and 5.3 work on any field, not just the real numbers. We would like to see a derivation of Theorem 5.2 as a consequence of Theorem 5.3 plus the assumption of determinism, instead of having to reprove it from scratch.

5.5 Learning PDFA

5.5.1 An Oracle Algorithm for Learning DWA

In this section we present a high-level algorithm for learning Deterministic Weighted Automaton assuming that we have *oracle* access to the function $f : \Sigma^\star \to \mathbb{R}$. This model differs from the one considered so far for probabilistic automata in that we receive the exact probabilities of strings instead of estimating them from a sample, and that learning must be exact and not approximate. Furthermore, this model allows us to discuss functions other than probability distributions, for which estimating function values from samples does not even make sense, and even functions mapping strings to finite fields. At the end of the subsection, we will discuss how to turn this high-level algorithm into a sample-based one for the case of PFA, and give specific (and more efficient) implementations in the following sections.

The high-level algorithm is given in Fig. 5.1. To clarify the implementation of line 3, initially set $X' = \{\lambda\}$ and repeatedly pick any $z \in X - X'$ such that $H[z, :]$ differs from all $H[x, :]$ with $x \in X'$, while such a z exists, and add it to X'. It is clear that the final X' is as desired. It is also clear that, excluding the time required to answer oracle queries, the algorithm runs polynomial time in $|\Sigma|$ and the sum of the lengths of strings in X and Y.

1. Choose any two finite sets $X, Y \subseteq \Sigma^\star$ with $\lambda \in X \cap Y$, in some way not specified by the algorithm;
2. Build the submatrix $H = H_f[X \cup X\Sigma, Y]$ of H_f by asking oracle queries on f;
3. Find a minimal $X' \subseteq X$ such that $\lambda \in X'$ and for every $z \in X$, $H[z,:]$ is a multiple of $H[x,:]$ for $x \in X'$;
4. Build a DWA from H, X, Y, X' as follows. Say $X' = \{x^1 = \lambda, x^2, \ldots, x^n\}$, then

$$\alpha_0^T = (1, 0, \ldots, 0)$$
$$\alpha_\infty^T = (f(x^1), \ldots, f(x^n))$$
$$T_a[i, j] = v \text{ if } H[x^i a, :] = vH[x^j, :], \text{ and 0 otherwise;}$$

Fig. 5.1 Learning DWA with an oracle

Following the proof of Theorem 5.2, it is now easy to prove that the algorithm is correct if it is given "right" sets X and Y.

Theorem 5.4 *If the cardinality of X' is at least the number of nonzero rows of H_f up to nonzero scalar factors, then the DWA generated by the algorithm computes f.*

We do not discuss here the question of how to find suitable sets X, Y that provide a large enough X' in the oracle model. We instead move to the problem of translating this algorithm to the setting in which f is a probability distribution and all the information about f is given by some sample S, a multiset of strings.

The obvious choice is to set $X = \mathit{prefixes}(S)$, $Y = \mathit{suffixes}(S)$, and then create an approximation $\hat{H}$ of H by $\hat{H}[x, y]$ = empirical probability of xy in S. Note that although X and Y are sets, the approximation $\hat{H}[x, y]$ is still computed taking into account the possible repetitions of xy in the multiset S. Now, the question "is $H[x, :]$ a multiple of $H[z, :]$?" becomes a statistical question on $\hat{H}$ for a finite S. If the answer is "no", it will become clear as S grows in size, but if the answer is "yes" no finite amount of information will prove it conclusively. Let the algorithm use a statistical test that will return the right yes/no answer with probability tending to 1 as the sample size tends to infinity. (Different statistical tests have been used in the literature for specific instantiations of this general algorithm.) For sufficiently large samples, representatives of all equivalence classes of rows will appear in the sample and furthermore the test will correctly answer all the finitely many row equivalences. It is now easy to argue that the algorithm above endowed with such a statistical test will identify the target probability distribution in the limit.

In the PAC paradigm, however, tests should be sufficiently reliable for samples of polynomial size *and* be computationally efficient. Unfortunately, such tests may not exist if the target machine complexity is defined as the number of states times the size of the alphabet. This is precisely what the negative results in [1, 30, 41] formalize. It may become possible if one restricts the class of target machines in some way, such as requiring a minimum L_∞-distinguishability among states; alternatively, letting the inverse of the distinguishability be also a parameter of complexity. This will be the case for, for example, the method by Clark and Thollard [16].

Most methods for learning PDFA described in the literature conceptually rely on the DWA construction described above, with a number of refinements towards practicality. For example, they tend to identify states not in one shot from the matrix H, but instead by iteratively splitting or merging simpler states. This has the advantage that once several strings (rows) have been merged, their statistical evidence accumulates, allowing for sounder decisions from then on. Additionally, they are able to come up with a PDFA when it is guaranteed that the target is a PDFA—while our construction here may return DWA which are not PDFA. Finally, they differ in the precise statistical tests for state identification and in the smoothing policies to account for unseen mass, both of which may make a large difference in practice.

5.5.2 State-Merging Learning Algorithms

State-merging algorithms form an important class of strategies of choice for the problem of inferring a regular language from samples. Basically, they try to discover the target automaton graph by successively applying tests in order to discover new states and merge them to previously existing ones according to some similarity criteria.

We show below two adaptations of the state-merging strategy to PDFA learning, the Red-Blue [14] and the Safe-Candidate [16] algorithms. Both infer gradually elements of the target graph, i.e. states and transitions, and estimate their corresponding probabilities. The Red-Blue algorithm starts by inferring a prefix tree acceptor and then merges equivalent states. The Safe-Candidate algorithm alternates between inferring and merging new elements. While the first one achieves learning in the limit, the second one has PAC guarantees whenever a polynomial dependence on an additional parameter— the target distinguishability— is accepted.

The usual description of these algorithms considers the learning from examples paradigm. In contrast, following [6, 17], the exposition below assumes a query framework: algorithms get information on the target by asking queries of possibly different kinds instead of analyzing a sample. Proceeding in this way, many details concerning probability approximation issues can be abstracted, and a more compact, clear and elegant presentation can be provided.

The query model we use allows two type of queries, so-called respectively *statistical* and L_∞-*queries* that can be solved, with high probability, by drawing a sample of the target distribution. Thus, learning algorithms in this query setting can be easily moved to standard learning from examples.

Given a distribution D on $\Sigma^\star$ a *statistical query for D* is a tuple (X, α) where X is an efficiently decidable subset of $\Sigma^\star$ and $0 < \alpha < 1$ is some *tolerance* parameter. The query $\mathsf{SQ}^D(X, \alpha)$ returns an α-approximation $\hat{p}$ of $D(X)$ such that $|\hat{p} - D(X)| < \alpha$.

Let X be a prefix-free subset of $\Sigma^\star$. Function $D^X(y) = \frac{D(Xy)}{D(X\Sigma^\star)}$ defines a probability distribution on $\Sigma^\star$ that corresponds to the distribution over suffixes conditioned on having a prefix in X. An L_∞-query is a tuple (X, Y, α, β) where X and Y are efficiently decidable, disjoint and prefix-free subsets of $\Sigma^\star$, and $0 < \alpha, \beta < 1$ are, respectively, the *tolerance* and *threshold* of the query. Query $\mathsf{DIFF}_\infty^D(X, Y, \alpha, \beta)$ is answered by the oracle according to the following rules:

1. If either $D(X\Sigma^\star) < \beta$ or $D(Y\Sigma^\star) < \beta$ it answers '?'.
2. If both $D(X\Sigma^\star) > 3\beta$ and $D(Y\Sigma^\star) > 3\beta$, it answers with some α-approximation $\hat{\mu}$ of $L_\infty(D^X, D^Y)$.
3. Otherwise, the oracle may either answer '?' or give an α-approximation of $L_\infty(D^X, D^Y)$.

Both query types can be easily simulated with high probability with a sample of distribution D. The following result holds.

Proposition 5.1 *For any probability distribution D, a statistical query* $\mathsf{SQ}^D(X, \alpha)$ *can be simulated using* $O(\alpha^{-2}\log(1/\delta))$ *examples from D. A* $\mathsf{DIFF}_\infty^D(X, Y, \alpha, \beta)$ *query can be solved using* $\tilde{O}(\alpha^{-2}\beta^{-2}\log(1/\delta))$ *examples. In both cases, the error probability is less than* δ.

5.5.2.1 The Red-Blue Algorithm

State merging algorithms were initially proposed for the regular language inference problem. Gold [27] had shown that regular languages are identifiable in the limit from positive and negative data, but consistency with the input sample was not guaranteed when the sample does not contain some crucial information—the so-called *characteristic set*. Oncina and García [34] proposed a state merging algorithm for inferring regular languages that overcomes this drawback. The algorithm always returns a DFA that is data consistent and, provided a complete presentation is given, achieves identification in the limit.

Carrasco and Oncina in [14] adapted the DFA state merging learning algorithm to the stochastic setting. The new algorithm, so-called ALERGIA, was shown to identify in the limit any PDFA provided stochastic examples of the target are given. We present here a version of ALERGIA in the query model we have just introduced that considers statistical and L_∞-queries. We rename this version the Red-Blue algorithm.

Given an integer m as input, Red-Blue starts by building a prefix tree acceptor representing significant prefixes by calling the PTA function on parameters m and λ, see Figs. 5.2 and 5.3. On these values, PTA returns a prefix tree DFA whose leaves at level j correspond to prefixes of probability at least $2^j/m$. Thus, the resulting tree DFA has depth at most $\lceil \log m \rceil$. In order to decide whether a prefix has enough probability to be represented, PTA makes statistical queries.

Once the tree acceptor is built, Red-Blue colors the initial state as red and its direct descendants as blue, leaving other states uncolored, and starts merging compatible states. Two states are considered merge-compatible when a call to the DIFF_∞^D oracle returns a numerical value not exceeding $1/m$, meaning that there is not strong evidence that they define different distributions. Specifically, we define for each state q a prefix free subset $H[q]$ consisting of words x such that $\tau(q_0, x) = q$ and for any prefix y of

```
input: integer m
output: PDFA H
algorithm Red-Blue
    H ← PTA(m, λ)
    Red ← {q_λ};
    Blue ← {τ(q_λ, σ), σ ∈ Σ}
    Let α ← 1/m
    while Blue ≠ ∅
        pick some state b ∈ Blue
        if there is r ∈ Red with DIFF^D_∞(H[r], H[b], α, α) ≤ α
            H ← Merge(r, b, H)
        else Red ← Red ∪ {b}
        Blue ← Blue − {b} ∪ {τ(b, σ) | σ ∈ Σ and τ(b, σ) is uncolored}
    for q ∈ H
        γ(q, ξ) ← SQ^D(H[q], α)/SQ^D(H[q]Σ*, α)
        for σ ∈ Σ such that τ(q, σ) is defined
            γ(q, σ) ← SQ^D(H[q]σΣ*, α)/SQ^D(H[q]Σ*, α)
return H
```

Fig. 5.2 Red-Blue algorithm

Fig. 5.3 PTA algorithm

```
input: integer m and string w ∈ Σ*
output: tree DFA H
algorithm PTA
    Set a initial state q_w of H
    Let α ← 1/m
    for each σ ∈ Σ
        if SQ^D(wσΣ*, α) > 3α
            H_σ ← PTA(⌈m/2⌉, wσ)
            Set a new transition τ(q_w, σ) = q_wσ
return H
```

x it holds $x = y$ or $\tau(q_0, y) \neq q$. Asking the query $\mathsf{DIFF}^D_\infty(H[q_1], H[q_2], \alpha, \alpha)$ we get information about how different the distributions defined by states q_1 and q_2 are.

The merging flow proceeds as follows. For an arbitrarily chosen blue state, either there is a red state that is merge-compatible with it or no red state is. In the first case, both states are merged and the result colored red. This may introduce nondeterministic choices, which are eliminated by further merging in the merge procedure. Note that every merge always involves a blue state. On the other hand, if there is no merge-compatible red state for this blue state, it is promoted to red and new blue states are generated from it. When the set of blue states is empty, Red-Blue stops merging and a PDFA is returned by setting transition probabilities of H according to prefix probability approximations obtained by issuing statistical queries.

```
input: DFA H and states q and q' of H
output: DFA
algorithm Merge
    Replace each occurrence q' in the description of H by q
    while H contains a nondeterministic transition
        Let p and p' be target states of a nondeterministic choice
        Merge(p, p', H)
    return H
```

Fig. 5.4 Merge function

Figures 5.2 and 5.4 show the Red-Blue and the merge algorithms. Let H be the resulting automaton after some iterations of Red-Blue. Red states of H and transitions between them represent the part of the graph we trust as correct, based on processed information. An invariant of the algorithm is that non-red states are always roots of trees in H and blue states are always direct successors of a red state.

Assuming unit time for oracle calls, the complexity of Red-Blue is $O(m^2)$. Moreover, if m is large enough that every state or transition of D has a counterpart in the tree DFA returned by the PTA algorithm and such that $2/m$ is less than the distinguishability of the target machine, the algorithm on input m learns a PDFA hypothesis whose graph agrees with the target graph. As probability estimations of states and transitions will improve as m is larger, we have the following.

Theorem 5.5 *The Red-Blue algorithm learns every PDFA in the identification in the limit paradigm.*

This result appears (for the equivalent ALERGIA algorithm) in [13, 14]. But from this point it is easy to argue that Red-Blue also learns in the PAC model if the complexity of the target PDFA is taken to be the number of states times the alphabet size times the inverse of the distinguishability.

5.5.2.2 The Safe-Candidate Algorithm

We describe a variant of the algorithm in [16], which was the first one to claim any formal PAC guarantee. Our version fits the query framework in [6, 17], which makes the exposition easier and simplifies correctness arguments.

The main differences of the presentation below with respect to the description in [16] are two. First, as said, the algorithm gets information on the target distribution asking statistical and L_∞-queries defined above instead of analyzing a sample. Second, it guarantees a good L_1 approximation, a weaker requirement than the good relative entropy approximation guaranteed in [16]. The latter choice avoids some rather subtle points in [16] by the fact that L_1 is a true distance unlike KL.

The Safe-Candidate algorithm works in two stages. In the first one, it builds a transition graph that is isomorphic to the target subgraph formed by important elements— these are states and transitions whose probability of being visited while generating a random string is above a threshold defined by the input parameters. The construction uses both statistical and L_∞-queries. The second stage consists of converting the learned graph into a PDFA by estimating the transition and stopping probabilities corresponding to each state. This is similar to the estimation step in the Red-Blue algorithm performed once the set of blue states is empty, see Fig. 5.2. In this stage, only statistical queries are necessary.

Figure 5.5 shows the code for the graph construction stage. The algorithm keeps two sets of nodes, the set of safe nodes and the set of candidates. Safe nodes and transitions between them are known to be correct. Initially, there is only a safe state q_λ corresponding to the empty word and one candidate q_λ^σ for each $\sigma \in \Sigma$. Each candidate represents a still unknown transition in the graph H. In each iteration, statistical queries are performed to choose the most informative candidate q and, after that, L_∞-queries are issued in order to decide if the candidate is either insignificant at all, it is an already known safe node q' or it is a new safe one. In the latter case, when a candidate is promoted to safe, new candidates are considered representing undefined transitions leaving the new safe node. The algorithm finishes when there

input: $n, \mu, \Sigma, \varepsilon$, oracles DIFF_∞^D and SQ^D
output: A graph H
algorithm Safe-Candidate
$\alpha \leftarrow \mu/2$; $\beta \leftarrow \varepsilon/24n|\Sigma|$
initialize the graph H with a safe q_λ and candidates q_λ^σ for $\sigma \in \Sigma$
while H has candidates
  choose a candidate q maximizing $\mathsf{SQ}^D(H[q]\Sigma^\star, \beta)$
  foreach safe q'
    make a call to the oracle $\mathsf{DIFF}_\infty^D(H[q], H[q'], \alpha, \beta)$
    if the answer is '?'
      remove q from the candidate list
      break
    if the answer $\hat{\mu} < \mu/2$
      merge q and q'
      remove q from the candidate list
      break
  if q is still a candidate
    promote q to safe
    add candidates q^σ for each $\sigma \in \Sigma$

Fig. 5.5 Safe-Candidate graph construction

are no candidates left. An inductive reasoning proves that the resulting graph is isomorphic to the target subgraph containing significant states and transitions.

The following theorem summarizes the performance of Safe-Candidate.

Theorem 5.6 *Let M denote an n-state PDFA computing a distribution D over $\Sigma^\star$, L denote the expected length of D, and π be the smallest nonzero stopping probability in M. Then the execution of Safe-Candidate on a sample from D satisfies:*

1. *It runs in time $\mathsf{poly}(n, |\Sigma|, \log(1/\pi))$.*
2. *It asks $O(n^2|\Sigma|^2 \log(n/\pi))$ statistical queries with tolerance $\tilde{\Omega}(\varepsilon^3\pi^2/n^3|\Sigma|^3 L)$.*
3. *It asks $O(n^2|\Sigma|)$ L_∞-query with tolerance $\Omega(\mu)$ and threshold $\Omega(\varepsilon/n|\Sigma|)$.*
4. *It returns a PDFA H such that $\mathrm{L}_1(D, H) \leq \varepsilon$.*

A short and complete proof of this theorem is in [17]. A similar theorem without any dependence on L and π is shown in [35] but the proof is more complex. The proof in [16] that shows PAC learnability under the KL-divergence measure is much longer.

5.6 Learning PFA

In this section we discuss algorithms having formal guarantees of learning the whole class of PFA. Similarly to the PDFA case, we first give an algorithm that has access to the target function as oracle, and can exactly learn the class of Weighted Automata when the right set of prefixes and suffixes is provided. Then, we specialize the algorithm to PFA and the case in which the input is a randomly drawn finite sample. We discuss the solutions given by Denis et al. [22] on the one hand and, on the other, by Mossel and Roch [33], Hsu et al. [29], and Bailly et al. [4]. The latter leads to the spectral method, which we expose in more detail following mostly the presentation in [9]. We finally mention a few of the most recent works extending the spectral method in several directions.

5.6.1 An Oracle Algorithm for Learning WA

The algorithm in Fig. 5.6 encapsulates the main idea for learning general Weighted Automata in the oracle model. It resembles the algorithm given by Beimel et al. [11] that learned instead from Evaluation and Equivalence queries.

There are two nondeterministic steps in this algorithm: One is the choice of sets X and Y; like for PDFA, we do not discuss how to choose them in the oracle model. The other one is the choice of a factoring QR of H', which represents the choice of an arbitrary basis for the rows of the Hankel matrix. The construction of the WA in the proof of Theorem 5.3 is the special case in which $R = H'$ and Q is the identity, but it is easy to see that the correctness of the special case implies the correctness of

1. Choose any two finite sets $X,Y \subseteq \Sigma^\star$ with $\lambda \in X \cap Y$, in some way not specified by the algorithm;
2. Build the submatrix $H = H_f[X \cup X\Sigma, Y]$ of H_f by asking oracle queries to f;
3. Find two minimal subsets $X' \subseteq X$, $Y' \subseteq Y$ such that $\lambda \in X'$ and $H' = H[X',Y']$ has the same rank as H; note that $|X'| = |Y'|$, say n, and H' has full rank;
4. Let $Q,R \in \mathbb{R}^{n\times n}$ be any two matrices factoring H', i.e., $H' = QR$;
5. For each symbol a, let $H'_a = H_f[X'a,Y'] = H_f[X',aY']$;
6. Build a WA from Q, R, and $\{T_a\}_a$ as follows.

$$\begin{aligned}\alpha_0^T &= H'[\lambda,:]R^{-1}\\ \alpha_\infty &= Q^{-1}H'[:,\lambda]\\ T_a &= Q^{-1}H'_aR^{-1}\end{aligned}$$

Fig. 5.6 Learning WA with an oracle

the general QR case. Indeed, the value of the automaton for any word w is

$$\begin{aligned}\alpha_0^T T_w\alpha_\infty &= (H'[\lambda,:]R^{-1})(Q^{-1}H'_{w_1}R^{-1})(Q^{-1}H'_{w_2}R^{-1})\ldots(Q^{-1}H'_{w_m}R^{-1})Q^{-1}H'[:,\lambda]\\ &= H'[\lambda,:]H'^{-1}H'_{w_1}H'^{-1}H'_{w_2}\ldots H'_{w_m}H'^{-1}H'[:,\lambda]\end{aligned}$$

which is the value computed for $R = H'$ and Q the identity matrix. The reason for the more general presentation will be clear later. It is also clear that the algorithm runs in time polynomial in Σ and the sums of lengths of strings in $X \cup Y$. Following the proof of Theorem 5.3 we have:

Theorem 5.7 *Let f be computed by some WA. If the cardinality of X' is at least the rank of H_f then the WA built in this way computes f.*

Proof The algorithm defines matrices T_a by $H'_a = T_aH'$; they are uniquely defined as H' has full rank. Furthermore, since the rank of H' is that of the whole Hankel matrix H of f, these matrices must also satisfy $H_a = T_aH$. Therefore, the automaton constructed by the algorithm is exactly the one built in the proof of Theorem 5.3, which by the theorem computes f. □

Consider now the adaptation of the algorithm to the case in which f is a probability distribution and we are given a finite sample S. As in the PDFA case, we take $X = \mathit{prefixes}(S)$, $Y = \mathit{suffixes}(S)$, and then create an approximation $\hat{H}$ of H by $\hat{H}[x,y]$ = empirical probability of xy in S. We know that the unperturbed H has rank at most n, the number of states of the smallest WA for f but, because rank is so fragile under perturbations, $\hat{H}$ will probably have maximal rank, even if $|S|$ is large.

The solution taken by [22] can be intuitively described as follows: Compute a subset $X' \subseteq X$ such that $|X'| = n$ and every row of $\hat{H}[X,:]$ is "close to" a linear combination of rows indexed by X'. For sufficiently small choice of "close to" (depending on the distribution), X' will be as in the algorithm above, and will lead to a correct solution.

The solution taken by, for example, [4, 29, 33], which leads to the spectral method, is less combinatorial and more algebraic, less local and more global, and can be phrased as follows: Let us instead find a matrix H' that (1) is easy to compute, (2) has the same dimensions as H, but rank at most n, and (3) is "as close as possible" to $\hat{H}$ under some metric, with this rank constraint. One particular way of computing such H' (among, perhaps, other possibilities) is the spectral method described next.

5.6.2 *The Spectral Method*

An important tool for the spectral method is the *Singular Value Decomposition* theorem; see, for example, [40].

Theorem 5.8 (SVD theorem) *Let $H \in \mathbb{R}^{p\times q}$. There are matrices $U \in \mathbb{R}^{p\times p}$, $D \in \mathbb{R}^{p\times q}$ and $V \in \mathbb{R}^{q\times q}$ such that:*

- $H = UDV^T$
- *U and V are orthonormal: $U^TU = I \in \mathbb{R}^{p\times p}$ and $V^TV = I \in \mathbb{R}^{q\times q}$*
- *D is a diagonal matrix of non-negative real numbers.*

The diagonal values of D, denoted $\sigma_1, \sigma_2, \ldots$, are the *singular values* of H, and column vectors of U are its *left singular vectors*. It follows that $\text{rank}(A) = \text{rank}(D)$ is the number of non-zero singular values. W.l.o.g. by rearranging rows and columns, the diagonal values in D are nondecreasing, i.e. $\sigma_1 \geq \sigma_2 \geq \cdots$. The SVD decomposition can be computed in time $O(pq^2)$.

The Frobenius norm of a matrix H is

$$\|H\|_F = \left(\sum_{i,j} H[i,j]^2\right)^{1/2}.$$

As this norm is invariant by unitary products, the square of the Frobenius norm of H is the sum of the squares of its singular values. It follows that from the singular value decomposition $H = UDV$ of H, we can compute a low rank approximation: Fix $n \leq \text{rank}(H)$, and define H_n' as

$$H_n' = \left(\begin{array}{c|c|c|c} u_1 & \ldots & u_n & \mathbf{0} \end{array}\right) \left(\begin{array}{ccc|c} \sigma_1 & & & \\ & \ddots & & \\ & & \sigma_n & \\ \hline & & & \mathbf{0} \end{array}\right) \left(\begin{array}{c} v_1 \\ \hline \vdots \\ \hline v_n \\ \hline \mathbf{0} \end{array}\right).$$

The following is the crucial fact.

Fact H'_n has rank n and minimizes $\|H - G\|_F$ among all rank-n matrices G.

Now we would like to use H'_n to find a full rank submatrix H' of H, since H'_n will not in general have full rank and cannot be inverted. Alternatively, there is a notion of *pseudoinverse* matrix that satisfies what we need for the algorithm, and is easily computable from the SVD decomposition.

The *Moore–Penrose pseudoinverse* of A, denoted A^+, admits many different definitions. The most algorithmic one is perhaps:

- If $A \in \mathbb{R}^{p\times q}$ is a diagonal matrix, $A^+ \in \mathbb{R}^{q\times p}$ is formed by transposing A, and taking the inverse of each non-zero element.
- In the general case, if $A = UDV^T$ then $A^+ = VD^+U^T$.

Some of its many interesting properties are:

- In general, $AA^+ \neq I$ and $A^+A \neq I$.
- But if A is invertible, $A^+ = A^{-1}$.
- If columns of A are independent, $A^+A = I \in \mathbb{R}^{q\times q}$.
- If rows of A are independent, $AA^+ = I \in \mathbb{R}^{p\times p}$.

With this artillery in place, the spectral method for learning probability distributions generated by WA is described in Fig. 5.7. The following PAC result was shown in [29] and reformulated in [9].

Theorem 5.9 [9, 29] *Let S be a sample of a probability distribution D such that H_D has rank n. Let σ_n be the nth largest singular value of H_D, and M the WA produced by the algorithm in Fig. 5.7 on input S. There is a polynomial p such that if $|S| \geq p(n, |\Sigma|, 1/\sigma_n, 1/\varepsilon, \log(1/\delta))$, with probability at least $1-\delta$:*

$$\sum_{|x|=t} |D(x) - M(x)| < \varepsilon.$$

1. get n and sample S;
2. $X = \mathit{preffixes}(S)$; $Y = \mathit{suffixes}(S)$;
3. define $H[X,Y] \in \mathbb{R}^{p\times q}$ and set $H[x,y] =$ empirical probability of xy;
4. define $H_a[X,Y] \in \mathbb{R}^{p\times q}$ and set $H_a[x,y] =$ empirical probability of xay;
5. Let QR be a rank-n factorization of H, that is:
   - $Q \in \mathbb{R}^{p\times n}$ and $R \in \mathbb{R}^{n\times q}$, both having rank n,
   - $H = QR$,

   for instance, take Q to be the first n left singular vectors of H;
6. output the WA M such that

$$\alpha_0^T = H[\lambda,:]R^+, \quad \alpha_\infty = Q^+H[:,\lambda], \quad T_a = Q^+H_aR^+$$

Fig. 5.7 Spectral learning of PFA

Observe that $\sigma_n \neq 0$ if and only if $\text{rank}(H_P) \geq n$, so $1/\sigma_n$ makes sense as a complexity parameter. Observe also that the output of the algorithm is not necessarily a PFA, even under the assumption that P is a probability distribution. It may assign negative values to some strings, and not add up to exactly 1. However, by the theorem, such oddities tend to disappear as the sample size grows. Converting a WA known to compute (or approximate) a probability distribution to a PFA is, in general, uncomputable. The problem is discussed in detail in [22]. Both [2, 33] give partial solutions by considering somewhat restricted machine models.

5.6.3 Variations and Implementation of the Spectral Method

Several variations of the spectral method have been introduced in order to improve its sample-efficiency or to extend it to wider settings. We point out below a couple of recent proposals.

Let D be a distribution on $\Sigma^\star$. Derived from D, we define function D_p assigning to each string x the probability of the set $x\Sigma^\star$, i.e. $D_p(x) = \sum_y D(xy)$. Similarly, we also consider function D_s that on input x evaluates to the expected number of times x appears in a random string w. It turns out that if any of these three functions has WA, then all three functions have WA. Moreover, WA descriptions can be obtained from original WA parameters of one of them. The following lemma is shown in [17]:

Lemma 5.1 *Let* $\langle \{T_\sigma\}_{\sigma \in \Sigma^\star}, \alpha_0, \alpha_\infty \rangle$ *be a WA and define* $S = \sum_\sigma T_\sigma$, $\tilde{\alpha}_0 = \alpha_0^T (I - S)^{-1}$ *and* $\tilde{\alpha}_\infty = (I - S)^{-1} \alpha_\infty$. *Then the following are equivalent*

1. $\langle \{T_\sigma\}_{\sigma \in \Sigma^\star}, \alpha_0, \alpha_\infty \rangle$ *computes* D.
2. $\langle \{T_\sigma\}_{\sigma \in \Sigma^\star}, \alpha_0, \tilde{\alpha}_\infty \rangle$ *computes* D_p.
3. $\langle \{T_\sigma\}_{\sigma \in \Sigma^\star}, \tilde{\alpha}_0, \tilde{\alpha}_\infty \rangle$ *computes* D_s .

Thus, besides using statistics of full strings in order to approximate the Hankel matrix from a sample we can also try to use statistics from prefixes and substrings in order to learn functions D_p and D_s. Experimentally, this yields more sample-efficient algorithms.

The use of statistics on prefixes instead of full strings in the spectral method was already proposed by Hsu et al. [29]. Later, Luque et al. [32] take advantage of substring statistics when applying the spectral method to learn non-deterministic split head-automata grammars, a hidden-state formalism for dependency parsing.

Recently, the spectral method in combination with matrix completion techniques has been also applied to a more general learning setting [8]. Here, the learning problem is to infer a weighted automaton from a sample of labeled examples but, in contrast with the standard paradigm, the sample is provided according to an arbitrary unknown distribution. Note that, for this type of learning problem, it is not guaranteed that a full approximation of a convenient Hankel submatrix can be achieved. This is because the input sample can lack information for many submatrix entries and now it can not be assumed that the WA function value must be close to 0 there, as one can in the standard probability setting. In [8], matrix completion techniques are used to fill the gaps in the Hankel submatrix and then the spectral method is used to infer the target WA. They prove formal learning guarantees for some of the completion techniques under mild conditions on the distribution.

5.7 Future Work

Let us mention a few open questions or future lines of research. Concerning the spectral method, it is clearly in an early stage and further extensions and applications will keep appearing. Making it generally practical and competitive is certainly of interest. Most spectral methods produce weighted automata with negative transition values when learning PFA; this may somewhat hinder their application in contexts where interpretability of the learned model is important.

At a more theoretical level, we would like to find some geometric or algebraic interpretation of PDFA distinguishability; this might explain its role in PDFA learning as a particular case of the spectral values that come up in learning PFA.

As mentioned, it is known that PDFA cannot exactly compute all distributions computed by PFA [22]. But, to our knowledge, it is not known whether PDFA can reasonably approximate distributions computed by PFA, say with a number of states polynomial in $1/\varepsilon$ for the desired approximation ε in the L_1 distance (a rather direct result for L_∞ is given in [26], which extends to every L_p for $p > 1$). If such approximability is true, then it may be possible to transfer PAC learnability results for PDFA to PFA with a polynomial overhead.

Acknowledgments This work is partially supported by MICINN projects TIN2011-27479-C04-03 (BASMATI) and TIN-2007-66523 (FORMALISM), by SGR2009-1428 (LARCA). We thank the chairs of ICGI 2012 for the invitation to present a preliminary version of this work as tutorial. We particularly thank the reviewer of this version for thorough and useful work.

Appendix: The Baum–Welch Method

The Baum–Welch algorithm [10] is one of the most popular methods to infer a hidden Markov model from observed data. Despite the fact that there are no bounds on convergence time and that it may get trapped in local optima, it is intuitively attractive since it has a clear focus—maximizing the observed data likelihood—and performs a simple step-by-step hill climbing progress to this goal. However, we think that as new techniques based on spectral methods progress and gain popularity, Baum–Welch may lose its status as first option. Theoretically, spectral methods will obtain global optima, come with performance guarantees in time and accuracy, and tend to work faster at least on large samples.

input: data observation $x = x_1 \dots x_m$ and some initial HMM guess $H = \langle \{T_a\}_{a\in\Sigma}, \alpha_0 \rangle$
output: updated H, locally maximizing sample likelihood
algorithm Baum-Welch
repeat
compute backward β_t and forward ϕ_t probability vectors for $t = 0 \dots m$
compute a posteriori state visit and transition probabilities α_0' and $\{T_a'\}_{a\in\Sigma}$
$H \leftarrow \langle \{T_a'\}_{a\in\Sigma}, \alpha_0' \rangle$
until stopping condition

Fig. 5.8 The Baum–Welch algorithm

The Baum–Welch algorithm starts by guessing some hidden Markov model; this is frequently done using problem-specific heuristics. Provided with a sequence of data observations $x = x_1 \dots x_m$, the algorithm continues by iterating a process where parameters of the last built model H are updated according to a posteriori state and transition probability values. Iteration finishes either when a parameter convergence criterion is achieved or there is a loss of prediction accuracy. The learning algorithm is shown in Fig. 5.8. Specifically, let $H = \langle \{T_a\}_{a\in\Sigma}, \alpha_0 \rangle$ the last hypothesis built. Backward and forward probability vectors for $t = 0 \dots m$ are, respectively:

$$\beta_t^T = \alpha_0^T T_{x_1} \dots T_{x_t},$$
$$\phi_t = T_{x_{t+1}} \dots T_{x_m} \mathbf{1}.$$

Component j of the backward vector β_t is the probability of generating prefix $x_1 \dots x_t$ and being in state j just after emitting x_t, i.e. $\beta_t[j] = \Pr[x_1 \dots x_t \wedge S(t) = j]$. On the other hand, component j of the forward vector ϕ_t is the probability of emitting suffix $x_{t+1} \dots x_m$ from state j, $\phi_t[j] = \Pr[x_{t+1} \dots x_m | S(t) = j]$. Let $S(t)$ denote the state at time t and $O(t)$ the observation at time (whose realization is thus x_t). Given data $x = x_1 \dots x_m$, the a posteriori state visit probability of state j is:

$$\alpha_0'[j] = \frac{1}{m+1} \sum_{t=0}^{m} \Pr[S(t) = j | x] = \frac{1}{m+1} \sum_{t=0}^{m} \frac{\beta_t[j]\phi_t[j]}{\beta_t^T \phi_t} = \frac{\sum_{t=0}^{m} \beta_t[j]\phi_t[j]}{(m+1)(\alpha_0 T_x \mathbf{1})}. \tag{5.1}$$

Similarly, the a posteriori probability of transition a from state i to j is:

$$T_a'[i,j] = \frac{\sum_{t|x_t=a} \xi_t^{i,j}}{\sum_{t=1}^{m} \xi_t^{i,j}} \tag{5.2}$$

where

$$\xi_t^{i,j} = \Pr[S(t-1) = i \wedge S(t) = j \wedge O(t) = x_t | x] = \frac{\beta_{t-1}[i] T_{x_t}[i,j] \phi_t[j]}{\alpha_0 T_x \mathbf{1}}.$$

Thus, $\xi_t^{i,j}$ denotes the probability that provided observed data x, state i is reached just after processing length $t-1$ prefix of x and x_t moves from state i to state j.

The following theorem shows that the iterated updating procedure in the Baum–Welch algorithm either increases the sample likelihood or, at local maxima, keeps it unchanged. We follow the presentation in [47].

Theorem 5.10 [10, 47] *Let $x = x_1 \dots x_m$ be a sample and let $H = \langle \alpha_0, \{T_a\}_{a\in\Sigma}\rangle$ and $H' = \langle \alpha'_0, \{T'_a\}_{a\in\Sigma}\rangle$ be the hidden Markov models defined above. Then, $H'(x) \geq H(x)$ with equality at local maxima.*

Proof (sketch) Let $s = s_0 \dots s_m$ be a sequence of states in machine H and consider conditional distributions $H(s|x)$ and $H'(s|x)$. Starting from the KL divergence formula and expanding conditional probabilities, it is easy to derive the relations

$$\begin{aligned} 0 &\leq \mathrm{KL}(H(\cdot|x), H'(\cdot|x)) \\ &= \sum_s H(s|x) \log \frac{H(s|x)}{H'(s|x)} = \log \frac{H'(x)}{H(x)} + \sum_s \frac{H(x \wedge s)}{H(x)} \log \frac{H(x \wedge s)}{H'(x \wedge s)}. \end{aligned}$$

Defining function Q as

$$Q(x, H, H') \doteq \sum_s H(x \wedge s) \log H'(x \wedge s),$$

the last inequality can be rearranged to show that

$$\frac{Q(x, H, H') - Q(x, H, H)}{H(x)} \leq \log \frac{H'(x)}{H(x)}.$$

Thus, $H'(x) > H(x)$ when $Q(x, H, H') > Q(x, H, H)$. We obtain a hill climbing procedure by finding H' maximizing the $Q(x, H, \cdot)$ function. Using Lagrange's method to find critical points of Q subject to stochastic constraints (H' must define a probability function) results in values for α'_0 and $\{T'_a\}_{a\in\Sigma}$ defining H' as the ones displayed in Eqs. (5.1) and (5.2). □

References

1. Naoki Abe and Manfred K. Warmuth. On the computational complexity of approximating distributions by probabilistic automata. *Machine Learning*, 9:205–260, 1992.
2. Animashree Anandkumar, Daniel Hsu, and Sham M. Kakade. A method of moments for mixture models and hidden Markov models. In *25th Annual Conference on Learning Theory (COLT); Journal of Machine Learning Research - Proceedings Track - COLT 2012*, volume 23, pages 33.1–33.34, 2012.

3. Dana Angluin. Queries and concept learning. *Machine Learning*, 2(4):319–342, 1987.
4. Raphaël Bailly, François Denis, and Liva Ralaivola. Grammatical inference as a principal component analysis problem. In *26th Intl. Conf. on Machine Learning (ICML)*, page 5. ACM, 2009.
5. Borja Balle, Jorge Castro, and Ricard Gavaldà. Learning PDFA with asynchronous transitions. In *10th Intl. Coll. on Grammatical Inference (ICGI)*, volume 6339 of *Lecture Notes in Computer Science*, pages 271–275. Springer, 2010.
6. Borja Balle, Jorge Castro, and Ricard Gavaldà. A lower bound for learning distributions generated by probabilistic automata. In *21st Intl. Conf. on Algorithmic Learning Theory (ALT)*, volume 6331 of *Lecture Notes in Computer Science*, pages 179–193. Springer, 2010.
7. Borja Balle, Jorge Castro, and Ricard Gavaldà. Bootstrapping and learning PDFA in data streams. In *11th Intl. Conf. on Grammatical Inference (ICGI)*, volume 21 of *JMLR Workshop and Conf. Proceedings*, pages 34–48, 2012.
8. Borja Balle and Mehryar Mohri. Spectral learning of general weighted automata via constrained matrix completion. In *Neural Information Processing Systems (NIPS)*, 2012.
9. Borja Balle, Ariadna Quattoni, and Xavier Carreras. Local loss optimization in operator models: A new insight into spectral learning. In *29th Intl. Conf. on Machine Learning (ICML)*, 2012.
10. L. E. Baum, T. Petrie, G. Soules, and N. Weiss. A maximization technique occurring in the statistical analysis of probabilistic functions of Markov chains. *Ann. Math. Statist.*, 41(1): 164–171, 1970.
11. Amos Beimel, Francesco Bergadano, Nader H. Bshouty, Eyal Kushilevitz, and Stefano Varricchio. Learning functions represented as multiplicity automata. *J. ACM*, 47(3):506–530, 2000.
12. J.W. Carlyle and A. Paz. Realization by stochastic finite automaton. *J. Comput. Syst. Sci.*, 5:26–40, 1971.
13. Rafael C. Carrasco and José Oncina. Learning stochastic regular grammars by means of a state merging method. In *2nd Intl. Coll. on Grammatical Inference (ICGI)*, volume 862 of *Lecture Notes in Computer Science*, pages 139–152. Springer, 1994.
14. Rafael C. Carrasco and José Oncina. Learning deterministic regular grammars from stochastic samples in polynomial time. *Informatique Théorique et Applications*, 33(1):1–20, 1999.
15. Jorge Castro and Ricard Gavaldà. Towards feasible PAC-learning of probabilistic deterministic finite automata. In *9th Intl. Coll. on Grammatical Inference (ICGI)*, volume 5278 of *Lecture Notes in Computer Science*, pages 163–174. Springer, 2008.
16. Alexander Clark and Franck Thollard. PAC-learnability of probabilistic deterministic finite state automata. *Journal of Machine Learning Research*, 5:473–497, 2004.
17. Borja de Balle Pigem. *Learning Finite-State Machines. Algorithmic and Statistical Aspects.* PhD thesis, Universitat Politècnica de Catalunya, 2013.
18. Colin de la Higuera. *Grammatical Inference Learning Automata and Grammars*. Cambridge University Press, 2010.
19. Colin de la Higuera and José Oncina. Learning stochastic finite automata. In *7th Intl. Coll. on Grammatical Inference (ICGI)*, volume 3264 of *Lecture Notes in Computer Science*, pages 175–186. Springer, 2004.
20. Colin de la Higuera and Franck Thollard. Identification in the limit with probability one of stochastic deterministic finite automata. In *5th Intl. Coll. on Grammatical Inference (ICGI)*, volume 1891 of *Lecture Notes in Computer Science*, pages 141–156. Springer, 2000.
21. François Denis and Yann Esposito. Learning classes of probabilistic automata. In *17th Annual Conference on Learning Theory (COLT)*, volume 3120 of *Lecture Notes in Computer Science*, pages 124–139. Springer, 2004.
22. François Denis, Yann Esposito, and Amaury Habrard. Learning rational stochastic languages. In *19th Annual Conference on Learning Theory (COLT)*, volume 4005 of *Lecture Notes in Computer Science*, pages 274–288. Springer, 2006.
23. Pierre Dupont and Juan-Carlos Amengual. Smoothing probabilistic automata: An error-correcting approach. In *5th Intl. Coll. on Grammatical Inference (ICGI 2000)*, volume 1891 of *Lecture Notes in Computer Science*, pages 51–64. Springer, 2000.

24. Pierre Dupont, François Denis, and Yann Esposito. Links between probabilistic automata and hidden Markov models: probability distributions, learning models and induction algorithms. *Pattern Recognition*, 38(9):1349–1371, 2005.
25. M. Fliess. Matrices de Hankel. *J. Math. Pures Appl.*, 53:197–222, 1974. (Erratum in vol. 54, 1975.).
26. Ricard Gavaldà, Philipp W. Keller, Joelle Pineau, and Doina Precup. PAC-learning of Markov models with hidden state. In *17th European Conf. on Machine Learning (ECML)*, volume 4212 of *Lecture Notes in Computer Science*, pages 150–161. Springer, 2006.
27. E. Mark Gold. Language identification in the limit. *Information and Control*, 10(5):447–474, 1967.
28. Omri Guttman, S. V. N. Vishwanathan, and Robert C. Williamson. Learnability of probabilistic automata via oracles. In *16th Intl. Conf. on Algorithmic Learning Theory (ALT)*, pages 171–182, 2005.
29. Daniel Hsu, Sham M. Kakade, and Tong Zhang. A spectral algorithm for learning hidden Markov models. In *22nd Annual Conference on Learning Theory (COLT)*. ACM, 2009.
30. Michael J. Kearns, Yishay Mansour, Dana Ron, Ronitt Rubinfeld, Robert E. Schapire, and Linda Sellie. On the learnability of discrete distributions. In *26th Annual ACM Symp. on Theory of Computing (STOC)*, pages 273–282. ACM, 1994.
31. Christopher Kermorvant and Pierre Dupont. Stochastic grammatical inference with multinomial tests. In *6th Intl. Coll. on Grammatical Inference (ICGI)*, volume 2484 of *Lecture Notes in Computer Science*, pages 149–160. Springer, 2002.
32. Franco M. Luque, Ariadna Quattoni, Borja Balle, and Xavier Carreras. Spectral learning for non-deterministic dependency parsing. In *13th Conf. of the European Chapter of the Association for Computational Linguistics (EACL)*, pages 409–419. The Association for Computer Linguistics, 2012.
33. Elchanan Mossel and Sébastien Roch. Learning nonsingular phylogenies and hidden Markov models. In *37th Annual ACM Symp. on Theory of Computing (STOC)*, pages 366–375. ACM, 2005.
34. José Oncina and Pedro García. Identifying regular languages in polynomial. In *Advances in Structural and Syntactic Pattern Recognition*, pages 99–108. World Scientific, 1992.
35. Nick Palmer and Paul W. Goldberg. PAC-learnability of probabilistic deterministic finite state automata in terms of variation distance. *Theor. Comput. Sci.*, 387(1):18–31, 2007.
36. Dana Ron, Yoram Singer, and Naftali Tishby. The power of amnesia: Learning probabilistic automata with variable memory length. *Machine Learning*, 25(2-3):117–149, 1996.
37. Steven Rudich. Inferring the structure of a Markov chain from its output. In *26th Annual Symp. on Foundations of Computer Science (FOCS)*, pages 321–326, 1985.
38. Marcel P. Schützenberger. On the definition of a family of automata. *Information and Control*, 4:245–270, 1961.
39. Cosma Rohilla Shalizi and Kristina Lisa Shalizi. Blind construction of optimal nonlinear recursive predictors for discrete sequences. In *20th Conf. on Uncertainty in Artificial Intelligence (UAI)*, pages 504–511, 2004.
40. Gilbert Strang. *Introduction to Linear Algebra, 4th edition*. Wellesley-Cambridge Press and SIAM, 2009.
41. Sebastiaan Terwijn. On the learnability of hidden Markov models. In *6th Intl. Coll. on Grammatical Inference (ICGI)*, volume 2484 of *Lecture Notes in Computer Science*, pages 261–268. Springer, 2002.
42. Franck Thollard. Improving probabilistic grammatical inference core algorithms with post-processing techniques. In *18th Intl. Conf. on Machine Learning (ICML 2001)*, pages 561–568, 2001.
43. Franck Thollard, Pierre Dupont, and Colin de la Higuera. Probabilistic DFA inference using Kullback-Leibler divergence and minimality. In *17th Intl. Conf. on Machine Learning (ICML)*, pages 975–982. Morgan Kaufmann, 2000.
44. Leslie G. Valiant. A theory of the learnable. *Commun. ACM*, 27(11):1134–1142, 1984.

45. Enrique Vidal, Franck Thollard, Colin de la Higuera, Francisco Casacuberta, and Rafael C. Carrasco. Probabilistic finite-state machines - part i. *IEEE Trans. Pattern Anal. Mach. Intell.*, 27(7):1013–1025, 2005.
46. Enrique Vidal, Franck Thollard, Colin de la Higuera, Francisco Casacuberta, and Rafael C. Carrasco. Probabilistic finite-state machines - part ii. *IEEE Trans. Pattern Anal. Mach. Intell.*, 27(7):1026–1039, 2005.
47. Lloyd R. Welch. Hidden Markov Models and the Baum–Welch Algorithm. *IEEE Information Theory Society Newsletter*, 53(4), 2003.

Chapter 6
Distributional Learning of Context-Free and Multiple Context-Free Grammars

Alexander Clark and Ryo Yoshinaka

Abstract This chapter reviews recent progress in distributional learning in grammatical inference as applied to learning context-free and multiple context-free grammars. We discuss the basic principles of distributional learning, and present two classes of representations, primal and dual, where primal approaches use nonterminals based on strings or sets of strings and dual approaches use nonterminals based on contexts or sets of contexts. We then present learning algorithms based on these two models using a variety of learning paradigms, and then discuss the natural extension to mildly context-sensitive formalisms, using multiple context-free grammars as a representative formalism.

6.1 Introduction

In this chapter we look at the problem of learning certain classes of phrase structure grammars from information about the language, a classic problem in grammatical inference. In particular we look at techniques using what is broadly called distributional learning, and which have been developed in recent years starting with Clark and Eyraud [16]. The term *distributional* as we use it has nothing to do with probability distributions or statistical learning, but rather concerns the linguistic notion of distribution: the set of contexts or environments in which strings or words can appear.

This is indeed a classic problem: Gold [23] suggests the following question as worthy of research:

> However, it would be useful to determine if there are interesting subclasses of context-free languages which can be identified in the limit by either of these approaches (i.e. by statistical approaches like distributional analysis or by approaches sensitive to order).

A. Clark (✉)
Department of Philosophy, King's College London, London, UK
e-mail: alexander.clark@kcl.ac.uk

R. Yoshinaka
Kyoto University, Kyoto, Japan
e-mail: ry@i.kyoto-u.ac.jp

J. Heinz and J.M. Sempere (eds.), *Topics in Grammatical Inference*,
DOI 10.1007/978-3-662-48395-4_6

Distributional learning itself has a long history, which we review briefly in Sect. 6.2 to provide some intellectual context. In this chapter we present a tutorial overview of modern approaches to distributional learning as applied to the inference of context-free grammars (CFGs) and multiple context-free grammars (MCFGs), which we take to be a representative mildly context-sensitive formalisms. We will focus on the general properties of these algorithms, and the representational ideas and the types of algorithms that exploit these representational assumptions. We will try to provide a full bibliography representing at least the current resurgence of interest in distributional learning, and including the key earlier papers, together with pointers into the rest of the literature. We are not trying to provide a complete survey of the inference of CFGs and MCFGs; we restrict ourselves to algorithms that are computationally efficient in some sense (and so exclude purely enumerative algorithms [53]), have some theoretical guarantees (as opposed to heuristic algorithms [33, 38]), and take as input only strings, or information about strings (as opposed to algorithms that take trees or partially bracketed strings as input [46]). Within these parameters, the only algorithms that we are aware of are distributional algorithms in the sense that we define below.

6.2 Distributional Learning: A Historical Note

Distributional learning has a long history. Beyond the well-known work of the American structuralists, most famously Harris [24] and Wells [61], structuralist linguistics had an autonomous history in Russia and Eastern Europe under the name of the Kulagina school, which has its origins in a seminar in mathematical linguistics initiated by Kolmogorov and which takes its name from Olga Kulagina's seminal 1957 paper [36]. While in the US, structuralist linguistics largely died out after Chomskyan linguistics became the dominant research paradigm, it continued for quite some time elsewhere. The most accessible introduction to this literature is either Marcus's book [41] or the two volume survey [57]. Important early papers are by Sestier [50] and Kunze [37]. None of this work has any real learning results—it merely uses distributional learning as an analytical tool. A lot of computational work also uses distributional learning explicitly or implicitly [1, 5, 34, 58], but we do not discuss this work here.

Distributional learning is also closely related to the context-free grammar formalism. The word 'context' after all appears in the term context-free and is also a foundational concept in distributional learning—this is not a coincidence. The context-free grammar formalism was originally devised to represent the outputs from distributional learning procedures. Chomsky [7, p. 172, fn.15] says:

> The concept of "phrase structure grammar" was explicitly designed to express the richest system that could reasonably be expected to result from the application of Harris-type procedures to a corpus.

Over the years since then many different learning procedures for context-free grammars have been devised based on the intuitions of distributional learning. Typically these algorithms are based on the justification that two strings derived from the same nonterminal will be distributionally similar; therefore one can try to reverse this process by finding clusters of distributionally similar strings and creating a grammar with nonterminals that generate these strings. The naive application of these heuristic approaches has been known for a long time to suffer from some serious problems; Chomsky was perhaps the first to articulate these problems explicitly in his doctoral thesis [6], and indeed much of the technical work that we describe can be seen as an attempt to either answer or avoid those problems, which at the time were taken to be compelling arguments against the very possibility of distributional learning.

6.3 Languages and Grammars

We start by defining some standard notation. We assume that we have a finite nonempty set called the *alphabet* which we denote by Σ. This set might consist of the words in a language, or the set of phonemes, a set of letters, or even DNA bases or amino acids, depending on the application. We write Σ^* for the set of all finite strings of elements of Σ. We write Σ^+ for the nonempty strings. We write λ for the empty string. A *(formal) language* is just a subset of Σ^*; if L is a language then $L \subseteq \Sigma^*$. In this chapter we consider this very restricted notion of a language: this might be for example the set of grammatical sentences in a language, or the set of phonotactically well formed words in a language or something else. We abstract away from the particular details and consider it just as a set of strings that is defined in some way.

The languages we are interested in are typically infinite, or even if finite are very large, so we need finite representations. In this chapter we look only at the class of multiple context-free grammars (MCFGs). Context-free grammars (CFGs) are a special case of MCFGs; we start by defining the standard class of CFGs, and in Sect. 6.8 we define the larger class of MCFGs.

A CFG over an alphabet Σ is a tuple which consists of a nonempty finite set of nonterminal symbols V, together with a set of productions of the form $N \rightarrow \alpha$ where $N \in V$ and $\alpha \in (V \cup \Sigma)^*$. We also have a finite set of initial symbols $I \subseteq V$, which in the standard definition consists of just one symbol S. The extension to multiple symbols does not change anything. We denote the standard derivation relation by $\overset{*}{\Rightarrow}_G$, and define the set of strings derivable from a nonterminal N to be $\mathscr{L}(G, N) = \{w \in \Sigma^* \mid N \overset{*}{\Rightarrow}_G w\}$. The language defined by the grammar is defined to be $\mathscr{L}(G) = \bigcup_{S \in I} \mathscr{L}(G, S)$.

6.3.1 Learning Models

We are interested in learning: we therefore assume that there is some language that we are trying to learn. We write L_* for the target language: i.e. the language that we are trying to learn. We consider a variety of learning models here, from ones where the information sources available to the learner are very limited to ones where they are quite rich; we always consider only information about the language (the set of strings generated by the grammar), and none about the grammar itself. Additionally we always require polynomial update time—the learner at each step can only use a polynomial amount of computation. This is not enough on its own to be truly restrictive; there is some technical detail which we omit here [44].

- The first model is the oldest: positive-only identification in the limit in the Gold style [23]. The learner receives a sequence of examples drawn from L_*, and must converge after a finite but unbounded time to an exactly correct hypothesis. Crucially there are no constraints on the sequence of examples other than the trivial ones that the examples are all in L_* and that every element of L_* must occur at least once somewhere in the sequence. In this model we cannot learn any superfinite classes of language; we obtain learnable classes by considering only languages which satisfy some language-theoretic closure properties. This model therefore places some significant restrictions on the classes of languages that can be learned. We require the existence of a polynomially bounded characteristic set and polynomial update time. There are some technical issues about the appropriate way of defining this for CFGs, since we can have grammars that define languages that have very long strings, and this needs to be taken into account when defining the appropriate bound on the characteristic set.
- The second model we consider is that of positive data and membership queries (MQs), the same as the previous model, but the learner can also ask MQs and find out whether a particular string is in the target language. This is the easiest model: the easiest model to learn under, but also the model that is easiest to understand and easiest to prove results in, and accordingly we will focus on this model. However, on its own it is not restrictive as it is possible to define vacuous enumerative algorithms that nonetheless can learn using various computational tricks. We do not use these tricks in the algorithms we present.
- Minimally adequate teacher (MAT) model[1] [2, 10, 54, 69]. Here the learner has two sources of information: it can ask MQs and equivalence queries (EQs). Here we allow extended EQs. The learner can construct any CFG and ask whether it is correct or not. The teacher either says yes or provides a counterexample in the symmetric difference of the hypothesis and the target. Note that this is not computable for all CFGs. This is a restrictive learning model, in that it is known that classes such as regular grammars and CFGs are not learnable in this model [3], whereas deterministic regular grammars and congruential CFGs are, as we shall see.

[1] See *Learning Grammars and Automata with Queries*, de la Higuera (Chap. 3).

- Finally, there are results that use stochastic data—we assume the learner has access to positive samples drawn independently and identically from a fixed distribution, where the distribution is generated by a probabilistic version of the target grammar [8, 40].

6.3.2 Contexts and Distributions

One of the most basic notions is that of a context which is just a pair of strings. In the original papers this is written as an ordered pair (l, r) where $l, r \in \Sigma^*$. Here we will use a slightly different notation $l\Box r$. This avoids confusion when we move to MCFGs, and makes it clearer that it represents a sentence with a hole. We therefore write the special empty context (λ, λ) as just $\Box$. In linguistics, a context is sometimes called an *environment*.

We can combine a context and a string using the 'wrap' operation, for which we use the symbol $\odot$. This combines a context with a string, by inserting the string into the gap in the context: we define this therefore as $(l\Box r) \odot u = lur$. The empty context thus does not change the string it is wrapped around: $\Box \odot u = u$.

We extend this to sets of strings and contexts in the natural way, so

$$C \odot S = \{lur \mid l\Box r \in C, u \in S\}.$$

If we fix a language $L \subseteq \Sigma^*$ then we can talk about the relation between contexts and substrings given by $l\Box r \sim_L u$ iff $lur \in L$. If $lur \in L$ then we say that u *occurs in* the context $l\Box r$ in the language L.

We now define the notion of the distribution of a string in the language. Note that this has nothing to do with the notion of a probability distribution.

$$C_L(u) = \{l\Box r \mid lur \in L\}$$

If we can successfully model this distribution then we will have learned the language since $\Box \in C_L(u)$ iff $u \in L$. We also write this set $C_L(u)$ as $u^{\triangleright}$ when L is understood. Conversely for a context $l\Box r$ we define $(l\Box r)^{\triangleleft}$ to be $\{u \mid lur \in L\}$. Note that the residual languages $u^{-1}L = \{v \mid uv \in L\}$ are in this notation $(u\Box)^{\triangleright}$. We extend these to sets of contexts: $C^{\triangleleft} = \{u \mid \forall l\Box r \in C,\ lur \in L\}$. Alternatively and perhaps more intuitively:

$$C^{\triangleleft} = \{u \mid C \odot u \subseteq L\}.$$

Given a set of strings S we can also define

$$S^{\triangleright} = \{l\Box r \mid (l\Box r) \odot S \subseteq L\}.$$

For a set of strings D we define

$$\mathrm{Sub}(D) = \{u \in \Sigma^* \mid lur \in D \text{ for some } l, r \in \Sigma^*\},$$

$$\mathrm{Con}(D) = \{l\Box r \mid lur \in D \text{ for some } u \in \Sigma^*\}.$$

Distributional learning techniques are based on modeling the context-substring relation of a language. There are two technical details which we need to pay attention to: one is the case of the empty string, which as always in CFGs needs to be dealt with as a special case; the second is whether the CFG has more than two nonterminals on the right hand side of a rule. While every CFG can be put into Chomsky normal form, these learning algorithms depend on a correspondence between the nonterminals and sets of strings in the grammar that may not be preserved under binarisation.

6.3.3 Observation Tables

A natural way of visualising the relation between strings and contexts is through observation tables (OTs) [1, 2]. We show a simple example in Table 6.1. We assume a finite set of substrings that we call K; these form the rows of the table; we have a finite set of contexts, F, that we use as rows. In the entry corresponding to the row indexed by u and the column indexed by $l\Box r$ we put a 1 if $lur \in L_*$ and a 0 if $lur \notin L_*$. The table in the example contains a limited amount of information about the language: the language includes $\lambda, ab, aabb$, but does not include the strings $a, b, aa, bb, aaa, aab, abb, bbb, aaab, abbb$ or $aabbb$. The table does not contain any information about other strings, for example $abab$, which may or may not be in L_*. Thus there are a number of different CFLs that are compatible with this information, from the finite language consisting just of those three examples, $\lambda, ab, aabb$, that are certainly in the language, to the nonregular language $\{a^n b^n \mid n \geq 0\}$.

These approaches can be seen to be closely related to the classical techniques for regular inference which are based on modeling the relationship between prefixes and

Table 6.1 Example of an OT

	$\Box$	$a\Box$	$\Box b$	$\Box bb$
λ	1	0	0	0
a	0	0	1	0
b	0	1	0	0
aa	0	0	0	1
ab	1	0	0	0
aab	0	0	1	0

suffixes. In that model (e.g. [2]), an OT has rows indexed by prefixes, and columns indexed by suffixes, where a cell in the table has a 1 if and only if the concatenation of the corresponding prefix and suffix is in the language.

6.4 Context-Free Algorithms

All of these algorithms correspond to making a representational decision about what sets of strings the nonterminals should generate. We use the notation $[\![x]\!]$ for a nonterminal corresponding to some object x, typically a string or context, or set of strings or contexts. We need to decide what $\mathscr{L}(G, [\![x]\!])$ should be.

6.4.1 *Primal-Dual Distinction*

There is an important conceptual division that we want to discuss in general terms now—which is the division between primal methods and dual methods. In distributional learning, we have contexts and substrings and the relation between them. We have a choice—we can either take the substrings as being the primary objects and consider the contexts as being features, or we can swap the role of the contexts and substrings, and consider the contexts as being the primary objects and the substrings as being features.

Primal algorithms thus define sets of strings using one of the following schemes, where u is a single string and X is a set of strings:

$$\begin{aligned} [u] &= \{v \mid v^{\triangleright} = u^{\triangleright}\} \\ u^{\triangleright\triangleleft} &= \{v \mid v^{\triangleright} \supseteq u^{\triangleright}\} \\ X^{\triangleright\triangleleft} &= \{v \mid v^{\triangleright} \supseteq X^{\triangleright}\} \end{aligned}$$

Dual approaches on the other hand take a context $(l\Box r)$ or a finite set of contexts C and use the sets of strings defined as follows:

$$\begin{aligned} (l\Box r)^{\triangleleft} &= \{w \mid lwr \in L\} \\ C^{\triangleleft} &= \{w \mid C \odot w \subseteq L\} \end{aligned}$$

In the case of left regular grammars, these two are essentially similar: we consider only contexts of the form $\Box u$, and switching between primal and dual approaches is just equivalent to reversing the strings of the language. There is therefore no theoretically interesting difference between the primal and dual techniques. In the case of CFGs and MCFGs the two approaches differ radically in the types of languages that can be learned. One immediately obvious difference is that using a dual approach one can always define the language itself, as a set of strings, using the single context $\Box$ since $\Box^{\triangleleft} = L$.

6.5 Primal Algorithms

We now survey the major primal algorithms for CFG inference, not in chronological order. We will start by considering the most basic and mathematically tractable model, the congruential model. This model is also the closest to the models informally described by the American structuralists. If we consider a clustering model based on the distributions of strings, then the most fundamental model is one where the clusters are sets of distributionally identical sets of strings.

6.5.1 *Congruential Languages*

The most basic result is the MAT-learner result for what are called *congruential CFGs* [10]. The class of languages that can be learned using this algorithm is the class of congruential CFLs. Congruential CFGs are such that for all nonterminals N if $u, v \in \mathscr{L}(G, N)$ then they are *congruent* in the sense that $C_L(u) = C_L(v)$, which we will write $u \equiv_L v$. These are closely related to the *non-terminally separated (*NTS*) languages* [4, 49]. These form a proper subclass of CFLs that nonetheless include all regular languages.

We can construct a grammar directly from an OT: we will explain this case in full detail, as this is the simplest model and the basic ideas are reused several times later. Taking the example from Table 6.1, we construct the grammar on the lower part of the same table using the following procedure. Recall that the rows in the table are indexed by substrings of strings that are in the language: for each row in the table, corresponding to a substring u, we create a new nonterminal $[\![u]\!]$. We want this nonterminal to generate the string u and all other strings that are congruent to it. This gives us six nonterminals. First of all we note that the two strings λ and ab occur in the empty context $\Box$ and are therefore in the language: we accordingly pick the two symbols, $[\![\lambda]\!]$ and $[\![ab]\!]$, as being the start symbols. We now add productions of three types: lexical, branching, and chain (unary) productions. First of all if w is of length 1 or 0, that is to say $w = a$ for some letter $a \in \Sigma$ or is equal to λ, we add a rule of the form $[\![a]\!] \to a$ or $[\![\lambda]\!] \to \lambda$. Note that in this rule $[\![a]\!]$ is a nonterminal symbol, and a is a terminal symbol, which are different in a CFG. Next, for every string w which is of length at least two, we add all possible branching rules of the following form: We split w into two strings u, v each of length at least 1 that occur in the table, such that $w = uv$, and add a production for each of these splits of the form $[\![w]\!] \to [\![u]\!][\![v]\!]$. This is a binary production with two nonterminal symbols on the right hand side of the rule. In the example, we have two strings of length 2, which each have a unique split. We have one string of length 3, aab, which can be split in two different ways. We therefore have two branching productions with the symbol $[\![aab]\!]$ on the left hand side. In general, if we have a string of length n, where $n > 1$, then we will have $n - 1$ corresponding branching productions.

These productions on their own are rather trivial—if the grammar consisted only of these productions then a nonterminal $[\![u]\!]$ could generate only the string u and no other string. We also want each nonterminal to generate other strings that are distributionally identical to u; accordingly we add nonbranching rules between two nonterminals $[\![u]\!]$ and $[\![v]\!]$ whenever it appears that the substrings u and v are distributionally identical. Of course it is impossible in general to tell from a finite amount of information whether $u \equiv_L v$ since this is an infinitary property—the distributions $C_L(u)$ and $C_L(v)$ are often infinite sets, and thus we cannot expect to get an exact answer in a finite amount of time. We can however get an approximate answer: we can use the OT to get a finite approximation of the distribution of the two strings u, v. For example, in Table 6.1, the two strings a and aab appear to be distributionally identical, or at least in the table there is no evidence suggesting that they are not identical, as the two rows that they label are identical. If we use F to refer to the finite set of contexts that label the columns of the OT, then we are testing whether $C_L(u) \cap F = C_L(v) \cap F$, rather than whether $C_L(u) = C_L(v)$. We therefore add productions of the form $[\![u]\!] \to [\![v]\!]$ whenever u and v have the same rows in the table. Now the grammar is more interesting: the nonterminals like $[\![a]\!]$ generate an infinite set of strings. The generated grammar is shown in Table 6.2.

The final grammar then has six nonterminals, two of which are initial, three lexical rules, three branching rules and four unary rules. In this form it is hard to see what is happening, and so it is convenient to convert it into a more readable grammar by merging nonterminals that are linked via unary rules. This gives us a grammar which has four nonterminals, one of which is initial. We can relabel the nonterminals for legibility, with S being the nonterminal corresponding to the two original nonterminals $[\![\lambda]\!]$ and $[\![ab]\!]$; A being the nonterminal corresponding to the nonterminals $[\![a]\!]$ and $[\![aab]\!]$; B corresponding to $[\![b]\!]$; and X corresponding to $[\![aa]\!]$. We then have the following grammar:

- Nonterminals are $\{S, A, B, X\}$ with one start symbol S
- Lexical productions $A \to a$, $B \to b$, $S \to \lambda$
- Branching productions $S \to AB$, $A \to AS$, $A \to XB$, $X \to AA$

Table 6.2 The generated grammar based on Table 6.1

N	$N \in I$?	Lexical rules	Branching rules	Chain rules
$[\![\lambda]\!]$	Y	$[\![\lambda]\!] \to \lambda$		$[\![\lambda]\!] \to [\![ab]\!]$
$[\![a]\!]$		$[\![a]\!] \to a$		$[\![a]\!] \to [\![aab]\!]$
$[\![b]\!]$		$[\![b]\!] \to b$		
$[\![aa]\!]$			$[\![aa]\!] \to [\![a]\!][\![a]\!]$	
$[\![ab]\!]$	Y		$[\![ab]\!] \to [\![a]\!][\![b]\!]$	$[\![ab]\!] \to [\![\lambda]\!]$
$[\![aab]\!]$			$[\![aab]\!] \to [\![a]\!][\![ab]\!]$, $[\![aab]\!] \to [\![aa]\!][\![b]\!]$	$[\![aab]\!] \to [\![a]\!]$

This grammar generates an infinite nonregular language which is $\{\lambda,$
$ab,$
$aabb,$
$aababb,$
$\ldots\}.$

So given an OT, we can write down a set of nonterminals and productions; but this leaves unanswered a very important question: how do we pick the rows and columns of the OT? The construction procedure that we have just defined has two interesting properties that make it possible to answer this question. These are called the *monotonicity* properties.

First, if we increase the columns in the table, the language defined by the grammar generated from the table will always be smaller than or equal to the grammar generated from the original table. Table 6.3 shows a table with two more columns, which we have filled in using MQs on some hypothetical language. Now the resulting grammar, shown in Table 6.4, contains no unary rules, and each nonterminal only generates the single string that it is labeled with. This grammar therefore defines a small finite language which consists of just the two strings $\{\lambda, ab\}$; this is a proper subset of the language defined by the original grammar. It is easy to see why this will in general always be the case: if we add columns, the generated grammar will have the same set of nonterminals, lexical and binary productions, but may have fewer unary productions. Adding additional columns means that the approximate test for

Table 6.3 Example where we have added two more columns to Table 6.1

	$\square$	$a\square$	$\square b$	$\square bb$	$aa\square bb$	$\square abb$
λ	1	0	0	0	1	0
a	0	0	1	0	0	1
b	0	1	0	0	0	0
aa	0	0	0	1	0	0
ab	1	0	0	0	0	0
aab	0	0	1	0	0	0

$aa\square bb$ and $\square abb$. As a result the generated grammar no longer contains any unary or chain rules and just generates the finite language $\{\lambda, ab, aabb\}$

Table 6.4 The generated grammar no longer contains any unary or chain rules and just generates the finite language $\{\lambda, ab, aabb\}$

N	$N \in I$?	Lexical rules	Branching rules
$[\![\lambda]\!]$	Y	$[\![\lambda]\!] \rightarrow \lambda$	
$[\![a]\!]$		$[\![a]\!] \rightarrow a$	
$[\![b]\!]$		$[\![b]\!] \rightarrow b$	
$[\![aa]\!]$			$[\![aa]\!] \rightarrow [\![a]\!][\![a]\!]$
$[\![ab]\!]$	Y		$[\![ab]\!] \rightarrow [\![a]\!][\![b]\!]$
$[\![aab]\!]$			$[\![aab]\!] \rightarrow [\![a]\!][\![ab]\!]$, $[\![aab]\!] \rightarrow [\![aa]\!][\![b]\!]$

Table 6.5 Example where we have added one more row (bab) to Table 6.1

	$\square$	$a\square$	$\square b$	$\square bb$
λ	1	0	0	0
a	0	0	1	0
b	0	1	0	0
aa	0	0	0	1
ab	1	0	0	0
aab	0	0	1	0
bab	0	1	0	0

Table 6.6 The resulting grammar now has more nonterminals and productions than the original and as a result generates a larger language

N	$N \in I$?	Lexical rules	Branching rules	Chain rules
$[\![\lambda]\!]$	Y	$[\![\lambda]\!] \rightarrow \lambda$		$[\![\lambda]\!] \rightarrow [\![ab]\!]$
$[\![a]\!]$		$[\![a]\!] \rightarrow a$		$[\![a]\!] \rightarrow [\![aab]\!]$
$[\![b]\!]$		$[\![b]\!] \rightarrow b$		$[\![b]\!] \rightarrow [\![bab]\!]$
$[\![aa]\!]$			$[\![aa]\!] \rightarrow [\![a]\!][\![a]\!]$	
$[\![ab]\!]$	Y		$[\![ab]\!] \rightarrow [\![a]\!][\![b]\!]$	$[\![ab]\!] \rightarrow [\![\lambda]\!]$
$[\![aab]\!]$			$[\![aab]\!] \rightarrow [\![a]\!][\![ab]\!]$, $[\![aab]\!] \rightarrow [\![aa]\!][\![b]\!]$	$[\![aab]\!] \rightarrow [\![a]\!]$
$[\![bab]\!]$			$[\![bab]\!] \rightarrow [\![b]\!][\![ab]\!]$	$[\![bab]\!] \rightarrow [\![b]\!]$

congruence becomes more accurate and stringent, and as a result some unary rules will be removed. Thus the resulting grammar will in general generate a language which is a subset of the original, though it may remain the same.

We have a complementary result when we add one or more additional rows to Table 6.1. Table 6.5 gives a simple example: we add one more row, labeled with the string bab. The resulting grammar shown in Table 6.6 is now larger: the set of productions and nonterminals include the original productions and nonterminals and as a result the language defined is going to be larger. The grammar after merging nonterminals and relabeling is this:

- Nonterminals are $\{S, A, B, X\}$ with one start symbol S
- Lexical productions $A \rightarrow a$, $B \rightarrow b$, $S \rightarrow \lambda$
- Branching productions $S \rightarrow AB$, $A \rightarrow AS$, $A \rightarrow XB$, $X \rightarrow AA$, $B \rightarrow BS$

This grammar generates a nonregular language that is larger and includes the string $abab$.

The end result of these two monotonicity properties is that it is easy to construct a learning algorithm. We maintain an OT; if the grammar generates too small a language, then we can add some rows to reinforce the grammar, and if, on the other hand, the grammar overgenerates, then we can add columns in order to make the grammar more accurate. There are a number of different ways of doing this: if we observe some string w which is not generated by the current grammar, the most naive approach is simply to add every element of $\text{Sub}(w)$ as a row. This is guaranteed to make the grammar generate w and simplifies the convergence analysis. In this learning model we have an oracle that can be used to make MQs, and so we can fill in all of the spaces in the OT easily. In other learning models, we need to use other approaches.

The fundamental representational assumption though is that nonterminals are indexed by substrings u, and we want each nonterminal $[\![u]\!]$ to generate all strings that are distributionally identical to u. That is to say we want $\mathscr{L}(G, [\![u]\!]) = [u]$. This assumption is what distinguishes these congruential approaches from others that we examine later: this is the simplest primal approach.

The same representational assumption can be used to define an algorithm to learn related classes of languages in a stochastic setting [8, 40, 51]. In these models, we cannot ask queries but only have a randomly generated sequence of examples. In this case we can have an OT that stores counts rather than just a 0/1 value. In each cell of the table we store the number of times we have seen the string that corresponds to that cell. Congruence then can be replaced by its stochastic variant. The classes of languages that we can prove we can learn here are quite limited, and the assumptions quite strong and unrealistic; nevertheless, this shows that stochastic variants of these algorithms are possible.

6.5.2 Substitutable Languages

If we want to learn under the more stringent Gold paradigm, where we have neither queries nor any constraints on how the positive samples are being selected, then we need to use a slightly different algorithm that relies on a language-theoretic closure property in order to guarantee convergence. We maintain the same representational assumption as in the previous section—the nonterminals will generate congruence classes.

Given two nonempty strings u and v, we say that $u \doteq_L v$ if there is a context $l \Box r$ such that $lur \in L$ and $lvr \in L$. A language L is substitutable if $u \doteq_L v$ implies $u \equiv_L v$. This is a very strong condition, analogous to reversibility in the inference of regular languages [16]. Indeed substitutability implies reversibility; there are however languages which are substitutable but not context-free (see for example the language MIX which we define below). Languages that are substitutable include examples like $\{a^n cb^n \mid n > 0\}$ but are too strong to be of much practical interest: for example even the language $\{a^n b^n \mid n > 0\}$ is not substitutable. There are even finite languages that are not substitutable: $\{a, aa\}$ is a trivial example.

Clark and Eyraud [16] show that this class of languages can be learned from positive data alone using a Gold model; the algorithm has polynomial update time, and has a characteristic set with polynomial number of elements. When the algorithm observes two substrings of the data that occur in a single context then it assumes that they are congruent; the restriction of the class substitutes for the lack of MQs. This has been extended to k, l-substitutable CFGs [63], in a manner analogous to k-reversibility in the inference of regular languages.

Interestingly the criteria of substitutability is quite natural and was already noted in the early days of structuralist linguistics: Myhill in 1950 [42] gave an equivalent definition and suggested calling languages that satisfy that definition 'regular'!

6.5.3 *Finite Kernel Property*

In the case of congruential languages, we considered nonterminals that generate congruence classes. $\mathscr{L}(G, [\![u]\!]) = [u]_{L_*} = \{v \mid C_L(v) = C_L(u)\}$. A slightly different condition would be to consider the set: $u^{\triangleright\triangleleft} = \{v \mid C_L(v) \supseteq C_L(u)\}$. Now we allow sets of strings that include strings whose distribution properly includes the distribution of u. These strings will have an asymmetric substitution property: whenever we have a string like lur we can substitute any string $v \in u^{\triangleright\triangleleft}$ to get a string lvr but perhaps not in reverse. There are a number of cases where this could be useful; in natural languages we often have that an ambiguous word has a wider distribution than an unambiguous one. In this case we might want to have a nonterminal that generates not just the unambiguous words but also words that can have other lexical categories as well.

A further generalization of this is to consider the case where rather than considering the nonterminals to be generated by individual strings, we can consider them to be generated by small finite sets of strings. Given a bound k, we can consider sets of strings S such that $|S| \leq k$, and consider nonterminals that are indexed by these sets. In this case we want the nonterminal $[\![S]\!]$ to generate the set of strings that can occur in all of the contexts that are *shared* by the elements of S. This allows us to have nonterminals that correspond to clusters of strings that are distributionally similar but not identical.

More formally we say a CFG has the *k-Finite kernel property* (k-FKP) if every nonterminal N has a set of strings S_N, $|S_N| \leq k$, such that $\mathscr{L}(G, N)^{\triangleright} = S_N^{\triangleright}$. The class of all CFGs with k-FKP can be learned using examples and MQs [67].

6.6 Dual Algorithms

In dual algorithms we swap the roles of the substrings and the contexts: we index nonterminals by contexts or sets of contexts, and use substrings to eliminate the incorrect rules. The representational assumption is then quite different. We take a context or more generally a finite set of contexts C, and consider the sets of strings that can occur in all of the contexts: $C^{\triangleleft}$. We then define grammars where the nonterminals correspond to these sets of strings.

6.6.1 Context Deterministic Grammars

The first dual learning result is by Shirakawa and Yokomori who, in an important early paper [54], which in our opinion has not received enough attention, define the class of c-deterministic grammars as those grammars G such that whenever $S \overset{*}{\Rightarrow}_G lNr$ it is the case that $\mathscr{L}(G, N) = (l\Box r)^{\triangleleft}$. They then provide a MAT learning algorithm for this class. There is a small error in this paper: the paper claims that the class includes all regular languages, but the grammar construction is slightly too weak for this. A minor modification—allowing rules of the form $N \to Pa$ and $N \to aP$, where N, P are nonterminals and $a \in \Sigma$—is sufficient to correct this.

Having defined the nonterminals as corresponding to sets of strings that occur in a given context $l\Box r$ we then can use the strings to eliminate rules. Suppose we have three nonterminals that correspond to the three contexts $l_1\Box r_1, l_2\Box r_2, l_3\Box r_3$; reusing the earlier notation we can say the nonterminal symbols are $[\![l_1\Box r_1]\!]$, $[\![l_2\Box r_2]\!]$ and $[\![l_3\Box r_3]\!]$. We can consider the possible production $[\![l_1\Box r_1]\!] \to [\![l_2\Box r_2]\!][\![l_3\Box r_3]\!]$. If this rule is correct, then the result of concatenating any string that can occur in $l_2\Box r_2$ with any string that can occur in $l_3\Box r_3$ will be a string that can occur in $l_1\Box r_1$, or, using the notation we defined earlier, $(l_1\Box r_1)^{\triangleleft} \supseteq (l_2\Box r_2)^{\triangleleft}(l_3\Box r_3)^{\triangleleft}$.

Crucially, if this is *false*, then we can observe some strings u, v that show that it is false: if l_2ur_2 and l_3vr_3 are in L_* but l_1uvr_1 is *not*, then we will know that the production is incorrect in a certain sense. Thus, just as contexts were used to eliminate undesirable unary chain rules in the congruential case earlier, strings are used to eliminate undesirable binary rules in this c-deterministic case.

6.6.2 Finite Context Property

One can weaken this condition in two ways. One is by requiring only that there be some context $l\Box r$ such that $\mathscr{L}(G, N) = (l\Box r)^{\triangleleft}$; this is a weaker condition because the c-deterministic condition requires this to be true for *any* context of the nonterminal N. The second is that we allow more than one context. This leads us to the *k-finite context property* (k-FCP) [39, 67].[2] A CFG has the k-FCP if for every nonterminal we can find a set of contexts C, where $|C| \leq k$, such that $\mathscr{L}(G, N) = C^{\triangleleft}$. One can also modify this to a slightly weaker form as in [67]. A closely related idea, context-separability, is defined in [1]—this is equivalent to the 1-FCP.

[2]The original paper defining this [12] unfortunately contains some errors.

6.7 Combined Primal-Dual Methods

The primal and dual techniques can be combined to produce an algorithm which can learn classes where the nonterminals can be defined either primally or dually [68]; we can use these techniques to combine for example the congruential learning result and the c-deterministic learning result to get an algorithm which can MAT-learn a larger class. The class that is learnable using these combined methods will be strictly larger than the union of the learnable classes with either primal or dual on its own, as it will include languages where some of the nonterminals can only be defined primally and some can only be defined dually. The classes learned are stratified by three natural numbers: r, the maximum number of nonterminals used on the right hand side of a production; p, the maximum number of strings used to define a nonterminal primally; and q, the maximum number of contexts used to define a nonterminal dually. We denote by $\mathbb{G}(p, q, r)$ the class of CFGs that satisfy these bounds. Yoshinaka [68] shows that the class $\mathbb{G}(p, q, r)$ can be learned using positive data and MQs.

It is important to realise that there are CFLs that cannot be generated by any $\mathbb{G}(p, q, r)$ for any values of p, q, r. A simple example is the language $\{a^n b^m \mid n \neq m\}$. This is clearly a CFL but cannot be represented by any grammar in this class. This is because in order to represent this language we need nonterminals that will generate sets of strings like $\{a^n b^m \mid n > m\}$ and $\{a^n b^m \mid n < m\}$. Neither of these sets can be defined by any finite number of strings or contexts. Thus this language, and others like it, are not learnable using any of these distributional techniques. Figure 6.1 shows the relationship of the various learnable classes to the classes of regular languages and CFLs. It is possible to get a more integrated view of the representational assumptions of these algorithms by looking at the Syntactic Concept Lattice—the residuated lattice consisting of all distributionally definable sets of strings [9, 14, 62].

6.8 Multiple Context-Free Grammars

CFGs are fairly expressive for describing natural languages, yet the literature has found several natural language phenomena that cannot be described by CFGs. The example which definitively established that CFGs were not weakly adequate was the case of cross-serial dependencies in Swiss German [26, 52]. We present here the data in a form very close to the original presentation. In the particular dialect of Swiss German considered by Shieber, the data concerns a sequence of embedded clauses.

Let us abstract this a little bit and consider a formal language for this non-context-free fragment of Swiss German. We consider that we have the following words or word types: n_a, n_d, which are respectively accusative and dative noun phrases, v_a, v_d, which are verb phrases that require accusative and dative noun phrases respectively, and finally c, which is a complementizer which appears at the beginning of the clause. Thus the 'language' we are looking at consists of sequences like $cn_a v_a$ and $cn_d v_d$ and $cn_a n_a n_d v_a v_a v_d$, but crucially does not contain examples where the sequence of

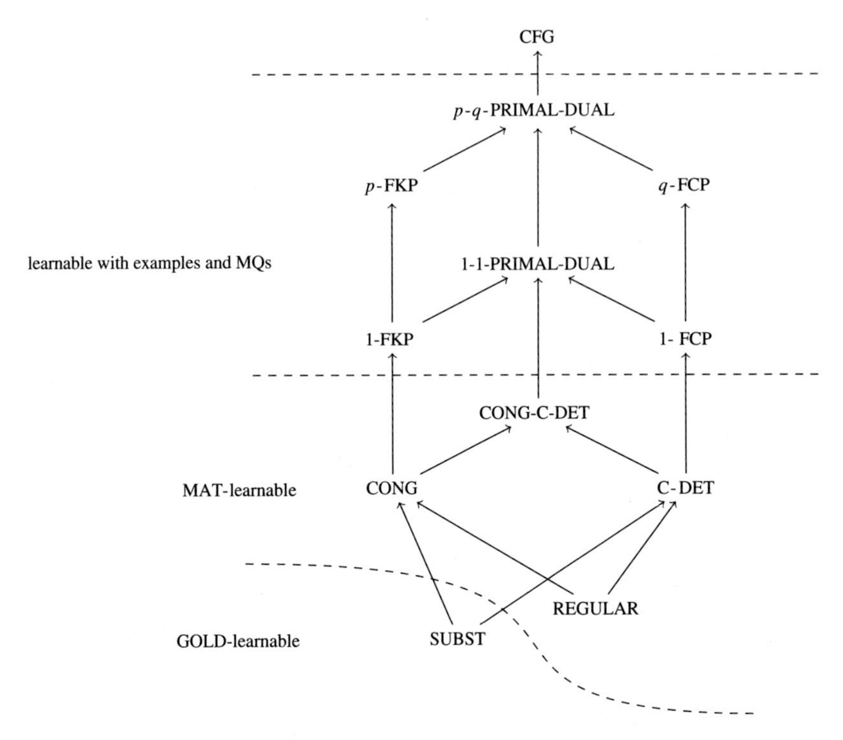

Fig. 6.1 Diagram showing the various classes of CFGs learnable using these techniques. All inclusions are strict. CONG is the class of congruential CFGs in Sect. 6.5.1, SUBST is the class of substitutable CFGs in Sect. 6.5.2, and C-DET is the class of context-deterministic grammars in Sect. 6.6. The dual techniques are on the right and the primal techniques are on the left; substitutable languages are both primal and dual

accusative/dative markings on the noun sequence is different from the sequence of requirements on the verbs. So it does not contain $cn_d v_a$, because the verb requires an accusative and it only has a dative. The sublanguage we are concerned with is the language $L_{\mathrm{SG}} = \{cn_a^i n_d^j v_a^i v_d^j \mid i, j \geq 1\}$. This language is defined through intersection of the original language with a suitable regular language and a homomorphism relabeling the strings. Since CFGs are closed under these operations, and L_{SG} is clearly not context-free, this establishes the non-context-freeness of the original language.

Joshi [27] proposed the notion of *mildly context-sensitive (*MCS*) grammars* to pursue a better formalism to describe natural languages. They suggested that an MCS family of languages should

1. include CFLs,
2. allow limited cross-serial dependencies,
3. have the constant-growth property,[3]
4. have polynomial-time parseability.

Here what "limited cross-serial dependencies" means is unclear, and various definitions of MCS formalisms have been proposed. These different grammar definitions have converged to three different language classes. The smallest class is the class defined by Linear Indexed Grammars, Combinatory Categorial Grammars and Tree Adjoining Grammars (TAGs) [59]. The largest class is defined by the formalism we use here, Multiple Context-Free Grammars (MCFG) [48], which are essentially equivalent to Linear Context-Free Rewriting Systems [60]; these are also equivalent[4] to the more linguistically motivated Minimalist Grammars [56] which came out of an attempt to formalise some ideas in contemporary syntactic theory. There is also an intermediate class of well-nested MCFGs [30], non-duplicating Macro Grammars [22] and Coupled CFGs [25], which form a proper subset of the class of all MCFGs, and a proper superset of the class of TAG-equivalent grammars, which are equivalent to well-nested MCFGs of dimension 2.

MCFGs are a very natural generalisation of CFGs. While nonterminals of a CFG generate strings, those in an MCFG generate tuples of strings; for example pairs of strings. These strings in a tuple need not be adjacent in a complete sentence in the language of the grammar. This allows MCFGs to generate languages which have cross-serial dependencies.

We will start this section by defining the MCFG formalism which may be unfamiliar to the reader. We will then discuss the learnability of these, using distributional methods. We structure this section somewhat differently from the CFG section; given that the reader is now familiar with the primal/dual distinction, we will start with the positive data-only learning model, and then move to the query-based learning models.

6.8.1 Definition

An MCFG is a quadruple $G = \langle \Sigma, V, R, I \rangle$ just like a CFG but each nonterminal symbol $N \in V$ is assigned a positive integer called the *dimension*, which we will denote by $\dim(N)$. By V_d we denote the subset of V whose elements have dimension d. Every start symbol has dimension 1: $I \subseteq V_1$. A nonterminal of dimension d generates d-tuples of strings. We write production rules of an MCFG in Horn clause

[3] An infinite language L is said to have the constant-growth property if $\exists k \in \mathbb{N}.\forall u \in L.\exists v \in L.\, 0 < |v| - |u| \leq k$.

[4] It is worth noting that MCFGs may be much larger than the smallest equivalent Minimalist Grammar.

notation,[5] which consists of a single literal on the left and a possible empty sequence of literals on the right, where a nonterminal of dimension d appears as a predicate of d-ary. Terminal symbols can only appear on the left hand side of the rule. A rule $N \to aPQ$ of a CFG, for example, is written in this notation as

$$N(axy) :- P(x), Q(y),$$

where x, y are *variables* and a a terminal symbol, or a constant. This rule is read as follows: if P derives a string x and Q derives y, N derives axy. In an MCFG, nonterminals may have dimension more than 1. For example, the rule

$$N(x_1 y_1, ax_2) :- P(x_1, x_2), Q(y_1)$$

means that if P derives a pair (x_1, x_2) and Q derives y_1, N derives the pair $(x_1 y_1, ax_2)$, where the dimensions of N, P and Q are 2, 2 and 1, respectively.

More formally, production rules have the following form in general:

$$N_0(\alpha_1, \dots, \alpha_{d_0}) :- N_1(x_{1,1}, \dots, x_{1,d_1}), \dots, N_k(x_{k,1}, \dots, x_{k,d_k})$$

where $N_0, N_1, \dots, N_k \in V$ for some $k \geq 0$, $d_i = \dim(N_i)$ for each $i \in \{0, \dots, k\}$; variables $x_{1,1}, \dots, x_{k,d_k}$ are pairwise distinct; and each $\alpha_1, \dots, \alpha_{d_0}$ are strings of terminals and variables such that all and only variables $x_{1,1}, \dots, x_{k,d_k}$ occur just once through $\alpha_1, \dots, \alpha_{d_0}$.[6] If $k = 0$ then the right-hand side is empty, and the production is of the form $N_0(\mathbf{v}) :-$ where $\mathbf{v} \in (\Sigma^*)^{d_0}$. If $x_{i,j}$ always occurs left of $x_{i,j+1}$ in $\alpha_1 \dots \alpha_m$ for $1 \leq i \leq k$ and $1 \leq j < d_i$, the rule is said to be *non-permuting*. An example of a rule that is *not* non-permuting is

$$P(x_1 y_2, x_2 y_1) :- Q(x_1, x_2), R(y_1, y_2),$$

as y_2 occurs left to y_1 in $P(x_1 y_2, x_2 y_1)$.

Example 6.1 We define an MCFG $G_{\text{SG}} = \langle \{c, n_a, n_d, v_a, v_d\}, V_1 \cup V_2, R, \{S\} \rangle$ where $V_1 = \{S\}$ and $V_2 = \{P, Q\}$. R consists of the rules

$$\begin{aligned} S(cx_1 y_1 x_2 y_2) &:- P(x_1, x_2), Q(y_1, y_2); \\ P(n_a x_1, v_a x_2) &:- P(x_1, x_2); \\ Q(n_d x_1, v_d x_2) &:- Q(x_1, x_2); \\ P(n_a, v_a) &:- ; \\ Q(n_d, v_d) &:- . \end{aligned}$$

All of the above rules are non-permuting.

[5] The notation adopted in this chapter follows Smullyan's elementary formal systems [55] rather than [48].

[6] We only consider non-deleting productions in this chapter.

The derivation process of an MCFG is formally defined with a substitution. A *substitution* θ is a map from variables to strings, which is extended to the homomorphism $\hat{\theta}$ such that $\hat{\theta}(y) = \theta(y)$ if y is in the domain of θ, and $\hat{\theta}(y) = y$ otherwise. We identify $\hat{\theta}$ and θ if no confusion arises. A substitution θ is often denoted as a suffix operator $[x_1 \mapsto \theta(x_1), \ldots, x_k \mapsto \theta(x_k)]$, or even as $[\theta(x_1), \ldots, \theta(x_k)]$ if the domain of θ is understood and ordered as $x_1, \ldots, x_k$.

We write $\vdash_G N(\mathbf{v})$ if $N(\mathbf{v})$:− is a rule in R. If we have $\vdash_G N_i(\mathbf{v}_i)$ for all $i = 1, \ldots, k$ and R has a rule $N_0(\alpha_1, \ldots, \alpha_{d_0})$:− $N_1(\mathbf{x_1}), \ldots, N_k(\mathbf{x_k})$, then we deduce

$$\vdash_G N_0(\theta(\alpha_1), \ldots, \theta(\alpha_{d_0}))$$

where $\theta(\mathbf{x_i}) = \mathbf{v}_i$ for all $i = 1, \ldots, k$. We will abbreviate this substitution θ as $[\mathbf{v}_1, \ldots, \mathbf{v}_k]$. The *language of* N is defined by

$$\mathscr{L}(G, N) = \{\mathbf{v} \in (\Sigma^*)^{\dim(N)} \mid \vdash_G N(\mathbf{v})\}.$$

The *language of* G is $\mathscr{L}(G) = \bigcup_{S \in I} \mathscr{L}(G, S)$.

Recall Example 6.1. It is easy to see that we have

$$\vdash_{G_{\mathrm{SG}}} P(n_a, v_a), \quad \vdash_{G_{\mathrm{SG}}} P(n_a n_a, v_a v_a), \quad \vdash_{G_{\mathrm{SG}}} P(n_a n_a n_a, v_a v_a v_a),$$
$$\vdash_{G_{\mathrm{SG}}} Q(n_d, v_d), \quad \vdash_{G_{\mathrm{SG}}} Q(n_d n_d, v_d v_d),$$

and

$$\vdash_{G_{\mathrm{SG}}} S(c n_a n_a n_a n_d n_d v_a v_a v_a v_d v_d),$$

for example. Figure 6.2 describes this derivation process in a tree form, where boxes emphasise the fact that the pair $\langle n_a n_a, v_a v_a \rangle$ generated by P appears as discontinuous strings in the final product $c n_a n_a n_a n_d n_d v_a v_a v_a v_d v_d$. It is easy to see that $\mathscr{L}(G_{\mathrm{SG}}) = L_{\mathrm{SG}} = \{c n_a^i n_d^j v_a^i v_d^j \mid i, j \geq 1\}$.

By MCFG(p, q) we denote the class of MCFGs such that

- every nonterminal has a dimension at most p,
- every rule has at most q occurrences of nonterminals on the right hand side.

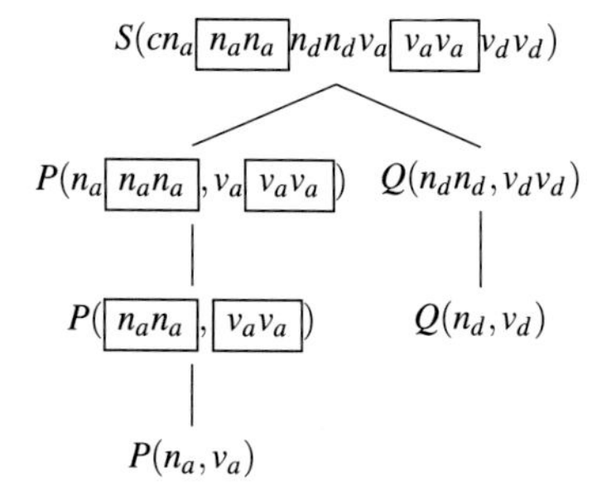

Fig. 6.2 Derivation tree of an MCFG

Thus G_{SG} belongs to MCFG(2, 2). Then we define $\text{MCFL}(p,q) = \{\mathscr{L}(G) \mid G \in \text{MCFG}(p,q)\}$. We also write $\text{MCFG}(p,*) = \bigcup_{q\in\mathbb{N}} \text{MCFG}(p,q)$ and $\text{MCFL}(p,*) = \bigcup_{q\in\mathbb{N}} \text{MCFL}(p,q)$. The class of CFGs is identified with $\text{MCFG}(1,*)$ and the CFGs in Chomsky normal form are all in MCFG(1, 2). Hence $\text{MCFL}(1,2) = \text{MCFL}(1,q)$.

Seki et al. [48] and Rambow and Satta [45] have investigated the hierarchy of MCFLs.

Proposition 6.1 (Seki et al. [48], Rambow and Satta [45]) *For* $p \geq 1$, $\text{MCFL}(p,*) \subsetneq \text{MCFL}(p+1,*)$.

For $p \geq 2$, $q \geq 1$, $\text{MCFL}(p,q) \subsetneq \text{MCFL}(p,q+1)$ *except for* $\text{MCFL}(2,2) = \text{MCFL}(2,3)$.

For $p \geq 1$, $q \geq 3$ *and* $1 \leq k \leq q-2$, $\text{MCFL}(p,q) \subseteq \text{MCFL}((k+1)p, q-k)$.

Hereafter we fix p and q to be small natural numbers. An important property of the class $\text{MCFG}(p,q)$ is the polynomial-time uniform parsability.

Theorem 6.1 (Seki et al. [48], Kaji et al. [29]) *Let p and q be fixed. It is decidable in* $O(\|G\|^2|w|^{p(q+1)})$ *time whether* $w \in \mathscr{L}(G)$ *for any* MCFG $G \in \text{MCFG}(p,q)$ *and* $w \in \Sigma^*$.

It is known that every MCFG in $\text{MCFG}(p,q)$ has an equivalent one in $\text{MCFG}(p,q)$ whose rules are all non-permuting [35], so we assume without loss of generality that all MCFG*s* are non-permuting in this chapter.

6.8.2 *Generalisation of Contexts and Substrings and Observation Tables*

Recall that classical algorithms for learning regular languages observe the relation between two strings p and s as prefixes and suffixes, respectively. That is, we have an OT whose rows are indexed by prefixes p and columns are by suffixes s and the entries show whether the concatenations ps belong to the target language L_*. The choice of those two types of objects corresponds to the fact that a nonterminal of a (right) regular grammar generates suffixes of members of the language: $S \overset{*}{\Rightarrow} pN \overset{*}{\Rightarrow} ps$. In the distributional learning of CFGs the two sorts of objects we choose are contexts $l\Box r$ and substrings v, which correspond to the fact that a nonterminal of a CFG generates substrings: $S \overset{*}{\Rightarrow} lNr \overset{*}{\Rightarrow} lvr$. In accordance with the fact that an MCFG generates discontinuous substrings, we now generalise the notion of a context to a *multi-context* and a substring to a *multi-word* and define the corresponding wrap operation in the natural way. We call a pair of strings $\langle u, v\rangle$ *2-word*. A *2-context* contains exactly two occurrences of the hole: thus a 2-context has the form $l\Box m\Box r$. The wrapping operation is accordingly generalised as

$$l\Box m\Box r \odot_2 \langle u, v\rangle = lumvr\,.$$

Similarly we can have 3-contexts, 3-words, and so on, if we target MCFGs with nonterminals of dimension 3 or more. We denote p-word and p-context by bold letters $\mathbf{u}, \mathbf{v}, \dots$ and sans-serif $\mathsf{u}, \mathsf{v}, \dots$, respectively, and the sets of p-words and p-contexts by $\mathscr{S}_p$ and $\mathscr{C}_p$, respectively. The wrapping operations $\odot_p$ for p-words and p-contexts are also defined accordingly. For a string set D, we define

$$\mathrm{Sub}_p(D) = \{\mathbf{v} \in \mathscr{S}_p \mid \mathsf{u} \odot_p \mathbf{v} \in D \text{ for some } \mathsf{u} \in \mathscr{C}_p\},$$
$$\mathrm{Con}_p(D) = \{\mathsf{u} \in \mathscr{C}_p \mid \mathsf{u} \odot_p \mathbf{v} \in D \text{ for some } \mathbf{v} \in \mathscr{S}_p\}.$$

Understanding the concerned language L, the polar maps $\rhd$ and $\lhd$ are also generalised for $\mathbf{S} \subseteq \mathscr{S}_p$ and $\mathsf{C} \subseteq \mathscr{C}_p$:

$$\mathbf{S}^{\rhd} = \{\mathsf{u} \in \mathscr{C}_p \mid \mathsf{u} \odot_p \mathbf{S} \subseteq L\},$$
$$\mathsf{C}^{\lhd} = \{\mathbf{v} \in \mathscr{S}_p \mid \mathsf{C} \odot_p \mathbf{v} \subseteq L\}.$$

For two p-words $\mathbf{u}$ and $\mathbf{v}$, we define

$$\mathbf{u} \equiv_L \mathbf{v} \text{ if and only if } \{\mathbf{u}\}^{\rhd} = \{\mathbf{v}\}^{\rhd}.$$

Distributional learning algorithms for CFGs use an OT to observe which combination of a context and a substring together forms a sentence in the concerned language. Since an MCFG may have nonterminals of different dimensions, we now have an OT for each dimension.[7]

Once we have obtained those generalisations, almost every technique in the distributional learning of CFGs can be translated for the learning of MCFGs straightforwardly as we will see in the following subsections.

6.8.3 *Substitutability*

The learnability result of substitutable CFGs presented in Sect. 6.5.2 can be translated to the MCFG learning [66]. For two 2-words $\langle u_1, u_2\rangle, \langle v_1, v_2\rangle \in \mathscr{S}_2$, let us write $\langle u_1, u_2\rangle \doteq_L \langle v_1, v_2\rangle$ if there is a 2-context $l\Box m\Box r$ such that $lu_1mu_2r \in L$ and $lv_1mv_2r \in L$. We say that a language L is 2D-*substitutable*[8] if $\langle u_1, u_2\rangle \doteq_L \langle v_1, v_2\rangle$ implies $\langle u_1, u_2\rangle \equiv_L \langle v_1, v_2\rangle$.

[7] Technically speaking, the OT for the highest dimension subsumes the other ones for lower dimensions, since m-contexts and m-words can be seen as special cases of n-contexts and n-words, respectively, for $m < n$: e.g., $l\Box r\Box \odot_2 \langle u, \lambda\rangle = l\Box r \odot u$. However, there are cases where it is more reasonable to exclude the empty string from consideration.

[8] The original definition [66] has a slightly weaker form, where strings m, u_1, u_2, v_1, v_2 are restricted to be non-empty strings. ab^*cd^*e is 2D-substitutable according to the original definition, but it is not the case in our simplified definition.

The learning algorithm for pD-substitutable MCFGs in MCFG(p, q) is essentially the same as the one for substitutable CFGs and it runs in polynomial time under the assumption that p and q are known to the learner. We note that the degree of the polynomial linearly depends on pq. The only difference is in the construction of the learner's conjecture grammar from a positive data set D. The nonterminal set is $V = \bigcup_{1 \le i \le p} V_i$, where

$$V_i = \{[\![\mathbf{v}]\!] \mid \mathbf{v} \in \mathrm{Sub}_i(D)\}$$

is the set of nonterminals of dimension i. What we would like $[\![\mathbf{v}]\!]$ to generate is $\{\mathbf{u} \mid \mathbf{u} \equiv_{L_*} \mathbf{v}\}$ where L_* is the learning target. The initial symbols are

$$I = \{[\![v]\!] \in V_1 \mid v \in D\}.$$

MCFGs do not have a simple and well-established normal form like the Chomsky normal form in CFGs. One may introduce a normal form for MCFG(p, q) but it should involve many different types of rules differently from the case of CFGs, where branching rules and lexical rules suffice. Instead we introduce rules of the conjecture grammar in a general form. We have a decomposition rule of the form

$$[\![\mathbf{v}]\!](\boldsymbol{\alpha}) :- [\![\mathbf{v}_1]\!](\mathbf{x}_1), \ldots, [\![\mathbf{v}_k]\!](\mathbf{x}_k)$$

if

- it is eligible for a rule of an MCFG in MCFG(p, q),
- $\boldsymbol{\alpha}[\mathbf{v}_1, \ldots, \mathbf{v}_k] = \mathbf{v}$.

Decomposition rules can be seen as a generalisation of branching and terminating rules in distributional learning of CFGs. We also have chain rules between two 'substitutable' k-words:

$$[\![\mathbf{v}_1]\!](\mathbf{x}) :- [\![\mathbf{v}_2]\!](\mathbf{x})$$

if there is $\mathbf{u}$ such that $\mathbf{u} \odot_k \mathbf{v}_1, \mathbf{u} \odot_k \mathbf{v}_2 \in D$.

For example, when $p = q = 2$, from $D = \{abcde, aabccdee\}$, we construct rules

$$\begin{aligned}
[\![aabccdee]\!](x_{1,1}x_{2,1}, x_{1,2}x_{2,2}) &:- [\![\langle a, cde\rangle]\!](x_{1,1}, x_{1,2}), [\![\langle abc, e\rangle]\!](x_{2,1}, x_{2,2})\,;\\
[\![\langle aa, ccdee\rangle]\!](x_{1,1}a, cx_{1,2}e) &:- [\![\langle a, cde\rangle]\!](x_{1,1}, x_{1,2})\,;\\
[\![\langle a, cde\rangle]\!](a, cde) &:- \,;\\
[\![\langle a, cde\rangle]\!](x_{1,1}, x_{1,2}) &:- [\![\langle aa, ccdee\rangle]\!](x_{1,1}, x_{1,2})
\end{aligned}$$

among others. The first three rules are decomposition rules, whereas the last one is a chain rule, which is induced from the fact $\Box b \Box \odot_2 \langle a, cde\rangle, \Box b \Box \odot_2 \langle aa, ccdee\rangle \in D$.

However, the property of pD-substitutability for $p \ge 2$ is far too strong a requirement to be useful. The flexibility of the decomposition of a sentence into 2-contexts and 2-words often makes many 2-words weakly substitutable, and thus the 2D-substitutability assumption causes too much generalisation. For example, some singleton language, say $\{aaabaaa\}$, is not 2D-substitutable. We have $a\Box a\Box a \odot_2 \langle aab, a\rangle = a\Box a\Box a \odot_2 \langle a, baa\rangle = aaabaaa$, which means $\langle aab, a\rangle \doteq_{\{aaabaaa\}} \langle a, baa\rangle$, but actually $\langle aab, a\rangle \not\equiv_{\{aaabaaa\}} \langle a, baa\rangle$ since $a\Box aa\Box \odot_2 \langle aab, a\rangle = aaabaaa$ and $a\Box aa\Box \odot_2 \langle a, baa\rangle = aaaabaa$. This argument also implies that $\{\, a^n ba^n \mid n \geq 1 \,\}$ which is still (1D-)substitutable, is not 2D-substitutable.

An interesting MCFL which is 2D-substitutable is MIX, the Bach language:

$$\mathsf{MIX} = \{w \in \{a, b, c\}^* \mid |w|_a = |w|_b = |w|_c\,\},$$

where $|w|_a$ denotes the number of occurrences of a in w. Salvati [47] showed that MIX is in MCFL(2, 2), while Kanazawa and Salvati [31] showed that it is not a tree-adjoining language. It is easy to see that MIX is indeed 2D-substitutable, since $\mathbf{u} \doteq_{\mathsf{MIX}} \mathbf{v}$ iff $\mathbf{u} \equiv_{\mathsf{MIX}} \mathbf{v}$ iff $|\mathbf{u}|_a - |\mathbf{u}|_b = |\mathbf{v}|_a - |\mathbf{v}|_b$ and $|\mathbf{u}|_b - |\mathbf{u}|_c = |\mathbf{v}|_b - |\mathbf{v}|_c$. Joshi et al. [28] suggested that MIX should be excluded from a family of MCS languages for its complete free word order of letters cannot be considered as 'limited' cross-serial dependencies. Yet MIX seems to be quite a simple language from the learnability point of view.

6.8.4 MAT Result

The other learning algorithms presented in Sects. 6.5 and 6.6 can also be translated in the same way as in the case of substitutable languages and we obtain diverse classes of learnable MCFGs. While the pD-substitutability with $p \geq 2$ is stronger than the original 1D-substitutability and accordingly the obtained learnable MCFGs are very much restricted, the subclasses of MCFGs defined by those properties are indeed larger than the corresponding classes of CFGs.

6.8.4.1 Congruential MCFGs

The MAT learnability result of congruential CFGs presented in Sect. 6.5.1 can be translated as follows [69]. We say that an MCFG G is *congruential* if for every nonterminal N and any elements $\mathbf{u}, \mathbf{v} \in \mathscr{L}(G, N)$, we have $\mathbf{u} \equiv_{\mathscr{L}(G)} \mathbf{v}$. By definition, every congruential CFG is a congruential MCFG.

The grammar G_{SG} of Example 6.1 is congruential. The languages of respective nonterminals are

$$\begin{aligned}
\mathscr{L}(G_{\mathrm{SG}}, S) &= \{cn_a^i n_d^j v_a^i v_d^j \mid i, j \geq 1\,\},\\
\mathscr{L}(G_{\mathrm{SG}}, P) &= \{(n_a^i, v_a^i) \mid i \geq 1\,\},\\
\mathscr{L}(G_{\mathrm{SG}}, Q) &= \{(n_d^j, v_d^j) \mid j \geq 1\,\},
\end{aligned}$$

and for all $cn_a^i n_d^j v_a^i v_d^j \in \mathscr{L}(G_{\mathrm{SG}}, S)$, $(n_a^i, v_a^i) \in \mathscr{L}(G_{\mathrm{SG}}, P)$ and $(n_d^j, v_d^j) \in \mathscr{L}(G_{\mathrm{SG}}, Q)$, their context sets are

$$\begin{aligned}(cn_a^i n_d^j v_a^i v_d^j)^{\triangleright} &= \{\Box\}\,,\\ (n_a^i, v_a^i)^{\triangleright} &= \{c\Box n_d^j \Box v_d^j \mid j \geq 1\}\,,\\ (n_d^j, v_d^j)^{\triangleright} &= \{cn_a^i \Box v_a^i \Box \mid i \geq 1\}\,.\end{aligned}$$

The MAT learner for congruential MCFGs constructs a grammar in a way similar to the one for pD-substitutable MCFGs, except for the condition for chain rules. Now we have membership oracle and all the entries of the OTs are fulfilled. Let K_d and F_d be the finite sets of d-words and d-contexts, which label the rows and columns of the OT for dimension d, respectively. The set of nonterminals of dimension d is $V_d = \{[\![\mathbf{v}]\!] \mid \mathbf{v} \in K_d\}$, where what we would like $[\![\mathbf{v}]\!]$ to generate are again $\mathbf{u}$ such that $\mathbf{u} \equiv_{L_*} \mathbf{v}$. Decomposition rules are constructed in exactly the same manner as before: we have $[\![\mathbf{v}]\!](\boldsymbol{\alpha}) :- [\![\mathbf{v}_1]\!](\mathbf{x}_1), \ldots, [\![\mathbf{v}_k]\!](\mathbf{x}_k)$ if it is eligible for a rule of an MCFG in MCFG(p, q), and $\boldsymbol{\alpha}[\mathbf{v}_1, \ldots, \mathbf{v}_k] = \mathbf{v}$. The criterion for a chain rule is different. For $\mathbf{u}, \mathbf{v} \in K_d$, we have the chain rule of the form $[\![\mathbf{u}]\!](\mathbf{x}) :- [\![\mathbf{v}]\!](\mathbf{x})$ if and only if $\mathbf{u}^{\triangleright} \cap F_d = \mathbf{v}^{\triangleright} \cap F_d$.

We again have the monotonicity properties. If we increase the rows in the table, no existing rules will be removed but new nonterminals and decomposition rules will be added. On the other hand, if we increase the columns in the table, no new rules will be added but possibly some chain rules will be removed.

For example, according to the OT in Table 6.7, we have a chain rule

$$[\![\langle n_d, v_d\rangle]\!](x_1, x_2) :- [\![\langle n_a n_d, v_a v_d\rangle]\!](x_1, x_2)$$

Table 6.7 Examples of OTs for $L_{\mathrm{SG}} = \{cn_a^i n_d^j v_a^i v_d^j \mid i, j \geq 1\}$

1D	$\Box$	$c\Box$	$c\Box v_a v_d$
λ	0	0	0
$cn_a v_a$	1	0	0
$n_a v_a$	0	1	0
$n_a n_d$	0	0	1
$n_a n_a$	0	0	0
2D	$cn_a\Box n_d v_a \Box v_d$	$c\Box n_a n_d \Box v_a v_d$	$cn_a\Box n_d\Box$
$\langle n_a, v_a\rangle$	1	1	0
$\langle n_d, v_d\rangle$	1	0	0
$\langle n_a n_d, v_a v_d\rangle$	1	0	0
$\langle n_a, n_d v_a v_d\rangle$	0	0	0
$\langle n_a, v_a v_a v_d\rangle$	0	0	1

Table 6.8 Expansion of the OT for dimension 2 in Table 6.7

2D	$cn_a\Box n_d v_a\Box v_d$	$c\Box n_a n_d\Box v_a v_d$	$cn_a\Box n_d\Box$	$cn_a n_d\Box v_a v_d\Box$
$\langle n_a, v_a\rangle$	1	1	0	0
$\langle n_d, v_d\rangle$	1	0	0	1
$\langle n_a n_d, v_a v_d\rangle$	1	0	0	0
$\langle n_a, n_d v_a v_d\rangle$	0	0	0	0
$\langle n_a, v_a v_a v_d\rangle$	0	0	1	0
$\langle n_d n_d, v_d v_d\rangle$	1	0	0	1

since $\langle n_d, v_d\rangle^{\triangleright} \cap F_2 = \langle n_a n_d, v_a v_d\rangle^{\triangleright} \cap F_2 = \{cn_a\Box n_d v_a\Box v_d\}$. If, however, $cn_a n_d\Box v_a v_d\Box$ is added to F_2, this chain rule should be discarded since $cn_a n_d\Box v_a v_d\Box \odot_2 \langle n_d, v_d\rangle \in L_{\mathrm{SG}}$ but $cn_a n_d\Box v_a v_d\Box \odot_2 \langle n_a n_d, v_a v_d\rangle \notin L_{\mathrm{SG}}$.

On the other hand if we add $\langle n_d n_d, v_d v_d\rangle$ to K_2, we obtain a new chain rule

$$[\![\langle n_d, v_d\rangle]\!](x_1, x_2) :- [\![\langle n_d n_d, v_d v_d\rangle]\!](x_1, x_2),$$

which will never been removed. See Table 6.8.

6.8.4.2 C-Deterministic MCFGs

One can easily define c-deterministic MCFGs as well according to the translation framework discussed in Sect. 6.8.2, though no preceding work has done it explicitly. A context of a nonterminal N of a CFG is defined through the top-down rewriting derivation process: $l\Box r$ is a context of N if $S \overset{*}{\Rightarrow} lNr$. In the case of an MCFG, we say that $\mathbf{u} \in \mathscr{C}_d$ is a *context* of a nonterminal $N \in V_d$ if it is generated from an initial symbol using a special rule $N(\underbrace{\Box, \dots, \Box}_{d \text{ times}}) :-$ just once [64]. We then say that an MCFG G is *c-deterministic* if for every nonterminal N of G, every context $\mathbf{u}$ of N characterises $\mathscr{L}(G, N)$, i.e., $\{\mathbf{u}\}^{\triangleleft} = \mathscr{L}(G, N)$ (or weakly $\{\mathbf{u}\}^{\triangleleft} \equiv_{\mathscr{L}(G)} \mathscr{L}(G, N)$). By generalising Shirakawa and Yokomori's algorithm for c-deterministic CFGs, one can obtain an analogous MAT algorithm for c-deterministic MCFGs in MCFG(p, q). The grammar G_{SG} itself is not c-deterministic, but a slight modification satisfies the definition:

$$\begin{aligned} S(cx_1 n_a n_d y_1 x_2 v_a v_d y_2) &:- P(x_1, x_2), Q(y_1, y_2); \\ P(n_a x_1, v_a x_2) &:- P(x_1, x_2); \\ Q(n_d x_1, v_d x_2) &:- Q(x_1, x_2); \\ P(\lambda, \lambda) &:- ; \\ Q(\lambda, \lambda) &:- . \end{aligned}$$

6.8.5 *Finite Kernel Property and Finite Context Property*

Yoshinaka [65] has given the MCFG counterpart of the learning of CFGs with the 1-FKP [17] and Clark and Yoshinaka's [18] result has established the learning of MCFGs with the k-FCP. The definitions of the FKP and FCP for MCFGs are now obvious. We say that an MCFG G has the k-FKP if every nonterminal N of dimension d admits a finite d-word set $\mathbf{S}_N \subseteq \mathscr{S}_d$ such that

- $|\mathbf{S}_N| \leq k$,
- $\mathbf{S}_N^{\triangleright} = \mathscr{L}(G, N)^{\triangleright}$.

We say that an MCFG G has the k-FCP if every nonterminal N of dimension d admits a finite d-context set $\mathsf{C}_N \subseteq \mathscr{C}_d$ such that

- $|\mathsf{C}_N| \leq k$,
- $\mathsf{C}_N^{\triangleleft} = \mathscr{L}(G, N)$.

Learners for MCFGs with the k-FKP in MCFG(p, q) and for those with the k-FCP are designed in a way similar to the ones for CFGs with the k-FKP and with the k-FCP, respectively, with the same straightforward translation technique presented in the previous subsections.

One can combine those primal and dual approaches, of course.

6.9 Discussion

6.9.1 *Distributional Learning Beyond MCFGs and CFGs*

Distributional learning has been applied recently to learning problems beyond the CFGs and MCFGs that we consider in this chapter; we briefly review some of these approaches here.

We also have the extension to Parallel Multiple Context Free Grammars (PMCFGs) [18, 19]; these grammars include a copying operation which allows them to represent some phenomena like reduplication. Two other extensions using nonstandard formalisms have also been proposed, Distributional Lattice Grammars [11] and Binary Feature Grammars [17]. These two formalisms use a limited form of conjunction; it thus seems possible to combine these results with the PMCFG formalism to have a formalism that includes copying and conjunction.

Synchronous CFGs that are used for modeling transduction can easily be modeled by MCFGs, but if we assume the transduction is a function this can simplify the learning problem. There is an extension to learning string transductions along the lines of the well-known OSTIA [43] algorithm using a very limited class of synchronous CFGs called Inversion Transduction Grammars [13]; these learn from input/output pairs.

All of these approaches have considered string languages: the objects being modeled are strings. These approaches have also been extended beyond string languages to learning languages over other types of objects such as trees [32], and they also have been applied to graphs with an intended application in computer vision [21], and to sentence/meaning pairs using Abstract Categorial Grammars [70].

Finally, all of these learning algorithms use only a weak notion of convergence: the learner must converge to a hypothesis that generates the right language considered as a set of strings. A stronger notion of convergence requires that the hypothesis be isomorphic to the target grammar: in other words, that the learner learn a grammar that generates not just the right set of strings but the right set of structures. Such an algorithm is presented in [15].

6.9.2 Conclusion

We have reviewed a wide spectrum of algorithms using distributional learning techniques: it is clear that the methods we have studied here do not exhaust the range of application of this approach. One important point is that from a learnability point of view, CFGs are just a special case of MCFGs. While there is a significant difference between regular inference and CFG inference, there seem to be no theoretically interesting differences between CFGs and MCFGs. Every learning result for CFGs can be converted into a corresponding result for MCFGs.

References

1. Adriaans, P.: Learning shallow context-free languages under simple distributions. Tech. Rep. ILLC Report PP-1999-13, Institute for Logic, Language and Computation, Amsterdam (1999)
2. Angluin, D.: Learning regular sets from queries and counterexamples. Information and Computation **75**(2), 87–106 (1987)
3. Angluin, D., Kharitonov, M.: When won't membership queries help? J. Comput. Syst. Sci. **50**, 336–355 (1995)
4. Boasson, L., Sénizergues, S.: NTS languages are deterministic and congruential. J. Comput. Syst. Sci. **31**(3), 332–342 (1985)
5. Brill, E., Magermann, D., Marcus, M., Santorini, B.: Deducing linguistic structure from the statistics of large corpora. In: Proceedings of the Third DARPA Workshop on Speech and Natural Language, pp. 275–282 (1990)
6. Chomsky, N.: The logical structure of linguistic theory. Ph.D. thesis, MIT (1955)
7. Chomsky, N.: Language and mind, 3rd edn. Cambridge University Press (2006)
8. Clark, A.: PAC-learning unambiguous NTS languages. In: Y. Sakakibara, S. Kobayashi, K. Sato, T. Nishino, E. Tomita (eds.) Grammatical Inference: Algorithms and Applications, *Lecture Notes in Computer Science*, vol. 4201, pp. 59–71. Springer Berlin Heidelberg (2006)
9. Clark, A.: A learnable representation for syntax using residuated lattices. In: Proceedings of the 14th Conference on Formal Grammar. Bordeaux, France (2009). http://www.papers/alexcFG2009.pdf

10. Clark, A.: Distributional learning of some context-free languages with a minimally adequate teacher. In: J. Sempere, P. García (eds.) Proceedings of ICGI, no. 6339 in LNCS, pp. 24–37. Springer (2010)
11. Clark, A.: Efficient, correct, unsupervised learning of context-sensitive languages. In: Proceedings of the Fourteenth Conference on Computational Natural Language Learning, pp. 28–37. Association for Computational Linguistics, Uppsala, Sweden (2010)
12. Clark, A.: Learning context free grammars with the syntactic concept lattice. In: J. Sempere, P. García (eds.) Grammatical Inference: Theoretical Results and Applications. Proceedings of the International Colloquium on Grammatical Inference, pp. 38–51. Springer (2010)
13. Clark, A.: Inference of inversion transduction grammars. In: Proceedings of ICML. Bellevue, Washington (2011)
14. Clark, A.: The syntactic concept lattice: Another algebraic theory of the context-free languages? Journal of Logic and Computation (2013). doi:10.1093/logcom/ext037
15. Clark, A.: Learning trees from strings: A strong learning algorithm for some context-free grammars. Journal of Machine Learning Research **14**, 3537–3559 (2014)
16. Clark, A., Eyraud, R.: Polynomial identification in the limit of substitutable context-free languages. Journal of Machine Learning Research **8**, 1725–1745 (2007)
17. Clark, A., Eyraud, R., Habrard, A.: Using contextual representations to efficiently learn context-free languages. Journal of Machine Learning Research **11**, 2707–2744 (2010)
18. Clark, A., Yoshinaka, R.: Beyond semilinearity: Distributional learning of parallel multiple context-free grammars. In: J. Heinz, C. de la Higuera, T. Oates (eds.) Proceedings of the Eleventh International Conference on Grammatical Inference, *JMLR Workshop and Conference Proceedings*, vol. 21, pp. 84–96 (2012)
19. Clark, A., Yoshinaka, R.: Distributional learning of parallel multiple context-free grammars. Machine Learning pp. 1–27 (2013). doi:10.1007/s10994-013-5403-2.
20. Dediu, A.H., Martín-Vide, C. (eds.): Language and Automata Theory and Applications - 6th International Conference, LATA 2012, A Coruña, Spain, March 5-9, 2012. Proceedings, *Lecture Notes in Computer Science*, vol. 7183. Springer (2012)
21. Eyraud, R., Janodet, J., Oates, T.: Learning substitutable binary plane graph grammars. In: Proceedings of ICGI, vol. 21, pp. 114–128 (2012)
22. Fisher, M.J.: Grammars with macro-like productions. Ph.D. thesis, Harvard University (1968)
23. Gold, E.M.: Language identification in the limit. Information and Computation **10**(5), 447–474 (1967)
24. Harris, Z.: Distributional structure. Word **10**(2-3), 146–62 (1954)
25. Hotz, G., Pitsch, G.: On parsing coupled-context-free languages. Theoretical Computer Science **161**(1&2), 205–233 (1996)
26. Huybrechts, R.A.C.: The weak inadequacy of context-free phrase structure grammars. In: G. de Haan, M. Trommelen, W. Zonneveld (eds.) Van Periferie naar Kern. Foris, Dordrecht, Holland (1984)
27. Joshi, A.K.: Tree adjoining grammars: how much context-sensitivity is required to provide reasonable structural descriptions? In: D.R. Dowty, L. Karttunen, A. Zwicky (eds.) Natural Language Parsing, pp. 206–250. Cambridge University Press, Cambridge, MA (1985)
28. Joshi, A.K., Vijay-Shanker, K., Weir, D.J.: The convergence of mildly context-sensitive grammar formalisms. In: P. Sells, S.M. Shieber, T. Wasow (eds.) Foundational Issues in Natural Language Processing, pp. 31–81. MIT Press, Cambridge, MA (1991)
29. Kaji, Y., Nakanishi, R., Seki, H., Kasami, T.: The universal recognition problems for parallel multiple context-free grammars and for their subclasses. IEICE Transaction on Information and Systems **E75-D**(7), 499–508 (1992)
30. Kanazawa, M., Salvati, S.: The copying power of well-nested multiple context-free grammars. In: Language and Automata Theory and Applications, pp. 344–355. Springer (2010)
31. Kanazawa, M., Salvati, S.: Mix is not a tree-adjoining language. In: ACL (1), pp. 666–674. The Association for Computer Linguistics (2012)
32. Kasprzik, A., Yoshinaka, R.: Distributional learning of simple context-free tree grammars. In: J. Kivinen, C. Szepesvári, E. Ukkonen, T. Zeugmann (eds.) Algorithmic Learning Theory, *Lecture Notes in Computer Science*, vol. 6925, pp. 398–412. Springer (2011)

33. Keller, B., Lutz, R.: Evolutionary induction of stochastic context free grammars. Pattern Recognition **38**(9), 1393–1406 (2005)
34. Klein, D., Manning, C.D.: A generative constituent-context model for improved grammar induction. In: Proceedings of the 40th Annual Meeting of the ACL (2002)
35. Kracht, M.: The Mathematics of Language, *Studies in Generative Grammar*, vol. 63, pp. 408–409. Mouton de Gruyter (2003)
36. Kulagina, O.S.: One method of defining grammatical concepts on the basis of set theory. Problemy Kiberneticy **1**, 203–214 (1958). (in Russian)
37. Kunze, J.: Versuch eines objektivierten Grammatikmodells I, II. Z. Zeitschriff Phonetik Sprachwiss. Kommunikat **20-21** (1967–1968)
38. Langley, P., Stromsten, S.: Learning context-free grammars with a simplicity bias. In: R. López de Mántaras, E. Plaza (eds.) Machine Learning: ECML 2000, *Lecture Notes in Computer Science*, vol. 1810, pp. 220–228. Springer Berlin Heidelberg (2000)
39. Leiss, H.: Learning CFGs with the finite context property: A note on A. Clark's algorithm (2012). Manuscript
40. Luque, F.M., Infante-Lopez, G.: PAC-learning unambiguous k,l-NTS$^{\leq}$ languages. In: J.M. Sempere, P. García (eds.) Grammatical Inference: Theoretical Results and Applications, *Lecture Notes in Computer Science*, vol. 6339, pp. 122–134. Springer Berlin Heidelberg (2010)
41. Marcus, S.: Algebraic Linguistics; Analytical Models. Academic Press, New York (1967)
42. Myhill, J.: Review of *On Syntactical Categories* by Yehoshua Bar-Hillel. The Journal of Symbolic Logic **15**(3), 220 (1950)
43. Oncina, J., García, P., Vidal, E.: Learning subsequential transducers for pattern recognition interpretation tasks. IEEE Transactions on Pattern Analysis and Machine Intelligence **15**, 448–458 (1993)
44. Pitt, L.: Inductive inference, DFAs, and computational complexity. In: Proceedings of 2nd Workshop on Analogical and Inductive Inference, *Lecture Notes in Computer Science*, vol. 397, pp. 18–44 (1989)
45. Rambow, O., Satta, G.: Independent parallelism in finite copying parallel rewriting systems. Theor. Comput. Sci. **223**(1-2), 87–120 (1999)
46. Sakakibara, Y.: Learning context-free grammars from structural data in polynomial time. Theoretical Computer Science **76**(2-3), 223–242 (1990)
47. Salvati, S.: MIX is a 2-MCFL and the word problem in $\mathbb{Z}^2$ is solved by a third-order collapsible pushdown automaton. Tech. Rep. Inria-00564552, version 1, INRIA (2011). URL http://hal.inria.fr/inria-00564552
48. Seki, H., Matsumura, T., Fujii, M., Kasami, T.: On multiple context-free grammars. Theoretical Computer Science **88**(2), 191–229 (1991)
49. Sénizergues, G.: The equivalence and inclusion problems for NTS languages. Journal of Computer and System Sciences **31**(3), 303–331 (1985)
50. Sestier, A.: Contribution à une théorie ensembliste des classifications linguistiques. In: Premier Congrès de l'Association Française de Calcul, pp. 293–305. Grenoble (1960)
51. Shibata, C., Yoshinaka, R.: PAC learning of some subclasses of context-free grammars with basic distributional properties from positive data. In: S. Jain, R. Munos, F. Stephan, T. Zeugmann (eds.) ALT, *Lecture Notes in Computer Science*, vol. 8139, pp. 143–157. Springer (2013)
52. Shieber, S.M.: Evidence against the context-freeness of natural language. Linguistics and Philosophy **8**, 333–343 (1985)
53. Shinohara, T.: Rich classes inferrable from positive data – length-bounded elementary formal systems. Information and computation **108**(2), 175–186 (1994)
54. Shirakawa, H., Yokomori, T.: Polynomial-time MAT learning of c-deterministic context-free grammars. Transactions of the Information Processing Society of Japan **34**, 380–390 (1993)
55. Smullyan, R.: Theory of Formal Systems. Princeton University Press (1961)
56. Stabler, E.: Derivational minimalism. In: C. Retoré (ed.) Logical aspects of computational linguistics (LACL 1996), pp. 68–95. Springer (1997)
57. van Helden, W.: Case and gender: Concept formation between morphology and syntax (II volumes). Studies in Slavic and General Linguistics. Rodopi, Amsterdam-Atlanta (1993)

58. van Zaanen, M.: ABL: Alignment-based learning. In: COLING 2000 - Proceedings of the 18th International Conference on Computational Linguistics, pp. 961–967 (2000)
59. Vijay-Shanker, K., Weir, D.J.: The equivalence of four extensions of context-free grammars. Mathematical Systems Theory **27**(6), 511–546 (1994)
60. Vijay-Shanker, K., Weir, D.J., Joshi, A.K.: Characterizing structural descriptions produced by various grammatical formalisms. In: Proceedings of the 25th annual meeting of Association for Computational Linguistics, pp. 104–111. Stanford (1987)
61. Wells, R.S.: Immediate constituents. Language **23**(2), 81–117 (1947)
62. Wurm, C.: Completeness of full Lambek calculus for syntactic concept lattices. In: Proceedings of the 17th conference on Formal Grammar 2012 (FG) (2012)
63. Yoshinaka, R.: Identification in the limit of k, l-substitutable context-free languages. In: A. Clark, F. Coste, L. Miclet (eds.) ICGI, *Lecture Notes in Computer Science*, vol. 5278, pp. 266–279. Springer (2008)
64. Yoshinaka, R.: Learning mildly context-sensitive languages with multidimensional substitutability from positive data. In: R. Gavaldà, G. Lugosi, T. Zeugmann, S. Zilles (eds.) ALT, *Lecture Notes in Computer Science*, vol. 5809, pp. 278–292. Springer (2009)
65. Yoshinaka, R.: Polynomial-time identification of multiple context-free languages from positive data and membership queries. In: J.M. Sempere, P. García (eds.) ICGI, pp. 230–244. Springer (2010)
66. Yoshinaka, R.: Efficient learning of multiple context-free languages with multidimensional substitutability from positive data. Theoretical Computer Science **412**(19), 1821–1831 (2011)
67. Yoshinaka, R.: Towards dual approaches for learning context-free grammars based on syntactic concept lattices. In: G. Mauri, A. Leporati (eds.) Developments in Language Theory, *Lecture Notes in Computer Science*, vol. 6795, pp. 429–440. Springer (2011)
68. Yoshinaka, R.: Integration of the dual approaches in the distributional learning of context-free grammars. In: Dediu and Martín-Vide [20], pp. 538–550
69. Yoshinaka, R., Clark, A.: Polynomial time learning of some multiple context-free languages with a minimally adequate teacher. In: P. de Groote, M.J. Nederhof (eds.) Formal Grammar: 15th and 16th International Conference on Formal Grammar, pp. 192–206. Springer (2012)
70. Yoshinaka, R., Kanazawa, M.: Distributional learning of abstract categorial grammars. In: S. Pogodalla, J.P. Prost (eds.) LACL, *Lecture Notes in Computer Science*, vol. 6736, pp. 251–266. Springer (2011)

Chapter 7
Learning Tree Languages

Johanna Björklund and Henning Fernau

Abstract Tree languages have proved to be a versatile and rewarding extension of the classical notion of string languages. Many nice applications have been established over the years, in areas such as Natural Language Processing, Information Extraction, and Computational Biology. Although some properties of string languages transfer easily to the tree case, in particular for regular languages, several computational aspects turn out to be harder. It is therefore both of theoretical and of practical interest to investigate how far and in what ways Grammatical Inference algorithms developed for the string case are applicable to trees. This chapter surveys known results in this direction. We begin by recalling the basics of tree language theory. Then, the most popular learning scenarios and algorithms are presented. Several applications of Grammatical Inference of tree languages are reviewed in some detail. We conclude by suggesting a number of directions for future research.

7.1 Introduction

As elaborated in other chapters of this book, several learning models have been developed in Grammatical Inference. These models describe how the learner can behave to reach its ends, as well as the other circumstances of the learning process, like the presence of a benevolent teacher or a hostile learning environment. Many models have first been suggested and discussed in Learning Theory and have then been adapted to Grammatical Inference.

There is another aspect of learning, namely that of the objects that are to be learned and also the way these objects are represented. This aspect was termed the

J. Björklund (✉)
Department of Computing Science, Universitet Umeå, 90750 Umeå, Sweden
e-mail: johanna@cs.umu.se

H. Fernau
FB IV—Abteilung Informatikwissenschaften, Universität Trier, 54286 Trier, Germany
e-mail: fernau@informatik.uni-trier.de

J. Heinz and J.M. Sempere (eds.), *Topics in Grammatical Inference*,
DOI 10.1007/978-3-662-48395-4_7

object axis by Kasprzik [107]. Notice that this axis is mostly neglected in classical Learning Theory, as objects like strings (also known as words in the context of Formal Language Theory) or trees can be easily encoded as natural numbers (and vice versa). Also grammars or automata can be viewed this way, but inherent structural properties are easily lost in the abstraction. Conversely, the area of Grammatical Inference has focused, at least in its very beginning, on learning devices designed to accept or generate strings, like classical deterministic finite automata that every student in Computer Science encounters in the first years of study. However, many formal descriptions of sets of objects others than strings have been developed over the last decades. For instance, the third volume of the famous *Handbook of Formal Languages* [176] is dedicated to the topic "Beyond Words". These objects can be many different things, like trees, graphs, or arrays, to name a few. This richness of object types is beyond the scope of this chapter.

Rather, we will revise known results of several of these settings in the specific situation of learning tree languages. The focus is on ordered ranked finite trees, but we discuss other models in passing, especially when they seem better suited for modeling real-life situations. For instance, ordered unranked finite trees are a convenient representation for XML documents.

We expect a readership coming from and having at least three different backgrounds: Formal Language Theory, Grammatical Inference (primarily of string languages) and application areas (like XML processing). We therefore chose a narrative style for presenting the main ideas, with many illustrative examples. This means that a mathematically detailed exposition of the most advanced technical results cannot be expected, but we try our best to point to the original sources.

Structure and Scope In Sect. 7.2, we provide the fundamentals on tree languages that support the technical results in later sections. Section 7.3 comprises the largest part of the survey; it collects the mathematical results. This section is organized according to different learning models, as typically different results show up for each of these basic models. The following section, Sect. 7.4, surveys applications of learning tree languages. It should be stressed that tree language learning is one of the areas of Grammatical Inference where the (potential) applications were an obvious driving force for theoretical developments. Section 7.5 summarizes the previous sections, outlines related research efforts, and gives an outlook on the future of the field.

Learning is a very rich area. In order to stay within a reasonable number of pages, we have to narrow down the (possible) scope of this survey. We already stated that we are going to restrict ourselves to the learning of trees. So, we do not consider inference of graph languages, even though corresponding results for trees are often obtained as special cases. Possibly more severely, we also restrict the learning models that we discuss. For instance, training of probabilistic language models or unsupervised learning such as clustering algorithms or Support Vector Machines is only mentioned by giving some references (but not explained in detail) in this chapter.

Moreover, we are going to concentrate on specific devices for describing tree languages. In this survey, we focus on automata that process their input trees bottom-up, but the reverse direction is also possible. Nondeterministic top-down tree automata characterize the regular (ordered, ranked) tree languages, whereas languages recognisable by deterministic top-down tree automata form a proper subclass of the regular tree languages. In the case of tree transducers, the top-down mode of operation is more common. The inference of transducers poses a difficult challenge already in the case of string languages, so there are relatively few papers on grammatical inference that look at tree transducers. (In a different setting, this is done for quite some time within so-called SMT systems; see Sect. 7.4.3.) At the same time, tree transducer inference has many practical applications, mainly in the area of XML processing. We will discuss some of these works in Sect. 7.4. Another topic not covered is stochastic automata. Again, this is quite an important area from a practical perspective. We will briefly sketch one bioinformatics application in Sect. 7.4.

Learning Models We refer here to the other chapters of this book, and also to the textbook of de la Higuera [96] on the inference of string languages. We concentrate on learning from positive examples (text learning), on learning from positive and negative examples (informant learning), and on active learning (also known as learning with query models; see Chap. 3, "Learning Grammars and Automata with Queries", C. de la Higuera).

7.2 Definitions for Tree Languages

There are quite a number of excellent books and surveys on tree languages. See for instance [54, 87, 134], and also the chapter written by Gécseg and Steinby in the *Handbook of Formal Languages* [177]. The technical details in these works differ, so we give a brief theoretical overview in order to fix notions and notations. The focus is on ordered ranked finite trees, but we explain in passing how the learnability results may be transfered to the unranked case.

7.2.1 Trees

A *ranked alphabet* V is a finite set of symbols together with a finite *rank relation* $r_V \subset V \times \mathbb{N}$. Define $V_n := \{f \in V \mid (f, n) \in r_V\}$. Since elements in V_n are often considered as *function symbols* (representing functions of *arity* n), elements in V_0 are also called *constant symbols*. As shown further below, ranked alphabets facilitate a compact string representation of the tree. As usual, given some (alphabet) set A, A^* refers to the set of strings with letters in A (or, more technically, to the free monoid generated by A, with neutral element λ). In the context of trees, also the infinite alphabet $\mathbb{N}$ of the natural numbers is important. Given some order $<$ on the alphabet A, this easily extends to a lexicographic ordering on A^* that we also denote by $<$.

A *tree domain* is a finite subset Δ_t of $\mathbb{N}^*$ such that (1) if $x \in \Delta_t$ and y is a prefix of x, then $y \in \Delta_t$; (2) if $y \cdot i \in \Delta_t$, $i \in \mathbb{N}$, then $y \cdot j \in \Delta_t$ for $1 \leq j \leq i$. A *tree over V* is a mapping $t : \Delta_t \to V$, where the Δ_t is a tree domain and for every $x \in \Delta_t$, $t(x) \in V_n$ if $\{i \in \mathbb{N} \mid x \cdot i \in \Delta_t\} = \{1, \ldots, n\}$. An element of Δ_t is also called a *node* of t, the node λ is the *root* of t, and for every node $x \in \Delta_t$, $t(x)$ is the *label* of x. A tree is finite if its domain is finite. Notice that the Hasse diagram of the order induced by $<$ on Δ_t coincides with the usual drawing of the corresponding tree; see Fig. 7.1.

A *frontier node* (or *leaf*) in t is a node $y \in \Delta_t$ such there is no $x \in \Delta_t$ with $y < x$. If $y \in \Delta_t$ is not a frontier node, it is an *interior node*. If $y \in \Delta_t$ is an interior node, then every node $y \cdot i \in \Delta_t$ for $i \in \mathbb{N}$ is called a *child* of y. The *depth* of a tree t is defined as $\text{depth}(t) = \max\{|x| \mid x \in \Delta_t\}$, whereas the *size* of t is given by $|\Delta_t|$. Letters will be viewed as trees of size 0 and depth 1.

Let $V^{\mathtt{t}}$ denote the set of all finite trees over V. By this definition, trees are rooted, directed, acyclic graphs in which every node except the root has one predecessor and the direct successors of any node are linearly ordered from left to right. Any subset of $V^{\mathtt{t}}$ is a *tree language*.

Interpreting V as a set of function symbols, $V^{\mathtt{t}}$ can be identified with the well-formed terms over V. This yields a compact string denotation of trees. Let us explain these notions with a little example.

Example 7.1 Consider Fig. 7.1. Here, we have $V = \{a, b, c, 2\}$. The rank relation is given by $V_0 = \{b, c, 2\}$, $V_1 = \{b\}$, $V_2 = \{a\}$, and $V_3 = \{a\}$. The tree on the left-hand side of Fig. 7.1 complies with these arities, and its domain is depicted on the right-hand side. Trees can be also interpreted as expressions of terms over some given (universal) algebra, and vice versa [54, 98]. For instance, the tree in Fig. 7.1 corresponds to the expression $a(a(b, a(b, c)), 2, a(2, b(2)))$. Hence, the labels of interior nodes can be seen as operators, while the labels of leaves are the operands. This also explains the special rôle of leaf labels, as further discussed below. For convenience, we will sometimes use this *term notation* for trees.

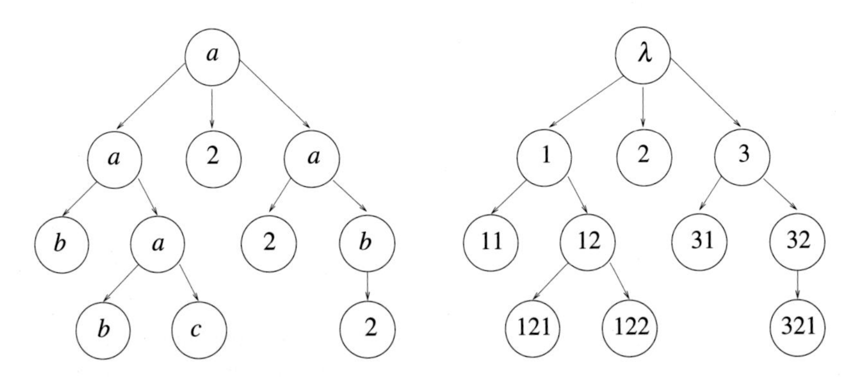

Fig. 7.1 An example of a tree, with the node labeling scheme to the right

7.2.2 Operations on Trees

We are going to define a catenation on trees, based on the notion of replacement. For $t \in V^{\mathrm{t}}$ and $x \in \Delta_t$, the *subtree* of t at x, denoted by t/x, is defined by $t/x(y) = t(x \cdot y)$ for any $y \in \Delta_{t/x}$, where $\Delta_{t/x} := \{y \mid x \cdot y \in \Delta_t\}$. $\mathrm{ST}(T) := \{t/x \mid t \in T \wedge x \in \Delta_t\}$ is the set of subtrees of trees from $T \subseteq V^{\mathrm{t}}$. The *replacement* of the subtree t/x with $s \in V^{\mathrm{t}}$ is defined as:

$$t(x \leftarrow s)(y) = \begin{cases} t(y), \text{ if } y \in \Delta_t \setminus (\{x\} \cdot \Delta_{t/x}), \\ s(z), \text{ if } y = x \cdot z \wedge z \in \Delta_s. \end{cases}$$

Let $S \subseteq V^{\mathrm{t}}$. We denote by $V(S)$ the set of trees

$$\{f(t_1, \ldots, t_n) \mid (f, n) \in V \text{ and } t_i \in S \text{ for every } i \in \{1, \ldots, n\}\}.$$

Let \$ be a new symbol, i.e., $\$ \notin V$, and let the arity of \$ be 0. Let $V_{\$}^{\mathrm{t}}$ denote the set of all trees over $V \cup \{\$\}$ which contain exactly one occurrence of label \$. We call these trees *contexts*. By definition, only frontier nodes can carry the label \$. For trees $u \in V_{\$}^{\mathrm{t}}$ and $s \in (V^{\mathrm{t}} \cup V_{\$}^{\mathrm{t}})$, we define an operation # to replace the frontier node labelled with \$ of u by s according to $u\#s = u(x \leftarrow s)$, where $x \in \Delta_u$ with $u(x) = \$$. This is the *catenation operation* that we were going to define. Notice that the classical catenation of strings (mostly denoted by $\cdot$) can be viewed as a special case if one considers strings as trees where all interior nodes have only one child.

If $U \subseteq V_{\$}^{\mathrm{t}}$ and $T \subseteq (V^{\mathrm{t}} \cup V_{\$}^{\mathrm{t}})$, then $U\#T := \{u\#t \mid u \in U \wedge t \in T\}$. Furthermore, for any $t \in V^{\mathrm{t}}$ and any tree language $T \subseteq V^{\mathrm{t}}$, the *quotient* of T and t is defined as:

$$U_T(t) := \begin{cases} \{u \in V_{\$}^{\mathrm{t}} \mid u\#t \in T\}, & \text{if } t \in V^{\mathrm{t}} \setminus V_0, \\ \{t\}, & \text{if } t \in V_0. \end{cases}$$

An equivalence relation $\equiv$ on V^{t} is *subtree invariant* if, for each $t \in V^{\mathrm{t}}$ and each $x \in \Delta_t$, $t_1 \equiv t_2$ implies that $t(x \leftarrow t_1) \equiv t(x \leftarrow t_2)$ for any trees $t_1, t_2 \in V^{\mathrm{t}}$.

7.2.3 Finite Tree Automata

The following definition of finite tree automata is made possible by the compact functional representation of trees. Let V be a ranked alphabet and m be the maximum arity of the symbols in V. A *(finite bottom-up) tree automaton* over V is a quadruple $A = (Q, V, \delta, F)$ such that Q is a finite state alphabet (disjoint with V_0), $F \subseteq Q$ is a set of final states, and $\delta = (\delta_0, \ldots, \delta_m)$ is an $m + 1$-tuple of state transition functions, where $\delta_k : V_k \times Q^k \to 2^Q$ for $k \in \{0, \dots, m\}$. Now, a transition relation (also denoted by δ) can be recursively defined on V^{t} (for $k \geq 0$) by letting

$$\delta(f(t_1, \dots, t_k)) := \bigcup_{q_i \in \delta(t_i), i=1,\dots,k} \delta_k(f, q_1, \dots, q_k).$$

A tree $t \in V^{\mathrm{t}}$ is accepted by A if $\delta(t) \cap F \neq \emptyset$. The tree language accepted by A is denoted by $T(A)$. A is *total* if each of the functions δ_k maps each possible argument to a set of cardinality of at least 1. A is *deterministic* if each of the functions δ_k maps each possible argument to a set of cardinality of at most 1. Similarly to deterministic finite-state acceptors recognizing strings, deterministic tree automata can be viewed as algorithms for labelling the nodes of a tree with states. Analogously to the string case, it can be shown that nondeterministic and deterministic finite tree automata accept the same class of tree languages, namely the *regular tree languages*, at the expense of a possibly exponential state explosion.

Many properties known for regular string languages are easily transferred to the tree domain. Most importantly, finite tree automata permit efficient membership and emptiness tests.

Example 7.2 Let us have a look at a simple example of a deterministic tree automaton. Let V consist of $V_0 = \{0, 1\}$ and $V_2 = \{\vee\}$. Let $Q = \{q_0, q_1\}$ be the set of states and $F = \{q_1\}$ be the set of final states. Let

$$\delta_0(x) = q_x \quad \text{and} \quad \delta_2(\vee, (q_x, q_y)) := q_{\max\{x,y\}}$$

for $x, y \in V_0$. This automaton accepts all Boolean terms (with $\vee$ as the only operation) of height 1 or more that evaluate to 1, that is, all terms that contain at least one leaf with label 1. For instance, the term (tree) $\vee(0, \vee(1, 0))$ is accepted, because

$$\begin{aligned}\delta(\vee(0, \vee(1, 0))) &= \delta_2(\vee, \delta(0), \delta(\vee(1, 0))) = \delta_2(\vee, q_0, \delta_2(\vee, \delta(1), \delta(0))) \\ &= \delta_2(\vee, q_0, \delta_2(\vee, q_1, q_0)) = q_{\max\{0,\max\{1,0\}\}} = q_1.\end{aligned}$$

The notions of *isomorphic* automata and (state subset induced) *subautomata* can be easily carried over from the well-known string case to the tree case. A state q of a deterministic tree automaton A is *useful* if there exists a tree t and some node $x \in \Delta_t$ such that $\delta(t/x) = q$ and $\delta(t) \in F$. A state that is not useful is called *useless*. A deterministic automaton containing only useful states is called *stripped*.

Remark 7.1 A couple of different definitions of tree automata can be found in the literature. We tried to give one of the more usual ones. However, in the context of learning, the one of Sakakibara [182] that considers elements of V_0 as if they were states is quite widespread.

7.2.4 Weighted Tree Automata

Finite tree automata accept crisp languages: either a tree is admitted as a member, or it is not. To allow for probabilities, and to model quantitative properties, it is useful to consider weighted devices.

Weighted tree automata (WTA) are obtained from regular tree automata by assigning weights, typically taken from some commutative semiring R, to all transitions and final states [8, 22, 30, 121]. More precisely, a WTA A is a TA in which the transition function has been replaced by a function $\delta_k : V_k \times Q^k \times Q \to R$ for $k \in \{0, \ldots, m\}$, and the final states by a function $F : Q \to R$. The tree $t = f[t_1, \ldots, t_k]$ is mapped to the state q with weight

$$A_q(t) = \sum_{(q_1,\ldots,q_k)\in Q^k} \delta_k(f, (q_1, \ldots, q_k), q) \cdot A_{q_1}(t_1) \cdot \ldots \cdot A_{q_k}(t_k).$$

The weight assigned to t by A is then given by

$$A(t) = \sum_{q\in Q} A_q(t) \cdot F(q).$$

A WTA A is *deterministic* if, for every number $k \in \{0, \ldots, m\}$, symbol $f \in V_k$, and sequence of states $(q_1, \ldots, q_k) \in Q^k$, there is at most one $q \in Q$ such that $\delta_k(f, (q_1, \ldots, q_k), q)$ is non-zero (a syntactic property). It is *all-accepting* if it assigns a non-zero weight to every input tree (a semantic property). If A is deterministic, then the additive operation is not needed, so it suffices to work with a monoid containing a 'zero' (i.e., an absorbing element) and a single multiplicative operation.

Weighted tree automata generalize finite tree automata, since the latter are WTA over the Boolean semiring: Any function $\delta_k : V_k \times Q^k \to 2^Q$ corresponds to a function $\delta'_k : V_k \times Q^k \times Q \to \{0, 1\}$ and vice versa. Having seen this, it should be clear that $\delta(t) \cap F \neq \emptyset$ iff $A(t) = 1$, where A is defined via δ'_k over the Boolean semiring.

More generally, we can associate to each WTA A a formal power series, namely $\sum_{t\in V^{\mathrm{t}}} A(t)\, t$. For the Boolean semiring, there is again a natural correspondence between these tree series and tree languages. Otherwise, weighted devices sometimes behave differently from their unweighted counterparts. For example, the classes of tree series computed by deterministic and nondeterministic WTA over the same semiring R do not always coincide. An informative example is the height function, which can be computed by a nondeterministic but not by a deterministic WTA over the tropical semiring, where the ground set is $\mathbb{N}$, the additive operation is max and the multiplicative operation is $+$; see [30].

7.2.5 Constructions for Tree Automata

We need some special constructions of tree automata in our exposition:

Base tree automaton Firstly, we define the analogue of the well-known prefix-tree acceptor in the string case: Let I_+ be a finite tree language over V. The *base tree automaton* (sometimes also called *subtree automaton*) for I_+, denoted by $Bs(I_+) = (Q, V, \delta, F)$, is defined as follows: $Q = \mathrm{ST}(I_+)$, $F = I_+$,

$$\delta_k(f, u_1, \ldots, u_k) = f(u_1, \ldots, u_k)$$

whenever $u_1, \ldots, u_k \in Q$, $k > 0$, and δ_0 maps any symbol a from V_0 to the one-node tree labeled a. Obviously, $T(Bs(I_+)) = I_+$.

Example 7.3 Consider the term (tree) $\vee(0, \vee(1, 0))$ as in the previous example. Possible subtrees are described by the following trees: $ST(\vee(0, \vee(1, 0))) = \{0, 1, \vee(1, 0), \vee(0, \vee(1, 0))\}$. $BS(\{\vee(0, \vee(1, 0))\})$ has two states apart from 0 and 1: $q_0 = \vee(1, 0)$ and $q_1 = \vee(0, \vee(1, 0))$. The only non-trivial state transition is $\delta_2(\vee, 0, q_0) = q_1$.

Canonical tree automaton Secondly, we transfer the notion of canonical automaton to the tree case: Let T be a regular tree language over V. The *canonical tree automaton* for T, denoted by $C(T) = (Q, V, \delta, F)$, is defined by: $Q = \{U_T(s) \mid s \in \mathrm{ST}(T)\}$, $F = \{U_T(t) \mid t \in T\}$, $\delta_k(f, U_T(s_1), \ldots, U_T(s_k)) = U_T(f(s_1, \ldots, s_k))$ if $f(s_1, \ldots, s_k)$ is in $\mathrm{ST}(T)$. Observe that $C(T)$ is a deterministic stripped automaton which is formed completely analogously to the minimal deterministic string automaton. As in the string case, to each regular tree language, there is a canonical tree automaton accepting that language. Moreover, the well-known characterization of Myhill-Nerode transfers to the tree case: A tree language T is regular if and only if it is of finite index, i.e., T is the union of equivalence classes of some subtree-invariant equivalence relation. This also explains why $C(T)$ is a finite-state automaton if T is regular.

Quotient automaton Finally, we define quotient automata. A *partition* of a set S is a collection of pairwise disjoint nonempty subsets of S whose union is S. If π is a partition of S, then, for any element $s \in S$, there is a unique element of π containing s, which we denote by $B(s, \pi)$ and call the *block* of π containing s. A partition π is said to *refine* another partition π' iff every block of π' is a union of blocks of π. If π is any partition of the state set Q of the automaton $A = (Q, V, \delta, F)$, then the *quotient automaton* $\pi^{-1}A = (\pi^{-1}Q, V, \delta', \pi^{-1}F)$ is given by $\pi^{-1}P = \{B(q, \pi) \mid q \in P\}$ (for $P \subseteq Q$) and, for $B_1, \ldots, B_k \in \pi^{-1}Q$, $f \in V_k$, $B \in \delta'_k(f, B_1, \ldots, B_k)$ whenever there exist $q \in B$ and $q_i \in B_i \in \pi^{-1}Q$ for $1 \leq i \leq k$ such that $q \in \delta_k(f, q_1, \ldots, q_k)$.

An alternative (and more illustrative) way to quotient automata is via the idea of *state merging*: $\pi^{-1}A$ is obtained from A by merging all states in each block into one state.

State merging can change the accepted language dramatically. Reconsider Example 7.3. The base tree automaton $A := BS(\{\vee(0, \vee(1,0))\})$ only accepts a single tree, that is, $T(A) = \{\vee(0, \vee(1,0))\}$. Merging the two states $q_0 = \vee(1,0)$ and $q_1 = \vee(0, \vee(1,0))$ into a single one, call it q, we arrive at the state transitions $\delta_2(\vee, 1, 0) = q$ and $\delta_2(\vee, 0, q) = q$. Hence, we find that $T := T(\{\{q_0, q_1\}\}^{-1}A)$ is infinite: $T = \{\vee(1,0), \vee(0, \vee(1,0)), \vee(0, \vee(0, \vee(1,0))), \vee(0, \vee(0, \vee(0, \vee(1,0)))), \ldots\}$. This makes it clear that automata properties like determinism can be lost by the quotient operation. However, state merging is the main idea behind many learning algorithms that start out with a simple automata presentation of given data, mostly via the base tree automaton.

In the case of unranked trees, the symbols that label tree nodes lose their arities. The usual finite automaton model yielding a notion of regular unranked tree languages is termed *hedge automaton*. Details can be found in [38, 134]. Regularity of unranked tree languages can be characterized by ranked tree languages in (at least) two ways, via currification and via a first-child-next-sibling encoding.

An obvious connection between string and tree languages is that strings can be represented as trees. There are in fact several different ways of doing so. A convenient representation for comparing and transferring results is to consider a string $w = a_1 \ldots a_n$ of length n over the alphabet Σ as a tree t on n nodes with domain $\Delta_t = \{1^j \mid 0 \leq j < n\}$, such that $t(1^j) = a_{n-j}$ for $0 \leq j < n$. The labels of t are taken from a ranked version of Σ, in which every symbol has arity at most 1. Notice that w is, in a sense, spelled out backwards by t, since string automata read input strings from left to right, whereas tree automata (usually) read trees from the leaves to the root, so that the root corresponds to the end of the string.

Much of the early interest in tree languages stemmed from the fact that the context-free string languages are the yields of the regular tree languages [200]. In fact, for this characterization, it is sufficient that all internal nodes are binary (recall the Chomsky normal form for context-free languages) and hence the trees of interest look just the same as those obtained by curryfying unranked trees.

The theory of tree languages is rich, even when restricted to the regular case: Just as with regular string languages, regular tree languages can be characterized by a particular type of logic, by regular expressions, by grammars, and by algebraic properties. Again, details must be omitted for lack of space, and we refer instead to the above-mentioned survey articles. There are also more specific surveys, e.g., [120]. The indicated connections to logic might be the most fruitful path, also for learning such devices, although this has been barely explored so far. In this context, it should be also mentioned that such kind of formalisms also exist for the case of weighted (tree) automata; see [71, 72, 123].

Moreover, the regular tree languages are not only characterized by nondeterministic bottom-up tree automata, as defined above, but also (via the well-known subset construction) by deterministic bottom-up tree automata, as well as by nondeterministic top-down tree automata (whose formal definition can be seen in [54], for instance). This formalism is, however, different from the bottom-up variety insofar as deterministic top-down tree automata are strictly weaker than nondeterministic ones regarding their descriptive power.

It is also worth mentioning that the class of regular tree languages is quite a robust class regarding its closure properties. For instance, it is closed under Boolean operations like union, under complementation, and under intersection. The mentioned expression characterization relies on the operations "tree substitution" and "iteration," whose formal definition can be found in [54].

There also exist notions of finite tree transducers, which are basically devices that work on trees but not only accept or reject, but again produce another tree as an output. These are particularly interesting in connection with the processing of XML documents, although they have various other applications. But with this particular application in mind, we defer further discussion to Sect. 7.4.1.

7.3 Mathematical Results

7.3.1 Gold-Style Learning

In his seminal paper [88], Gold considered the (passive) learning of a language by a computer program that receives an infinite stream of information with which it must converge to the learning target. Soon, it was recognized that it matters whether the learner only receives positive examples (this scenario is also known as *learning from text*) or both positive and negative examples (this scenario is called *informant learning* and will be described more in detail in Sect. 7.3.1.1).

It was then discovered that no language class with *infinite elasticity* is learnable in the limit from text [11]. The elasticity of a class C is the length of the longest chain of inclusions $L_1 \subsetneq L_2 \subsetneq L_3 \subsetneq \cdots$, where $L_i \in C$ for every $i \in \mathbb{N}$. If the learner is presented with a sequence of examples $e_1, e_2, \ldots$ with $e_i \in L_i$ for every $i \in \mathbb{N}$, then it will never reach a point when it can say with certainty what the target language is. There are however three different non-trivial classes of (string) languages that can be learned from positive data only: pattern languages [10], reversible (regular) languages [12] and languages akin to the k-gram approach known from practical lexical analysis, leading to different notions of so-called testability; see [85]. All three ideas have been transferred to the tree case, as we will show in the following. Text learners formally follow the protocol that they receive an infinite stream of positive data, but should stabilize towards one hypothesis after some finite number of steps. Mostly, it is easier to present these learners as receiving a finite language (of trees in our case) and then producing a hypothesis.

Moreover, it was well known that regular string languages can be learned from informants. This has then been generalized to the tree case, as we will exhibit next.

7.3.1.1 Informant Learning of Regular Tree Languages

For the string case, several learning algorithms that infer finite automata have been proposed, the most famous ones are those presented by Gold [89] and by García and

Oncina, which has been generalized to trees shortly after its presentation for strings [84]. We will try to convey the gist of the latter algorithm in the following.

Recall that a learning algorithm for informant learning receives an infinite input of positive and negative examples. This means, more formally and assuming that the target language is a tree language over the ranked alphabet V, that the learner receives a sequence of correctly labeled examples (t_i, l_i), i being a positive integer, so that $t_i \in V^{\mathrm{t}}$ and $l_i \in \{+, -\}$ for any i. The target language is $T = \{t_i \mid l_i = +\}$, and V^{t} equals $T \cup \{t_i \mid l_i = -\}$. At time step i, the learner receives (t_i, l_i) from the informant. As a reaction to this, the learner forms a hypothesis, which is, in our case, a deterministic tree automaton A_i. To the sequence of hypothesis automata $A_1, A_2, \ldots,$ there corresponds a sequence of hypothesis languages $T(A_1), T(A_2), \ldots,$ and the requirement of successful learning then means that, from a certain point of time j on, $A_j = A_k$ for all $k \geq j$, and also $T(A_j) = T$.

Notice that up to time step i, the learner has seen the examples $(t_1, l_1), \ldots, (t_i, l_i)$. Although this is not required from the learning model, all practical algorithms also satisfy that the hypothesis A_i formed in step i obeys $\{t_j \mid 1 \leq j \leq i, l_j = +\} \subseteq T(A_i)$ and $T(A_i) \cap \{t_j \mid 1 \leq j \leq i, l_j = -\} = \emptyset$, a property called *global-consistency* [28], although this is quite a restriction from the point of view of Learning Theory. Therefore, the task of forming a hypothesis automaton can be also seen in the following setting: To the hypothesizing algorithm, we input two disjoint sets of trees X^+ and X^- over the alphabet V, and the algorithm outputs a hypothesis automaton $A(X^+, X^-)$ whose language $T(A)$ satisfies $X^+ \subseteq T(A)$ and $X^- \cap T(A) = \emptyset$.

It should be observed that learnability as such is not a question, because regular tree languages are recursive, so that general results from learning theory apply [88]. This general approach is of purely enumerative nature: Such an algorithm could systematically enumerate all deterministic finite tree automata and stick with the last one enumerated as long as no inconsistencies with the datum (X^+, X^-) are discovered. If an inconsistency is discovered, the algorithm moves to the next automaton in the enumeration and tests it against the datum. But such an algorithm could take arbitrarily long for its computations if we happen to encounter an enumeration of the automata that is "bad" for the current datum, as it is not clear when a consistent automaton will be enumerated. It is only clear that it will show up sooner or later.

There have been two main suggestions for overcoming this problem, one by Gold [89] and one by García and his colleagues. We are going to describe the latter one, also known as RPNI (regular positive and negative inference), but we also like to mention that Gold's approach shares data structures like observation tables with the famous LSTAR algorithm explained below in the context of active learning in Sect. 7.3.2.3. As worked out by Kasprzik [105, 107], RPNI can be also described using observation tables, but this is a minor technical issue here.

The RPNI algorithm starts by building the base tree automaton for the positive samples X^+. This gives the first potential hypothesis PH_0. It then constructs a sequence of potential hypotheses PH_s, where PH_s is obtained from PH_{s-1} by merging states. Intuitively, state mergings are done as long as they do not introduce inconsistencies with respect to the negative sample X^-.

The algorithm maintains two sets of states, usually called "red states" and "blue states". The red states represent states in the hypothesis automaton, and the blue states represent transitions. After PH_0 has been computed, one of the automaton's states is selected at random and placed in the red set, and the remainder is put in the blue set. The algorithm repeatedly selects some blue state and tentatively merges it with some red state. If the merger does not cause trees in X^- to be accepted, then the merger is confirmed and leads to the next potential hypothesis. If every merger of the chosen blue state with a red state fails to respect X^-, then the blue state becomes a red state (this is usually called "promotion"), but the hypothesis automaton is not changed. The algorithm terminates if all originally blue states have been checked in the way described. In particular, if $X^- = \emptyset$, then all states will eventually be merged into a single accepting one. Clearly, the algorithm's running time is bounded by a polynomial in the size of $X^+ \cup X^-$.

Example 7.4 Assume that we have the alphabet $V = V_0 \cup V_2$ with $V_0 = \{0, 1\}$ and $V_2 = \{\vee\}$, and let $X^+ = \{\vee(1, 0), \vee(0, \vee(1, 0))\}$ and $X^- = \{\vee(0, 0)\}$. As $BS(X^+)$ has no transition for the case that is interesting in the negative sample, its two states can be merged to yield a hypothesis with the transition function $\delta_2(\vee, 0, p) = \delta_2(\vee, 1, 0) = p$. This changes if we take $X^+ = \{\vee(1, 0), \vee(0, \vee(1, 0)), \vee(1, \vee(0, 0))\}$ instead (for instance, after seeing the next labeled example). The automaton $BS(X^+)$ now has four states, $p = \vee(1, 0)$, $q = \vee(0, p)$, $r = \vee(0, 0)$ and $s = \vee(1, r)$. RPNI will discover that p, q, s can be merged into one state z without violating X^-, but s will remain separate. This yields the hypothesis with a transition function featuring $\delta_2(\vee, 1, 0) = \delta_2(\vee, 0, z) = \delta_2(\vee, 1, s) = z$ and $\delta_2(\vee, 0, 0) = s$. Assuming that these examples were drawn from the language from Example 7.2, we can see that this hypothesized automaton is already "close" to the target automaton.

As mentioned in the introduction, we do not cover probabilistic or stochastic extensions of tree languages, but only refer to some relevant papers [37, 183, 185]. Informant learning is in many ways similar to query learning [105, 107], a subject which we return to in Sect. 7.3.2.

We now turn our attention to learning from positive examples only, and consider a number of learning strategies and target language classes. Notice that this mode of learning can be formalized analogously to informant learning. For practical purposes, we present these learners as algorithms receiving a finite language (of trees in our case) and in response outputting a hypothesis.

7.3.1.2 Text Learning of Tree Patterns

One way of explaining tree patterns is by generalizing the catenation operation # towards a parallel operation. To this end, we extend the ranked alphabet V by an alphabet X of variables (of rank 0), and assume a linear order $<$ on X. A *tree pattern* P is a tree over the ranked alphabet $V \cup X$. To each tree pattern P, we associate with the help of the operator oc the k-tuple $oc(P) = (x_1, \dots, x_k)$ of variables in X that occur in P, such that: (a) Whenever there is a frontier node in P that carries the label $x \in X$, then there is an index i, $1 \leq i \leq k$, such that $x = x_i$, and (b) $x_1 < x_2 < \cdots < x_k$ according to the linear order $<$ on X. The language associated to P is the language of trees that can be obtained from P, with $oc(P) = (x_1, \ldots, x_k)$, by choosing a k-tuple of trees $(t_1, \ldots, t_k)$ where each $t_i \in V^{\mathrm{t}}$, and consistently replacing each occurrence of x_i at some frontier node by the tree t_i. A tree language (over V) is a *tree pattern language* if it is a language associated to some tree pattern P.

For instance, if $V = \{(a, 2), (a, 0)\}$, then all binary trees (with label a in each node) comprise V^{t}. The tree pattern language associated to the binary tree of depth 1, where both frontier nodes are labeled with variable x, however, does not consist of all binary trees (with all nodes labeled a) of depth at least 1, but rather of trees that can be obtained from two copies of some arbitrary binary tree, connected by a new root node.

It is shown in [124] that the class of tree pattern languages has finite thickness. Actually, this result is already contained in [171], and as a consequence, tree pattern languages can be learned from text. Due to Wright [160, 214], this implies further that also finite unions of tree pattern languages have finite elasticity and are text-learnable. For more on this subject, see [19, 20].

More general ideas of tree pattern languages are explored in [153, 197, 199]. There, variables can also label inner nodes of the tree pattern, enabling more complex replacements. From the point of view of (string) pattern languages, this looks like the right generalization to trees, but the replacements of variables must be carried out with care, taking into account the arities of the symbols involved. Also, height constraints have been investigated due to practical motivations in the context of discovering common characteristics in semistructured documents [198]. Notice that in the context of XML, a different notion of (tree) patterns was established; see [164].

We should also mention here research on the learnability of *unordered* tree pattern languages [192]. This is more like learning special types of graph languages, and the research has been moved into that direction, mostly driven by Shoudai and his co-authors; we only refer to [215, 217], but there are more papers in that research area. There have also been investigations of a purely graph-theoretical nature, for instance, the inference and recovery of edge-colored trees from sequences of edge colors obtained from walking a tree [150].

From a theoretical point of view, it may be interesting to see how elementary formal systems translate to the tree case, as they are believed to form a basis for designing text-learnable language classes; see [16, 157].

7.3.1.3 Text Learning of Testable Tree Languages

For every natural number k, there are only finitely many k-testable languages, so the text learnability of these language classes was easily established. Since then, many variants have been investigated [85, 86, 178], some of them in the Ph.D. thesis of Ruiz [167]. It is worth noticing that the yields of tree languages which are 2-testable in the strict sense, are exactly the context free languages [175].

The basic setting was generalized to tree languages by Knuutila and by García in [83, 114, 115], but also variants were discussed, for example, in [139]. We should also mention precursor papers like [81] that propose a method called k-followers akin to the k-tail tree inference method of Brayer and Fu [31]. In this context, also two papers of Levine deserve to be mentioned [129, 130]. Further results in this direction are contained in [174]. These ideas were also generalized beyond trees, for instance, see [141, 195].

Let us explain the idea in its simplest form. To this end, fix some natural number $k \geq 1$. For trees $t \in V^{\mathrm{t}}$, define the tuple $\mathrm{test}_k(t) = (r_{k-1}(t), p_k(t), \ell_{k-1}(t))$, the components of which refer to the *root region*, the set of *tree parts* and the set of *leaf regions*, respectively. Formally, $r_k(t)$ is the restriction of t to the domain $\Delta_t \cap \{x \in \mathbb{N}^* \mid |x| \leq k\}$, while $\ell_k(t)$ is the set of subtrees of t whose domain is a subset of $\{x \in \mathbb{N}^* \mid |x| \leq k\}$. Finally, $p_k(t)$ collects all trees $r_k(t')$, where $t' \in \mathrm{ST}(t)$ and t' has depth at least $k-1$. Clearly, $p_k(t) = \emptyset$ if the depth of t is smaller than $k-1$. The reader familiar with k-testable string languages will easily see the similarities, as we are also extracting the information on the beginning, the middle parts and the ends of the tree. A tree language T is *k-testable in a strict sense* if there exist three finite sets of trees R, P, L such that $t \in T$ if and only if $r_{k-1}(t) \in R$, $p_k(t) \subseteq P$ and $\ell_{k-1}(t) \subseteq L$. The learning algorithm proposed in [83] (also see [140]) basically deduces from a given positive sample the sets of root regions, the possible tree parts, and the union of the sets of leaf regions that are encountered in the sample, and takes this as a hypothesis (R, P, L). This hypothesis can be also interpreted as a deterministic finite tree automaton $A = (Q, V, \delta, F)$ with $Q = R \cup L \cup p_{k-1}(P)$, $F = R$, $\delta(t) = t$ for $t \in L$, and $\delta_n(\sigma, t_1, \ldots, t_n) = r_{k-1}(\sigma(t_1, \ldots, t_n))$ for $\sigma(t_1, \ldots, t_n) \in P$.

7.3.1.4 Text Learning of Reversible Tree Languages

One of the first non-trivial examples of a subclass of the regular languages that can be learned from text were the k-reversible languages [12]. Sakakibara [182] continued this line of work by proving the text-learnability of zero-reversible tree languages. More precisely, that context-free (string) languages are learnable from positive structural information. The property of k-reversibility was later generalized to *functional distinguishability*, first in the context of string languages [75] and then of tree languages [76]. The concept of k-reversibility was also independently extended towards trees in [140].

We only provide some further details for zero-reversible tree languages. Sakakibara called a tree automaton $A = (Q, V, \delta, F)$ *reset-free* if there are no two distinct states p, q with

$$\delta_n(\sigma, q_1, \ldots, q_{i-1}, p, q_{i+1}, \ldots, q_n) = \delta_n(\sigma, q_1, \ldots, q_{i-1}, q, q_{i+1}, \ldots, q_n).$$

A deterministic finite tree automaton is *zero-reversible* if it is reset-free and has at most one final state.

The learner begins by constructing the base tree automaton for the finite sample language. This yields the hypothesis PH_0. Inductively, at stage s, the learner begins by determining whether PH_s is zero-reversible. If it is not, then PH_s has more than one accepting state, is not deterministic, or is not reset-free. Each of these problems can be resolved by merging states. Doing such a (forced) state merge results in the automaton PH_{s+1} for the next stage. Upon termination, the learner outputs a hypothesis that is zero-reversible. It can be shown that this algorithm is a valid text learner for zero-reversible tree languages. Moreover, the hypothesis will always be a canonical tree automaton. This is also the case with k-reversible tree languages, while the (more general) function-distinguishable tree languages require custom canonical automata. Notice that the first step in Example 7.4 is the same as that of the learner for zero-reversible tree languages.

The idea of using reversibility (or functional distinguishability) was also used for defining text-learnable language classes from deterministic top-down tree automata, although the results are mostly given in terms of special context-free string languages. The most general published results in this direction are contained in [196], building on ideas from [125, 146].

7.3.2 *Active Learning*

In active learning (query learning, optimal experimental learning), the learning algorithm iteratively refines an hypothesis by querying an oracle for information (see Chap. 3, *Learning Grammars and Automata with Queries*, de la Higuera). Its success in deducing the target language depends on the type of queries that the oracle can answer. The complexity of inferring a class of languages is typically expressed as a function of the size of the canonical representation for the target language, and the number of different types of queries needed. A range of querying strategies have been proposed, such as asking about the data points that at the moment seem hardest to classify [132], or which have highest entropy [190]. Goldman and Kearns have taken the opposite view and considered the complexity of *teaching* various classes of languages [90].

7.3.2.1 Learning from Membership Queries

A natural form of active learning is learning from membership queries. A *membership query* consists in asking the oracle to label an unknown object in the learning domain as inside or outside the target language. Under the additional assumption that at least one positive example is available, Matsumoto and Shoudai give a polynomial-time algorithm that infers trees with *height-bounded variables*, that is, with a set of internal variables $\{x_i\}_{i\in\mathbb{N}}$ such that x_i can be substituted by any tree of height at most i [152].

Besombes and Marion [24] provide an inference algorithm for regular tree languages that learns from membership queries and representative samples. (Some corrections to this algorithm are given by Kasprzik [108].) A representative sample S for a language L is a finite subset of L such that each transition of the canonical automaton A_L for L is used to produce at least one tree in S. It is easy to see that there is always a representative sample of size exponential in that of A_L. Given such a representative sample, the algorithm uses membership queries to derive A_L in polynomial time by following the general approach of the LSTAR algorithm by Angluin (see Sect. 7.3.2.2).

The same learning model was adopted by Kasprzik for learning residual finite-state tree automata (RFTA) [106, 108]. These devices are a generalisation of residual finite-state automata [63], which in turn are a restricted form of NFA where each state represents a residual language of the language recognized. The RFTA offer a canonical and often compact form of representation for regular tree languages.

7.3.2.2 The Minimal Adequate Teacher Model MAT

Once it had been established that the regular languages (which have infinite elasticity) cannot be learnt from text [88], a learning model was sought that was strong enough to allow their inference, but sufficiently simple to have nice mathematical properties. This led to the *minimal adequate teacher* (MAT) model, an oracle capable of answering membership queries as well as a second type of query called equivalence query [13]. With an *equivalence query*, the learner asks the teacher to verify its current hypothesis, and to provide a counterexample that disproves it, if the hypothesis is indeed wrong. When equivalence queries are allowed, the learner cannot be fooled by a malevolent teacher to stall at the same mistaken hypothesis for an unbounded number of inference steps.

A survey of MAT learners for weighted and unweighted tree languages was compiled by Drewes [65]. According to this source, Sakakibara was the first to generalize the LSTAR algorithm to trees [181]. He considered skeletal languages, that is, the parse trees of context-free languages in which all internal nodes are unlabeled [131]. Drewes and Högberg considered regular tree languages in general [66], and proved that it is possible to avoid useless states in the inference process [67]. By doing so, the automaton A becomes a language-equivalent partial automaton with potentially exponentially fewer states. Notice that while most MAT learning algorithms aim at learning deterministic finite-state tree automata, there is also one that has residual finite-state tree automata as its target structure, published by Kasprzik [108].

As for the weighted case, Denis and Habrard considered the problem of learning a probability distribution P over a set of trees from an independent sample [62], given that P can be computed by a weighted tree automaton (WTA) over a suitable semiring [82]. Drewes and Vogler gave a MAT learner for deterministic all-accepting WTA [68], and Maletti one for deterministic WTA over commutative semifields [147]. Under the stronger assumption that weights are taken from a field (allowing both multiplicative and additive inverses), it is possible to learn also non-deterministic WTA within the MAT model [95].

In [69], key properties of observation tables are captured in an abstract data type (ADT) for learning deterministic WTA. The authors show that the ADT can be realized as an observation tree in the sense of Kearns and Vazirani [110], which may reduce the algorithm's complexity considerably.

Arimura et al. give an algorithm that learns the union of k tree pattern languages using $O(k \cdot n)$ equivalence queries and $O(k^2 \cdot n \cdot \max(\{k, n\}))$ membership queries [17], where n is the size of the largest counterexample provided by teacher. Amoth et al. show that unordered tree patterns are not learnable from equivalence and subset queries. Since subset queries are strictly more powerful than membership queries, this result also holds for the MAT model [9].

7.3.2.3 The LSTAR Algorithm for Regular Tree Languages

Let us look at a simple version of the LSTAR algorithm for regular tree languages. To find the minimal deterministic automaton A for a target language L, the algorithm maintains a set of trees S and a set of *contexts* C. Recall that a context is a tree in which exactly one of the leaves has been replaced by a special symbol \$, which could be also interpreted as a variable. The algorithm will achieve its goal by collecting in S representatives of the congruence classes of L (or equivalently, the states of A), and the purpose of contexts is to provide evidence that the trees in S belong to different classes.

As it runs, the algorithm builds up an *observation table* T. The rows of T are indexed by trees in $S \cup V(S)$ and the columns by contexts in C. At position (s, c), the algorithm writes a $+$ if $c\#s \in L$, and a—otherwise. This information can be obtained by asking the teacher membership queries.

Let $row(s)$ denote the row of $s \in (S \cup V(S))$, represented as a string in $\{+, -\}^*$. This representation supposes a certain ordering on C, say, to the order in which elements are added to C by the algorithm. The table T is *closed* if for every $s \in V(S)$, there is an $s' \in S$ such that $row(s) = row(s')$. If the table is not closed, then it can be closed by adding s to S (and for technical reasons, all its subtrees). The intuition here is that the tree $f(s_1, \ldots, s_k) \in V(S)$ encodes a transition from the states $row(s_1), \ldots, row(s_k)$ on symbol f, and only if the table is closed does the target state of this transition exist.

The table T is *consistent* if, for every $s_1, \ldots, s_k, s'_1, \ldots s'_k \in S$ such that $row(s_i) = row(s'_i)$, we have $row(f(s_1, \ldots, s_k)) = row(f(s'_1, \ldots, s'_k))$. If the table is not consistent, then it can be made so by adding a new context to C, assembled from a $c \in C$, the label f, and $k - 1$ of the trees $s_1, \ldots, s_k, s'_1, \ldots, s'_k$. This time, the intuition is that when the trees $f(s_1, \ldots, s_k)$ and $f(s'_1, \ldots, s'_k)$ encode the same transition, then the target state should be uniquely defined.

If a table is both closed and consistent, then it represents a unique deterministic tree automaton with states $\{row(s) \mid s \in S\}$, transitions as given by $V(S)$, and the set of accepting states being all states from $\{+ \cdot w \mid w \in \{+, -\}^*\}$, where we assume that the first element from C is \$.

	$
0	−
1	−

	$
0	−
1	−
$\vee(0,1)$	+
$\vee(0,0)$	−
$\vee(1,0)$	+
$\vee(1,1)$	+
$\vee(0,\vee(0,1))$	+
$\vee(\vee(0,1),0)$	+
$\vee(1,\vee(0,1))$	+
$\vee(\vee(0,1),1)$	+
$\vee(\vee(0,1),\vee(0,1))$	+

	$	$\vee(\$,0)$
0	−	−
1	−	+
$\vee(0,1)$	+	+
$\vee(0,0)$	−	−
$\vee(1,0)$	+	+
$\vee(1,1)$	+	+
$\vee(0,\vee(0,1))$	+	+
$\vee(\vee(0,1),0)$	+	+
$\vee(1,\vee(0,1))$	+	+
$\vee(\vee(0,1),1)$	+	+
$\vee(\vee(0,1),\vee(0,1))$	+	+

Fig. 7.2 Three observation tables for Example 7.5

The LSTAR algorithm thus proceeds as follows. It starts with sets $S = \emptyset$ and $C = \{\$\}$ and expands them until T is closed and consistent. When this happens, it synthesises an automaton A_T from T and passes it through the teacher in the form of an equivalence query. If A_T accepts L, then the algorithm terminates and we know that A_T is the target automaton T. If it does not, then the teacher will return a counterexample t in the symmetric difference of $L(A_T)$ and L. The learner then adds $ST(T)$ to S and continues to expand S and C until it again has a closed and consistent table, at which point a new equivalence query is posed, and so forth.

Example 7.5 As an illustration, we consider a run of the LSTAR algorithm on the language of Boolean terms in Example 7.2. The algorithm starts with the sets $S = \emptyset$ and $C = \{\$\}$, and builds up the observation table shown leftmost in Fig. 7.2 through a series of membership queries. When we draw the tables, we separate the rows corresponding to S and $V(S) \setminus S$ by a horizontal line. Since the table is both closed and consistent, the learner asks the teacher whether the automaton $(\{-\}, \{0, 1, \vee\}, \emptyset, \emptyset)$ is what he or she had in mind. Since this is not the case, the teacher returns a counter-example, say $t = \vee(0, 1)$. After adding $ST(t)$ to S, the table now looks as shown in the middle of Fig. 7.2. Since $row(0) = row(1)$ but $row(\vee(0, 0)) \neq row(\vee(1, 0))$ the table is not consistent, so the algorithm will add $\vee(\$, 0)$ to C. After further membership queries, the table looks as shown rightmost in Fig. 7.2. Since T is again closed and consistent, the learner can synthesize the automaton $A_T = (\{--, -+, ++\}, \{0, 1, \vee\}, \delta, \{++\})$ with $\delta(0) = \delta(\vee, --, --) = --$, $\delta(1) = -+$, and $\delta(\vee, x, y) = \delta(\vee, y, x) = ++$ for all $x \in \{-+, ++\}$ and $y \in \{--, -+, ++\}$. Since $L(A_T) = L$, the teacher accepts the hypothesis and the algorithm terminates.

7.3.2.4 Alternative Types of Active Learning

So far, we have discussed generalisations of active learning algorithms to the domain of trees. There is also work in which the learning model itself has been adjusted, e.g., by considering alternative types of queries. Becerra-Bonache et al. argue that when the aim is to model human language acquisition, it is more plausible to consider what they call correction queries than membership queries. In this type of active learning, the learner makes a sequence of statements, and whenever it says something that is ungrammatical, the learner responds with a structurally similar but grammatically correct sentence [21]. A similar line is taken by Tirnăucă [202], who considers three types of corrections based on prefixes, length bounds, and edit distances.

Correction learning has also been applied to tree languages by Tîrnaucă and Tîrnaucă [201], using the following kind of feedback: Given a tree t, the oracle returns the smallest context c (with respect to a certain order) such that $c\#t$ is contained in the target language, or a special token if no such context exists. The authors show that regular tree languages are learnable from a combination of correction and equivalence queries, and that a subclass called injective languages can be inferred from correction queries alone.

A common critique against the MAT model is that equivalence queries are too difficult to answer in reality. This has led some researchers to replace the equivalence queries with subset inclusion queries, which can often be decided by over-approximating the current hypothesis with a simpler language. A good example is the application of MAT learning to system verification tasks by Chen et al. [39].

Matsumoto et al. infer languages represented by finite unions of regular tree patterns from membership queries and a restricted type of subset query to which the teacher answers yes or no, but is not obliged to provide a counterexample [154]. Another reference for using queries in relation with tree patterns is [99], which is motivated by complexity-theoretic problems concerning the consistency problem.

7.3.3 *Error-Correcting Grammatical Inference (ECGI)*

Coping with errors and noise within the learning process is important for many applications. The classical approach of doing this in a mathematically sound way is through Probably Approximately Correct (PAC) learning. However, the PAC model has received moderate interest in the context of tree language learning, which is why we only mention it in the Conclusions and Perpectives Section. Another way of handling errors is that of *error-correcting grammatical inference* (ECGI). This idea was introduced by Rulot [179, 180] for string languages and extended to tree languages by López and his co-authors [136–138]. In short, ECGI provides a learning model that is robust against certain errors. This technique has been applied for optical character-recognition (OCR) tasks as described in [187]. For this type of application, robustness is essential. Notice that the advantages of using syntactic features in OCR tasks have been described before, and also using quite different formalisms [78, 142]. Applications have been also reported in the context of bioinformatics; see Sect. 7.4.2.

7.4 Applications

Grammatical Inference has found many interesting applications. One of these is within the more applied ramifications of pattern recognition, e.g., OCR, as mentioned in Sect. 7.3.3. The applicability of Grammatical Inference theory to Pattern Recognition can also be seen from the fact that quite a number of Grammar Induction papers appear in the leading conferences and journals of the Pattern Recognition community. In this section, however, we focus on three application areas that are typical for tree language inference, namely hypertext/XML processing, bioinformatics, and computational linguistics. We should not forget to mention, however, that many other types of applications have been reported in the literature. As an example, we point to the visual language learner of Crimi et al. [61].

7.4.1 XML Learning

The Extensible Markup Language (XML) is a data format advocated by the World Wide Web Consortium (W3C) for storing and exchanging information.[1] Together with HTML and XHTML, XML is an example of a Standard Generalized Markup Language (SGML), a precursory ISO standard for sharing machine-readable documents in government, law, and industry. Theoretical properties of SGML have been addressed by Wood [213], who suggested Extended Context-Free Grammars (ECFGs) as a suitable model. In terms of descriptive power, ECFGs and CFGs are equivalent, but the syntax of the former grammar type is more similar to that of SGMLs and allows for convenient representation of number ranges.

Simply put, an XML document is made up of well-balanced opening and closing tags, sometimes carrying attributes such as text strings, numerical values, and links to external resources. Figure 7.3 (left) shows an excerpt from a library database in XML format. Due to the restricted nesting, XML documents are typically represented as unranked trees. See Fig. 7.3 (right) for an example. The wide use of XML has therefore motivated a new line of research in tree automata theory that focuses on devices such as hedge and tree-walking automata. Oftentimes, finite alphabets are used as an abstraction of potentially infinite sets of document elements such as attributes and text values. For a survey description on the relations between trees and XML, we refer you to [163, 186].

7.4.1.1 Learning XML Schema

As its name suggests, the definition of XML can be extended by different schemas to obtain special-purpose languages such as RSS, SOAP, and XHTML. The first

[1] http://www.w3.org/XML/.

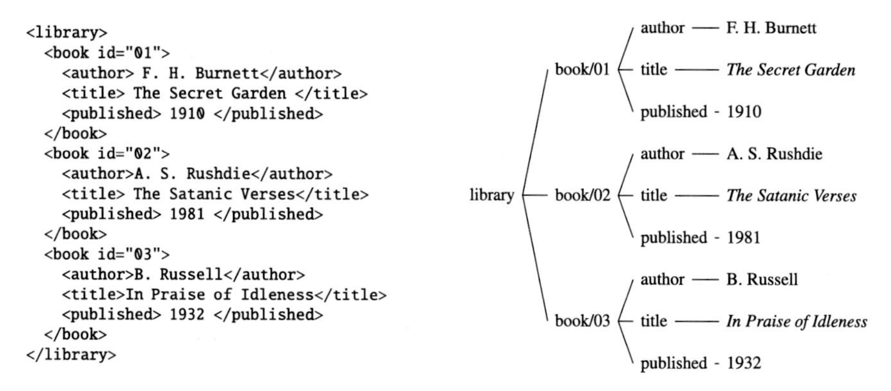

Fig. 7.3 A library database in XML format (*left*), and a tree representation of same database (*right*)

schema language for XML was the Document Type Definition (DTD). Ignoring attributes, DTDs can be seen as restricted context-free grammars and have been studied by Berstel and Boasson under the name XML grammars [23]. The application of grammatical inference techniques for learning XML grammars was investigated by Fernau [74]. The author showed that although the full class of XML languages is not learnable in the limit from positive samples, the subclass of f-distinguishable languages is Bex et al. later added the subclass of DTDs that can be described using the class of single-occurrence regular expressions in which every element name occurs at most once [25]. This study was further expanded in [80].

Chidlovskii proposed the use of regular tree automata to represent XML data, because they are more powerful than DTDs and are closed under algebraic operations such as union and negation [44]. Chidlovskii also did work on XML Schema inference based on extended CFGs [45] and probabilistic CFGs [48].

Kosala et al. learn local unranked tree automata with k-contextual expressions from positive samples of XML structured documents [116, 117].

A similarity measure for DTD-defined families of XML documents was suggested by Carrasco and Rico-Juan. They base their measure on a type of probabilistic tree automaton and evaluate their method on XML document sets for different authors in the Miquel de Cervantes Digital Library. According to [36], the results reflect genre variations among the writers.

A more recent alternative to DTD is the XML Schema Definition (XSD) language. Unlike DTD, XSD is a typed language, which allows the content model of an element to depend on the context in which it is used [149]. To manage the added complexity, Bex et al. divide XSD into an hierarchy of k-local XSDs, defined by the size of the context that is considered. They then continue to describe and experimentally evaluate a learning algorithm for single occurrence k-local XSDs [26].

7.4.1.2 Wrapper Induction

A *wrapper* for an XML-based language is a procedure for extracting data from documents written in the language [122]. There are numerous ways of formalising the notion of a wrapper for a particular language. Chidlowskii viewed wrappers as transducers from an input to an output structure and used transducer induction to learn potentially ambiguous wrappers [46]. In a later work by the same author, target document nodes are identified by the simple paths that reach them. In a simple path, no node except the root may have more than one child. Experimental results suggest that tree wrappers are more precise than string wrappers, which only regard the surface form, but that tree wrappers require in return longer processing time [47].

Later, Carme, Lemay, Niehren, and Gilleron [34, 127] used node-selecting tree transducers (NSTTs) to select target nodes from an XML document. Intuitively, an NSTT is a regular tree automaton that annotates the nodes of each input tree with Boolean values in an unambiguous way. The authors show how deterministic NSTTs can be inferred in polynomial time from annotated examples using a variation on the RPNI algorithm by Oncina and García [168], and that NSTTs capture the class of selection queries expressible in monadic second-order logic on trees [33]. Learning algorithms are also given for queries that select n-tuples of nodes [35]. This work has been further extended to cope with different pruning strategies [33, 166].

Lemay et al. give a Myhill-Nerode theorem for deterministic top-down tree transductions, which they then use to learn deterministic top-down tree transducers from characteristic samples. The work is motivated by automatic inference of XML–to–XML translations. Equivalence classes are represented by pairs of labelled paths, one in the input tree and one in the output tree, and canonical representations are obtained by considering the earliest transducers, i.e., transducers in which the output of symbols is pushed 'upwards' through the system of rewrite rules [128].

Another popular way of defining wrappers are through trees patterns. These are structured patterns with variables that can be used to select nodes. Ordered tree patterns with gaps are polynomial-time learnable in the MAT model of Angluin [18]. The inference of a simpler type of tree pattern called term trees was considered by Aikou et al. [6]. In [198], Suzuki et al. study methods for discovering maximally frequent tree patterns in Web documents.

Kosala et al. who did work on XML Schema inference also apply their techniques to wrapper induction and learn k-testable tree automaton from positive examples [118]. This idea was soon generalized in [173] towards (k, l)-contextual tree languages. Without giving details of this language's class(es) here, we only mention that the corresponding string language class(es) were also introduced before for reasons of automization of XML processing; see [3–5].

Wrappers can also be defined by query expressions in the XML Path Language (XPath). Given an XPath expression p and a tree t, we denote by $p(t)$ the set of nodes in t that are selected by p. Carme et al. discuss how active learning could be used to infer a target expression p from a MAT oracle [32]. In their interpretation of the learning model, answering equivalence query for a hypotheses expression q means verifying whether $p(t) = q(t)$ for all trees t. If q and p are not equivalent, then the teacher returns all nodes in the symmetric difference of $p(t)$ and $q(t)$, i.e., the set of nodes of t that belong to exactly one of the two sets. The advantage of this definition is that the question can be visualised for a human oracle by showing a document in which some elements have been highlighted in a color corresponding to q, and the human answers whether this is a fair representation of the document elements that he or she had in mind. Carme et al. argue that it is harder to find a natural visualisation of membership queries, and that it would for this reason be nice if XPath could be learned from equivalence queries alone. However, using Angluin's idea of approximate fingerprints [15], they show that this is not possible.

7.4.2 Bioinformatics

An early application of learning algorithms in Computational Biology is the investigation of genome sequences; see, for instance, [1, 156, 185]. Yokomori et al. [216] propose locally testable string languages for identifying protein α-chains. A nice and more recent overview has been written by Sakakibara [184] (see also Chap. 8, *Learning the Language of Biological Sequences*, Coste). For other aspects relevant in the context of tree languages, we refer you to [42, 43].

Recall that DNA and RNA strands correspond to strings over a four-letter alphabet. The *gene finding problem* consists in identifying coding or non-coding regions in the genome. Finite-automata and related learning techniques have been applied to the gene finding problem for about 20 years, mostly in the form of Hidden Markov Models (HMMs); the actual identification problem is then solved by the well-known Viterbi algorithm. Although they have their own notions and notations, it should be clear that HMMs are in essence stochastic finite automata. For details, we refer you to standard textbooks on bioinformatics, like [73].

The limitations of the HMM approach are described by Sakakibara et al. in [185]. The authors argue that viewing the DNA molecule as a one-dimensional object is an over-simplification, since it folds in a two-dimensional or even a three-dimensional fashion. As the interplay between parts of the molecule that seem far apart when viewed in one dimension can be important for its biological effect, good syntactical models that capture also the higher-dimension structure are sought. Interestingly, among the first proposals for such a model was the use of probabilistic context-free grammars (PCFG); see [185]. This idea was later refined by Sakakibara, using stochastic finite top-down tree automata in [183]. However, trees were recognized as an important modeling tool much earlier; see [126, 191, 212]. As the "parenthesis structure" (in the PCFG model) obviously corresponds to the folding of the molecule, this means that the structural information that is usually hidden in the form of a string is made explicit, so that we can also (and immediately) make use of this by employing trees. However, the internal node labels do not matter, so in the upcoming discussion, we simply label all internal nodes by σ, and let this symbol have arity 2 or 3 as appropriate.

Biologists have developed a classification scheme for folding patterns in the secondary structure of RNA. The RNA molecules, when seen as one-dimensional objects, are strings over the alphabet $\{A, U, G, C\}$, abbreviating the nucleobases **a**denin, **u**racil, **g**uanine, and **c**ytosine. As explained by Sakakibara, these can be readily described by certain patterns occurring in the corresponding (skeletal) tree. In the following, X and Y are symbols from $\{A, U, G, C\}$. For instance, a node of the form $\sigma(\sigma, \sigma)$ in the skeletal tree refers to a branching. A node of the form $\sigma(X, \sigma, Y)$ refers to base pairs, where the pairing reflects the secondary structure of the molecule. Nodes like $\sigma(X, \sigma)$ and $\sigma(\sigma, X)$ refer to unpaired bases, where the molecule looks and behaves like a string. Pictorially, the tree can be viewed as scanning the folded structure and showing links between different parts of the strand. This principle is illustrated in Fig. 7.4 with an example from [183]. It refers to the term

$$\sigma(\sigma(C, \sigma(C, \sigma(G, \sigma(A, \sigma(A, G)), C), G), G), \sigma(?, U)).$$

The corresponding strand is $CCGAAGCGGU$. We observe linear substructures (without base pairings in the secondary structure) like $\sigma(A, \sigma(A, G))$ and also base pairs like $\sigma(C, \sigma, G)$.

As with many models, there is a certain arbitrariness involved, for instance, the choice of the root of the tree. This is also the case if one considers alternative models like probabilistic context-free grammars.

The idea of using Grammatical Inference is to automatically find the secondary structure of RNA molecules. Due to possible errors in identifying the RNA, and also due to the variations that are common in RNA molecules of the same species, variations that do not produce problems in the physical functioning of RNA molecules as given by the secondary and tertiary structures, probabilistic and stochastic approaches seem to be most promising. In the context of learning trees, this approach has been

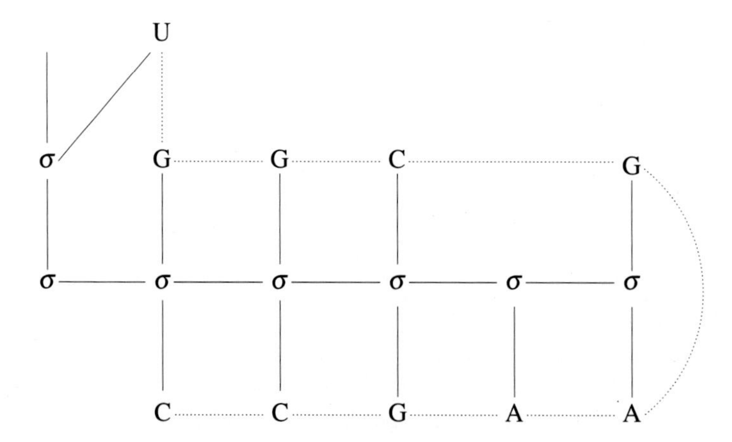

Fig. 7.4 An example of an RNA tree for the sequence $CCGAAGCGGU$; the *straight lines* show the tree structure, while the *dotted lines* indicate links within the strand

investigated by Sakakibara for automata working top-down. As we did not define these formally, we refrain from giving details here. We only mention that learning probabilistic finite automata can be considerably harder than learning deterministic ones, as already shown by Abe and Warmuth for the string case [2].

It is also worth mentioning that learning algorithms for grammatical formalisms that go beyond regular tree languages have been applied to RNA structure prediction. This was motivated by the observation that certain patterns, most notably related to so-called β-sheets, cannot be described by regular tree languages, not even by tree languages whose yields characterize the tree-adjoining string languages. The (stochastic) tree language learners are formally described in [1]. Here, tree-adjoining languages are used to describe pseudoknot structures [151]. Similar connections between generalizations of regular tree languages and non-context-free string languages have been described by Kasprzik [104, 107]. In the context of learning tree-adjoining string language, we would also like to refer to [59] and our Sect. 7.4.3 on computational linguistics.

As already indicated, the use of either tree automata or context-free string grammars for modeling various aspects of RNA molecules is rather a matter of taste, so readers interested in tree languages in the context of bioinformatics may also want to study papers dealing with context-free grammar modeling. A number of empirical studies have also been reported on; we refer here only to some of the papers of Unold and his colleagues [207, 208]. It might be interesting to notice that in these papers Grammatical Inference is seen as a method for classifying strings or trees, rather than as language description learners as such. The method used by Unold et al. is based on metaheuristics (genetic algorithms), and they employ it also in the context of natural language processing [209]. In [185], Sakakibara et al. use expectation-maximisation training of stochastic CFGs to separate between tRNA and non-tRNA.

There has also been research on bio-inspired computation. One of the sub-areas in this discipline is membrane computing. In short, the objects dealt with in that field are not languages of words, but rather multiset languages, in which the exact order of the letters does not matter. Sempere and López generalized learners for trees towards what they called tree multisets and applied this to learning within the area of membrane computing [188, 189]. As bio-inspired computing is believed to provide a model of future computers, this type of research has also potential applications for those future devices.

7.4.3 Computational Linguistics

Grammatical inference in natural language processing (NLP) is of practical interest because it offers tools for managing the complexity of natural languages. Most efforts in this direction address the inference of a language model from a corpus of sample sentences, but differ in the choice of language model and in the type of annotations available in the corpus: some consider only positive sample sentences, others both positive and negative examples, sometimes together with structural information such as parse trees. Grammatical inference in NLP can also be studied from a wish to model the human language acquisition process. In this case, different types of active learning seem more appropriate to describe how a human learner comes to possess a new language. The algorithms' performances are typically evaluated by simple inspection of the language models they output, through cross-validation, or by comparisons with target grammars.

One of the most studied language models in NLP are the context-free grammars. Sakakibara introduced reversible context-free grammars, a normal form for CFGs [182]. A context-free grammar is *reversible* if $A \to \alpha$ and $B \to \alpha$ imply that $A = B$, and $A \to \alpha B\beta$ and $A \to \alpha C\beta$ imply that $B = C$. The class of reversible CFGs has the attractive property that it is learnable in the limit from structural examples, unlabelled parse trees of strings in $L(G)$. For NLP applications, the fact that context-free languages can be characterized by categorical grammars is interesting, as learning the corresponding structures (different from regular tree languages); we mention [58, 60, 102, 103, 218] in this context, sometimes also surpassing the power of context-free string languages. Also, other formalisms (weakly) equivalent to context-free grammars have been investigated from the viewpoint of learnability, for instance, split head-automata grammars, that again lead to different types of derivation trees; see [145]. In each case, the task is to finally learn string languages from unlabelled examples, the trees being rather a hidden underlying structure. More on this topic can be found in Chap. 6, *Distributional Learning of Context-Free and Multiple Context-Free Grammars* by Clark and Yoshinaka.

Another way of formulating the learning task draws on the close relation between context-free string languages and regular tree languages. An epsilon-free string language L is context-free if and only if it is the yield of a regular tree language [200]. It therefore makes sense to learn the language of parse trees directly, restating the problem of finding a CFG for a target language L as finding a regular tree grammar or finite tree automaton that describes the parse trees of L. Then, the learner should be provided with labeled examples, making the parse trees explicit. The most natural theoretical model is then arguably learning from text. In [76], Fernau lifts his notion of distinguishability for string languages to tree languages. This leads to a source of identifiable subclasses of the regular tree languages which might be helpful for finding the ones most appropriate for the application at hand. These connections are also employed in so-called feature grammars, which extend PCFG by adding features to nonterminals, whose processing can be easily interpreted as the work of a tree automaton. Inferring such grammars has been discussed by Dreyer and Eisner [70]. Another aspect where trees naturally creep in into the discussion of learning PCFG is that of having some probability distributions on sets of trees, as suggested by Corrazza and Satta [55, 56].

Probabilistic tree-substitution and tree-adjoining grammars are extensions of PCFG in which nonterminals can be replaced by tree fragments [100, 101]. Advocates for these types of formalisms argue that they strike a good balance between expressive power and efficiency of parsing, which is still polynomial. However, learning tree-substitution grammars is a challenging task, partly because it is difficult to obtain structurally annotated training data.

In the case of unsupervised learning, where parse information is not required, Nesson et al. induce probabilistic synchronous tree-insertion grammars for machine translation from aligned multi-lingual corpora [162]. Cohn et al. consider a non-parametric Bayesian model for supervised and unsupervised induction of tree-substitution grammar and evaluate their algorithms experimentally by learning a target language consisting of dependency parses [53]. Tree transducers also offer a rich source of formalisms in this context [148] that have been also investigated from the viewpoint of Grammatical Inference [40, 41]. A linguistically relevant type of transducer is one translating trees into strings; see [94, 211] for one NLP-relevant discussion of (statistical) learning in this setting. The overview article of Lopez on Statistical Machine Translation [135] (SMT) is a recommendable source of information for the topic; a bit older but more relevant to the topic of this chapter is the survey of Knight and Graehl [113]. Other important papers include [133, 165, 172, 219]. This list is for sure not complete, but it should give a good starting point for finding other papers from the NLP area that deal with learning of tree languages and tree transducers, including the learning of string language concepts like tree-adjoining grammars, or translations like string-to-tree or tree-to-string transductions. A short but nice discussion of using tree versus string language learners for NLP purposes can be found in [155]. Recently, also ideas from active learning found their way into SMT systems; see [92]. Let us finally mention that another topic relevant for NLP, more precisely for the assistance of corpus construction, is that of tree annotation; see [170] for one approach on statistical learning in this context.

7.5 Conclusions and Perspectives

In this section, we present some directions of research that have yet to be fully covered by the tree learning community.

7.5.1 PAC Learning

A learning model that has received much attention in the string case is *probably approximately correct* (PAC) learning. The model was introduced by Valiant [210] and formalized by Angluin [14]. In this setting, the learning algorithm repeatedly draws a tree t_i from a universe V^{t} according to some probability distribution and produces an hypothesis L_i. The learner is a PAC learner for the target class of languages if the probability is at least $1 - \delta$ that there is an index k such that for every $j > k$, the probability that the tree t_j is in the symmetric difference of the target language L and L_j is at most ε. Here, ε is the accuracy of the learner, and δ its confidence. For tree languages, the interest in PAC learning seems surprisingly low compared to learning from, e.g., finite data or queries. Some reasons for this are given in [91].

7.5.2 Support Vector Machines

Support Vector Machines (SVM) have been a very successful methodology to classify and hence also to learn many different objects. They have been applied to various scenarios. Essentially, a method to compare two such objects is needed, implemented in a so-called kernel. Here, we only want to mention the works of Moschitti (and his colleagues); see [7, 49, 50, 158, 159] as starting points. This already shows the breadth of the application areas, from computational linguistics to biomedical applications, to mention two of them.

7.5.3 Unranked Trees

So far, we have only treated unranked trees in passing. However, many of the results for the ranked case can be transferred through an appropriate encoding, e.g., first-child-next-sibling (FCNS) or curryfication. For details, we refer you to [54]. The idea is to produce a sample of ranked trees from a sample of unranked trees, then to apply a learning algorithm for ranked trees, and finally to interpret the result by reversing the encodings. Note that the results thus accomplished can be quite different from those obtained by translating the learning algorithms themselves and computing everything in the unranked domain. For instance, tree patterns make perfect sense also in the unranked case, but with the standard construction, only replacements of variables at the frontier nodes are possible. However, after, say, curryfication, we might obtain a ranked tree pattern that encodes a way of replacing parts of a tree that look quite different, possibly allowing for replacements "in the middle" of an unranked tree. These effects deserve further study in the future, and we also refer to studies like [9, 199].

7.5.4 Infinite Trees

We have not touched at all the learning of languages consisting of infinite trees. For definitions, we refer you to [161]. To some extent, such languages of infinite trees correspond to ω-languages in the string case. Even for infinite strings, there are only a few learning algorithms, and most of them are based on inferring traditional languages of finite strings that characterize certain ω-languages, mostly due to topological properties [97]. A similar approach might be viable for infinite trees, as well.

7.5.5 Alternative Representations

While there are at least some suggestions that implement the idea of directly learning regular expressions from positive data for string languages [77, 80, 112], we are not aware of a single paper that does this for tree languages. This is an interesting problem in itself, as learning a finite automaton and then translating it into an equivalent regular expression usually produces something that is neither intuitive nor readable. For the sake of clarity and conciseness, it may therefore be better to develop direct learning algorithms.

7.5.6 Learning Within or Beyond Regularity?

We have seen that many ideas from string language inference transfer easily to the tree case, opening up new applications. Are further extensions possible? The right notions of regularity seem to be important here. To see this, it would be also good idea to look at this notion both from a more theoretical viewpoint and from the viewpoint of potential applications; see [27, 93, 107] for both approaches. However, as argued by Kasprzik [107], with the correct notion of regularity, most learning algorithms sketched in this survey translate directly into other settings, for instance to multi-dimensional trees. Another possible direction of extensions is given by the idea of *distributional learning*, mainly developed by Clark and Yoshinaka (see Chap. 6, *Distributional Learning of Context-Free and Multiple Context-Free Grammars*); in the context of learning non-regular tree languages (from positive data), we refer you to [51, 52, 109].

A caveat should be formulated in this place, however. Although in principle many important properties carry over from regular string languages to regular tree languages and beyond, as they are often relying on some sort of Myhill-Nerode theorem that is true in various object settings, this does not mean that the resulting algorithms are practical ones. The point is that several decidability questions and related tasks can become significantly harder for trees as compared to strings. For instance, useless states pose no problem for string processing finite automata, but they do so for trees; see [67]. Also, there are certain intricacies that make a simple translation from the string to the tree case sometimes an error-prone task, as certified by the learner for representative samples and membership queries as discussed in [24, 108]. Hence, it can become an important issue to choose the most appropriate formalism, balancing its expressibility against its complexity. The discussions of Maletti [148] regarding tree transducers can give some ideas thereof.

7.5.7 Shock Trees: A Potential Application

Apart from these applications that have been already used, we will now describe one potential application that (as we think) might give some ideas on future research.

There is a growing body of literature on shock graphs (sometimes also called shape graphs) and shock trees; see [64, 111, 143, 169, 193, 194, 203–205]. Moreover, the shock tree graphs described in [169] are used as standard test examples to validate new algorithms as emerging in the area of (multisub)graph matching; see [119]. In short, shock trees are a tree representation of two- or also three-dimensional shapes [57].

So far, learning algorithms for finding classes of shapes have been based on statistical properties or the Minimum Description Length principle [206]. It would be interesting to apply tree learning algorithms to this area of Pattern Recognition. The idea would be to learn a specific grammar or automaton for each kind of picture. This basic idea was already described in [144], there called "discovering shape categories". It would also be a nice testbed for applying syntactic methods to classification tasks, given the fact that several established shape datasets exist.

7.5.8 Beyond Trees

We mentioned at the very beginning of this chapter that we are not going to speak about the learning of structures beyond graphs. However, it should be finally mentioned that there are several ways to describe various structures with strings or with trees. With respect to any such fixed representation, learners for strings or trees can be also seen as learners for other structures. For instance, representations of graphs and multigraphs by trees or strings are described in [29, 79].

Acknowledgments We are grateful to the referee for many helpful suggestions. We are also thankful to Andreas Maletti and Markus L. Schmid for their careful reading of the manuscript.

References

1. N. Abe and H. Mamitsuka. Predicting protein secondary structure using stochastic tree grammars. *Machine Learning*, 29:275–301, 1997.
2. N. Abe and M. K. Warmuth. On the computational complexity of approximating distributions by probabilistic automata. *Machine Learning*, 9:205–260, 1992.
3. H. Ahonen. Automatic generation of SGML content models. *Electronic Publishing – Origination, Dissemination and Design*, 8:195–206, 1995.
4. H. Ahonen. *Generating grammars for structured documents using grammatical inference methods*. PhD thesis, Department of Computer Science, University of Helsinki, Finland, 1996.
5. H. Ahonen, H. Mannila, and E. Nikunen. Forming grammars for structured documents: an application of grammatical inference. In R. C. Carrasco and J. Oncina, editors, *Proceedings of the Second International Colloquium on Grammatical Inference (ICGI-94): Grammatical Inference and Applications*, volume 862 of *LNCS/LNAI*, pages 153–167. Springer, 1994.
6. K. Aikou, Y. Suzuki, T. Shoudai, and T. Miyahara. Automatic wrapper generation for metasearch using ordered tree structured patterns. In G. I. Webb and X. Yu, editors, *AI 2004: Advances in Artificial Intelligence, 17th Australian Joint Conference on Artificial Intelligence*, volume 3339 of *LNCS*, pages 1030–1035. Springer, 2004.
7. F. Aiolli, G. Da San Martino, A. Sperduti, and A. Moschitti. Efficient kernel-based learning for trees. In *Proceedings of the IEEE Symposium on Computational Intelligence and Data Mining, CIDM*, pages 308–315. IEEE, 2007.

8. A. Alexandrakis and S. Bozapalidis. Weighted grammars and Kleene's theorem. *Information Processing Letters*, 24(1):1–4, 1987.
9. T. R. Amoth, P. Cull, and P. Tadepalli. On exact learning of unordered tree patterns. *Machine Learning*, 44(3):211–243, 2001.
10. D. Angluin. Finding patterns common to a set of strings. *Journal of Computer and System Sciences*, 21:46–62, 1980.
11. D. Angluin. Inductive inference of formal languages from positive data. *Information and Control (now Information and Computation)*, 45:117–135, 1980.
12. D. Angluin. Inference of reversible languages. *Journal of the ACM*, 29(3):741–765, 1982.
13. D. Angluin. Learning regular sets from queries and counterexamples. *Information and Computation (formerly Information and Control)*, 75:87–106, 1987.
14. D. Angluin. Queries and concept learning. *Machine Learning*, 2:319–342, 1988.
15. D. Angluin. Negative results for equivalence queries. *Machine Learning*, 5:121–150, 1990.
16. S. Arikawa, T. Shinohara, and A. Yamamoto. Elementary formal systems as a unifying framework for language learning. Technical Report RIFIS-TR-CS-14, Research Institute of Fundamental Information Science, 1989.
17. H. Arimura, H. Ishizaka, and T. Shinohara. Learning unions of tree patterns using queries. In Klaus P. Jantke, Takeshi Shinohara, and Thomas Zeugmann, editors, *Algorithmic Learning Theory, 6th International Conference, ALT '95, Fukuoka, Japan, October 18-20, 1995*, volume 997 of *LNCS*, pages 66–79, Berlin, Heidelberg, 1995. Springer-Verlag.
18. H. Arimura, H. Sakamoto, and S. Arikawa. Efficient learning of semi-structured data from queries. In N. Abe, R. Khardon, and T. Zeugmann, editors, *Algorithmic Learning Theory, 12th International Conference, ALT*, volume 2225 of *LNCS*, pages 315–331. Springer, 2001.
19. H. Arimura, T. Shinohara, and S. Otsuki. A polynomial time algorithm for finding finite unions of tree pattern languages. In G. Brewka, K. P. Jantke, and P. H. Schmitt, editors, *Nonmonotonic and Inductive Logic, Second International Workshop, 1991*, volume 659 of *LNCS*, pages 118–131. Springer, 1993.
20. H. Arimura, T. Shinohara, and S. Otsuki. Finding minimal generalizations for unions of pattern languages and its application to inductive inference from positive data. In P. Enjalbert, E. W. Mayr, and K. W. Wagner, editors, *11th Annual Symposium on Theoretical Aspects of Computer Science, STACS*, volume 775 of *LNCS*, pages 649–660. Springer, 1994.
21. L. Becerra-Bonache, A. H. Dediu, and C. Tîrnaucă. Learning DFA from correction and equivalence queries. In Y. Sakakibara, S. Kobayashi, K. Sato, T. Nishino, and E. Tomita, editors, *Grammatical Inference: Algorithms and Applications, 8th International Colloquium, ICGI*, volume 4201 of *LNCS*, pages 281–292. Springer, 2006.
22. I. Berstel and C. Reutenauer. Recognizable formal power series on trees. *Theoretical Computer Science*, pages 115–148, 1982.
23. J. Berstel and L. Boasson. Formal properties of XML grammars and languages. *Acta Informatica*, 38(9):649–671, August 2002.
24. J. Besombes and J.-Y. Marion. Learning tree languages from positive examples and membership queries. *Theoretical Computer Science*, 382:183–197, 2007.
25. G. J. Bex, F. Neven, T. Schwentick, and K. Tuyls. Inference of concise DTDs from XML data. In U. Dayal, K.-Y. Whang, D. B. Lomet, G. Alonso, G. M. Lohman, M. L. Kersten, S. K. Cha, and Y.-K. Kim, editors, *32nd International Conference on Very Large Data Bases VLDB*, pages 115–126. ACM, 2006.
26. G. J. Bex, F. Neven, and S. Vansummeren. Inferring XML schema definitions from XML data. In C. Koch, J. Gehrke, M. N. Garofalakis, D. Srivastava, K. Aberer, A. Deshpande, D. Florescu, C. Yong Chan, V. Ganti, C.-C. Kanne, W. Klas, and E. J. Neuhold, editors, *Proceedings of the 33rd International Conference on Very Large Data Bases, VLDB*, pages 998–1009. ACM, 2007.
27. H. Björklund and T. Schwentick. On notions of regularity for data languages. *Theoretical Computer Science*, 411(4-5):702–715, 2010.
28. M. Blum and L. Blum. Towards a mathematical theory of inductive inference. *Information and Control*, 28:125–155, 1975.

29. C. Blume, H. J. S. Bruggink, M. Friedrich, and B. König. Treewidth, pathwidth and cospan decompositions with applications to graph-accepting tree automata. *Journal of Visual Languages & Computing*, 24(3):192–206, 2013.
30. B. Borchardt. *The Theory of Recognizable Tree Series*. Akademische Abhandlungen zur Informatik. Verlag für Wissenschaft und Forschung, 2005.
31. J. M. Brayer and K. S. Fu. A note on the k-tail method of tree grammar inference. *IEEE Transactions on Systems, Man and Cybernetics*, 7(4):293–300, 1977.
32. J. Carme, M. Ceresna, and M. Goebel. Query-based learning of XPath expressions. In Y. Sakakibara, S. Kobayashi, K. Sato, T. Nishino, and E. Tomita, editors, *Grammatical Inference: Algorithms and Applications, 8th International Colloquium, ICGI*, volume 4201 of *LNCS/LNAI*, pages 342–343. Springer, 2006.
33. J. Carme, R. Gilleron, A. Lemay, and J. Niehren. Interactive learning of node selecting tree transducer. *Machine Learning*, 66(1):33–67, 2007.
34. J. Carme, A. Lemay, and J. Niehren. Learning node selecting tree transducer from completely annotated examples. In G. Paliouras and Y. Sakakibara, editors, *International Colloquium on Grammatical Inference ICGI*, volume 3264 of *LNCS/LNAI*, pages 91–102. Springer, 2004.
35. J. Carme, A. Lemay, and J. Niehren. Learning node selecting tree transducer from completely annotated examples. In *7th International Colloquium on Grammatical Inference*, pages 91–102, Berlin, Heidelberg, 2004. Springer-Verlag.
36. R. C. Carrasco and J. R. Rico-Juan. A similarity between probabilistic tree languages: Application to XML document families. *Pattern Recognition*, 36(9):2197–2199, 2003.
37. R. C. Carrasco-Jiménez, J. Oncina, and J. Calera-Rubio. Stochastic inference of regular tree languages. *Machine Learning*, pages 185–197, 2001.
38. J. Champavère. *Induction de requêtes guidée par schéma (Schema-Guided Query Induction)*. PhD thesis, Univ. Lille 1, Science et technologies, Laboratoire d'informatique fondamentale de Lille, France, 2010.
39. Y.-F. Chen, A. Farzan, E. M. Clarke, Y.-K. Tsay, and B.-Y. Wang. Learning minimal separating DFA's for compositional verification. In S. Kowalewski and A. Philippou, editors, *Tools and Algorithms for the Construction and Analysis of Systems, 15th International Conference, TACAS*, volume 5505 of *LNCS*, pages 31–45. Springer, 2009.
40. D. Chiang. Hierarchical phrase-based translation. *Computational Linguistics*, 33(2):201–228, 2007.
41. D. Chiang. Learning to translate with source and target syntax. In J. Hajic, S. Carberry, and S. Clark, editors, *ACL 2010, Proceedings of the 48th Annual Meeting of the Association for Computational Linguistics*, pages 1443–1452. The Association for Computer Linguistics, 2010.
42. D. Chiang, A. K. Joshi, and K. A. Dill. A grammatical theory for the conformational changes of simple helix bundles. *Journal of Computational Biology*, 13(1):21–42, 2006.
43. D. Chiang, A. K. Joshi, and D. B. Searls. Grammatical representations of macromolecular structure. *Journal of Computational Biology*, 13(5):1077–1100, 2006.
44. B. Chidlovskii. Using regular tree automata as XML schemas. In *IEEE Advances in Digital Libraries Conference ADL*, pages 89–104, 2000.
45. B. Chidlovskii. Schema extraction from XML: A grammatical inference approach. In M. Lenzerini, D. Nardi, W. Nutt, and D. Suciu, editors, *8th International Workshop on Knowledge Representation meets Databases KRDB*, volume 45 of *CEUR Workshop Proceedings*. CEUR-WS.org, 2001.
46. B. Chidlovskii. Wrapping web information providers by transducer induction. In L. De Raedt and P. A. Flach, editors, *12th European Conference on Machine Learning ECML*, volume 2167 of *LNCS/LNAI*, pages 61–72. Springer, 2001.
47. B. Chidlovskii. Information extraction from tree documents by learning subtree delimiters. In Subbarao Kambhampati and Craig Knoblock, editors, *IIWeb*, pages 3–8, 2003.
48. B. Chidlovskii and J. Fuselier. A probabilistic learning method for XML annotation of documents. In L. Pack Kaelbling and A. Saffiotti, editors, *Proceedings of the Nineteenth International Joint Conference on Artificial Intelligence IJCAI*, pages 1016–1021. Professional Book Center, 2005.

49. Md. F. M. Chowdhury and A. Lavelli. Combining tree structures, flat features and patterns for biomedical relation extraction. In W. Daelemans, M. Lapata, and L. Màrquez, editors, *EACL 2012, 13th Conference of the European Chapter of the Association for Computational Linguistics*, pages 420–429. The Association for Computer Linguistics, 2012.
50. Md. F. M. Chowdhury, A. Lavelli, and A. Moschitti. A study on dependency tree kernels for automatic extraction of protein-protein interaction. In *Proceedings of the BioNLP 2011 Workshop*, pages 124–133. The Association for Computer Linguistics, 2011.
51. A. Clark. Three learnable models for the description of language. In A.-H. Dediu, H. Fernau, and C. Martín-Vide, editors, *Language and Automata Theory and Applications LATA*, volume 6031 of *LNCS*, pages 16–31. Springer, 2010.
52. A. Clark. Towards general algorithms for grammatical inference. In M. Hutter, F. Stephan, V. Vovk, and T. Zeugmann, editors, *Algorithmic Learning Theory, 21st International Conference, ALT*, volume 6331 of *LNCS/LNAI*, pages 11–30. Springer, 2010.
53. T. Cohn, P. Blunsom, and S. Goldwater. Inducing tree-substitution grammars. *Journal of Machine Learning Research*, 11:3053–3096, 2010.
54. H. Comon, M. Dauchet, R. Gilleron, F. Jacquemard, D. Lugiez, C. Löding, S. Tison, and M. Tomassi. *Tree Automata, Techniques and Applications*. 2007.
55. A. Corazza and G. Satta. Cross-entropy and estimation of probabilistic context-free grammars. In *Proceedings of the main conference on Human Language Technology Conference of the North American Chapter of the Association of Computational Linguistics, HLT-NAACL*, pages 335–342. Association for Computational Linguistics, 2006.
56. A. Corazza and G. Satta. Probabilistic context-free grammars estimated from infinite distributions. *IEEE Transactions on Pattern Analysis and Machine Intelligence*, 29(8):1379–1393, 2007.
57. L. P. Cordella, P. Foggia, C. Sansone, and M. Vento. Learning structural shape descriptions from examples. *Pattern Recognition Letters*, 23:1427–1437, 2002.
58. C. Costa Florêncio. *Learning Categorial Grammars*. PhD thesis, Universiteit Utrecht, The Netherlands, 2003.
59. C. Costa Florêncio. Learning tree adjoining grammars from structures and strings. In J. Heinz, C. de la Higuera, and T. Oates, editors, *Proceedings of the Eleventh International Conference on Grammatical Inference, ICGI*, volume 21 of *JMLR Workshop and Conference Proceedings*, pages 129–132. Journal of Machine Learning Research, 2012.
60. C. Costa Florêncio and H. Fernau. On families of categorial grammars of bounded value, their learnability and related complexity questions. *Theoretical Computer Science*, 452:21–38, 2012.
61. C. Crimi, A. Guercio, G. Pacini, G. Tortora, and M. Tucci. Automating visual language generation. *IEEE Transactions on Software Engineering*, 16(10):1122–1135, 1990.
62. F. Denis and A. Habrard. Learning rational stochastic tree languages. In M. Hutter, R. A. Servedio, and E. Takimoto, editors, *Algorithmic Learning Theory, 18th International Conference, ALT*, volume 4754 of *LNCS*, pages 242–256. Springer, 2007.
63. F. Denis, A. Lemay, and A. Terlutte. Residual finite state automata. *Fundamenta Informaticae*, 51(4):339–368, 2002. Conference version at STACS 2000.
64. P. Dimitrov, C. Phillips, and K. Siddiqi. Robust and efficient skeletal graphs. In *Conference on Computer Vision and Pattern Recognition CVPR 2000*, volume 1, pages 417–423. IEEE, 2000.
65. F. Drewes. MAT learners for recognizable tree languages and tree series. *Acta Cybernetica*, 19:249–274, 2009.
66. F. Drewes and J. Högberg. Learning a regular tree language from a teacher. In Z. Ésik and Z. Fülöp, editors, *Developments in Language Theory, 7th International Conference, DLT*, volume 2710 of *LNCS*, pages 279–291. Springer, 2003.
67. F. Drewes and J. Högberg. Query learning of regular tree languages: How to avoid dead states. *Theory of Computing Systems*, 40(2):163–185, 2007.
68. F. Drewes and H. Vogler. Learning deterministically recognizable tree series. *Journal of Automata, Languages and Combinatorics*, 12:333–354, 2007.

69. F. Drewes, J. Högberg, and A. Maletti. MAT learners for tree series: an abstract data type and two realizations. *Acta Informatica*, 48(3):165–189, 2011.
70. M. Dreyer and J. Eisner. Better informed training of latent syntactic features. In *Proceedings of the 2006 Conference on Empirical Methods in Natural Language Processing, EMNLP*, pages 317–326. Association for Computational Linguistics, 2006.
71. M. Droste and P. Gastin. Weighted automata and weighted logics. *Theoretical Computer Science*, 380(1–2):69–86, 2007.
72. M. Droste and H. Vogler. Weighted tree automata and weighted logics. *Theoretical Computer Science*, 366(3):228–247, 2006.
73. R. Durbin, S. Eddy, A. Krogh, and G. Mitchison. *Biological Sequence Analysis*. Cambridge University Press, Cambridge, UK, 1998.
74. H. Fernau. Learning XML grammars. In P. Perner, editor, *Machine Learning and Data Mining in Pattern Recognition MLDM'01*, volume 2123 of *LNCS/LNAI*, pages 73–87. Springer, 2001.
75. H. Fernau. Identification of function distinguishable languages. *Theoretical Computer Science*, 290:1679–1711, 2003.
76. H. Fernau. Learning tree languages from text. *RAIRO Informatique théorique*, 41:351–374, 2007.
77. H. Fernau. Algorithms for learning regular expressions from positive data. *Information and Computation*, 207:521–541, 2009.
78. H. Fernau and R. Freund. Bounded parallelism in array grammars used for character recognition. In P. Perner, P. Wang, and A. Rosenfeld, editors, *Advances in Structural and Syntactical Pattern Recognition (Proceedings of the SSPR'96)*, volume 1121 of *LNCS*, pages 40–49. Springer, 1996.
79. H. Fernau and M. Paramasivan. Formal language questions for Eulerian trails. In T. Neary and M. Cook, editors, *Machines, Computations and Universality, MCU*, volume 128 of *Electronic Proceedings in Theoretical Computer Science EPTCS*, pages 25–26. Open Publishing Association, 2013.
80. D. D. Freydenberger and T. Kötzing. Fast learning of restricted regular expressions and DTDs. In W.-C. Tan, G. Guerrini, B. Catania, and A. Gounaris, editors, *Joint 2013 EDBT/ICDT Conferences, ICDT '13 Proceedings*, pages 45–56. ACM, 2013.
81. H. Fukuda and K. Kamata. Inference of tree automata from sample set of trees. *International Journal of Computer and Information Sciences*, 13:177–196, 1984.
82. Z. Fülöp and H. Vogler. Weighted tree automata and tree transducers. In W. Kuich, M. Droste, and H. Vogler, editors, *Handbook of Weighted Automata*, pages 313–403. Springer, 2009.
83. P. García. Learning k-testable tree sets from positive data. Technical Report DSIC-II/46/1993, Departamento de Sistemas Informáticos y Computación, Universidad Politécnica de Valencia, http://www.dsic.upv.es/users/tlcc/tlcc.html, 1993.
84. P. García and J. Oncina. Inference of recognizable tree sets. Technical Report DSIC-II/47/93, Departamento de Sistemas Informáticos y Computación, Universidad Politécnica de Valencia, http://www.dsic.upv.es/users/tlcc/tlcc.html, 1993.
85. P. García and E. Vidal. Inference of k-testable languages in the strict sense and applications to syntactic pattern recognition. *IEEE Transactions on Pattern Analysis and Machine Intelligence*, 12:920–925, 1990.
86. P. García, E. Vidal, and J. Oncina. Learning locally testable languages in the strict sense. In *First International Workshop on Algorithmic Learning Theory ALT'90*, pages 325–328, 1990.
87. F. Gécseg and M: Steinby. *Tree Automata*. Akadémiai Kiadó, 1984.
88. E. M. Gold. Language identification in the limit. *Information and Control (now Information and Computation)*, 10:447–474, 1967.
89. E. M. Gold. Complexity of automaton identification from given data. *Information and Control (now Information and Computation)*, 37:302–320, 1978.
90. S. A. Goldman and M. J. Kearns. On the complexity of teaching. *Journal of Computer and System Sciences*, 50(1):20–31, 1995.
91. S. A. Goldman and S. Kwek. On learning unions of pattern languages and tree patterns in the mistake bound model. *Theoretical Computer Science*, 288(2):237–254, 2002.

92. J. González-Rubio, D. Ortiz-Martínez, and F. Casacuberta. Active learning for interactive machine translation. In W. Daelemans, M. Lapata, and L. Màrquez, editors, *EACL 2012, 13th Conference of the European Chapter of the Association for Computational Linguistics*, pages 245–254. The Association for Computer Linguistics, 2012.
93. G. Gottlob, P. G. Kolaitis, and T. Schwentick. Existential second-order logic over graphs: Charting the tractability frontier. *Journal of the ACM*, 51(2):312–362, 2004.
94. J. Graehl, K. Knight, and J. May. Training tree transducers. *Computational Linguistics*, 34(3):391–427, 2008.
95. A. Habrard and J. Oncina. Learning multiplicity tree automata. In Y. Sakakibara, S. Kobayashi, K. Sato, T. Nishino, and E. Tomita, editors, *Grammatical Inference: Algorithms and Applications, 8th International Colloquium, ICGI*, volume 4201 of *LNAI/LNCS*, pages 268–280. Springer, 2006.
96. C. de la Higuera. *Grammatical inference. Learning automata and grammars*. Cambridge University Press, 2010.
97. C. de la Higuera and J.-C. Janodet. Inferring omega languages from prefixes. In N. Abe, R. Khardon, and T. Zeugmann, editors, *Algorithmic Learning Theory ALT*, volume 2255 of *LNCS/LNAI*, pages 364–378. Springer, 2001.
98. T. Ihringer. *Allgemeine Algebra*. Stuttgart: Teubner, 1988.
99. H. Ishizaka, H. Arimura, and T. Shinohara. Finding tree patterns consistent with positive and negative examples using queries. *Annals of Mathematics and Artificial Intelligence*, 23(1-2):101–115, 1998.
100. A. Joshi. Tree adjoining grammars. In R. Mikkov, editor, *The Oxford Handbook of Computational Linguistics*, pages 483–501. Oxford University Press, 2003.
101. A. K. Joshi, L. S. Levy, and M. Takahashi. Tree adjunct grammars. *Journal of Computer and System Sciences*, 10:133–163, 1975.
102. M. Kanazawa. Identification in the limit of categorial grammars. *Journal of Logic, Language, and Information*, 5:115–155, 1996.
103. M. Kanazawa. *Learnable Classes of Categorial Grammars*. PhD, CSLI, 1998.
104. A. Kasprzik. Making finite-state methods applicable to languages beyond context-freeness via multi-dimensional trees. In J. Piskorski, B. Watson, and A. Yli-Jyrä, editors, *Post-proceedings of the 7th International Workshop on Finite-State Methods and Natural Language Processing*, pages 98–109. IOS Press, 2009.
105. A. Kasprzik. Generalizing over several learning settings. In J. M. Sempere and P. García, editors, *International Colloquium on Grammatical Inference ICGI*, volume 6339 of *LNCS*, pages 288–292. Springer, 2010.
106. A. Kasprzik. Inference of residual finite-state tree automata from membership queries and finite positive data. In G. Mauri and A. Leporati, editors, *Developments in Language Theory DLT*, volume 6795 of *LNCS*, pages 476–477. Springer, 2011.
107. A. Kasprzik. *Formal Tree Languages and Their Algorithmic Learnability*. PhD thesis, Fachbereich IV, Universität Trier, Germany, 2012.
108. A. Kasprzik. Four one-shot learners for regular tree languages and their polynomial characterizability. *Theoretical Computer Science*, 485C:85–106, 2013.
109. A. Kasprzik and R. Yoshinaka. Distributional learning of simple context-free tree grammars. In J. Kivinen, C. Szepesvári, E. Ukkonen, and T. Zeugmann, editors, *Algorithmic Learning Theory ALT*, volume 6925 of *LNCS*, pages 398–412. Springer, 2011.
110. M. J. Kearns and U. V. Vazirani. *An Introduction to Computational Learning Theory*. MIT Press, Cambridge, MA, USA, 1994.
111. B. B. Kimia, A. R. Tannenbaum, and S. W. Zucker. Shapes, shocks, and deformations, I. *International Journal of Computer Vision*, 15:189–224, 1995.
112. E. B. Kinber. On learning regular expressions and patterns via membership and correction queries. In A. Clark, F. Coste, and L. Miclet, editors, *Grammatical Inference: Algorithms and Applications, 9th International Colloquium, ICGI*, volume 5278 of *LNCS*, pages 125–138. Springer, 2008.

113. K. Knight and J. Graehl. An overview of probabilistic tree transducers for natural language processing. In A. F. Gelbukh, editor, *Computational Linguistics and Intelligent Text Processing, 6th International Conference, CICLing*, volume 3406 of *LNCS*, pages 1–24. Springer, 2005.
114. T. Knuutila. Inference of k-testable tree languages. In *Proc. IAPR Workshop on Structural and Syntactical Pattern Recognition*, pages 109–120. World Scientific, 1992.
115. T. Knuutila and M. Steinby. The inference of tree languages from finite samples: an algebraic approach. *Theoretical Computer Science*, 129:337–367, 1994.
116. R. Kosala, M. Bruynooghe, Jan Van Den Bussche, and H. Blockeel. Information extraction in structured documents using tree automata induction. In T. Elomaa, H. Mannila, and H. Toivonen, editors, *Proceedings of the 6th European Conference on Principles of Data Mining and Knowledge Discovery, PKDD*, volume 2431 of *LNCS*, pages 299–310. Springer, 2002.
117. R. Kosala, M. Bruynooghe, J. Van Den Bussche, and H. Blockeel. Information extraction from web documents based on local unranked tree automaton inference. In *Proceedings of the 18th International Joint Conference on Artificial Intelligence, IJCAI*, pages 403–408. Morgan Kaufmann, 2003.
118. R. Kosala, M. Bruynooghe, and J. Van den Bussche, H. Blockeel. Information extraction from structured documents using k-testable tree automaton inference. *Data & Knowledge Engineering*, 58(2):129–158, 2006.
119. S. Kosinov and T. Caelli. Inexact multisubgraph matching. In T. Caelli, A. Amin, R. P. W. Duin, M. Kamel, and D. de Ridder, editors, *Structural, Syntactic, and Statistical Pattern Recognition SSPR and SPR 2002*, volume 2396 of *LNCS*, pages 133–142. Springer, 2002.
120. M. Kuhlmann and J. Niehren. Logics and automata for totally ordered trees. In A. Voronkov, editor, *Rewriting Techniques and Applications, RTA*, volume 5117 of *LNCS*, pages 217–231. Springer, 2008.
121. W. Kuich. Formal power series over trees. In Symeon Bozapalidis, editor, *Proceedings of the 3rd International Conference Developments in Language Theory, DLT 1997, Thessaloniki, Greece, July 20-23, 1997*, pages 61–101. Aristotle University of Thessaloniki, 1997.
122. N. Kushmerick, D. S. Weld, and R. B. Doorenbos. Wrapper induction for information extraction. In *Proceedings of the Fifteenth International Joint Conference on Artificial Intelligence, IJCAI (1)*, pages 729–737. Morgan Kaufmann, 1997.
123. N. Labai and J. A. Makowsky. Weighted automata and monadic second order logic. In G. Puppis and T. Villa, editors, *Proceedings Fourth International Symposium on Games, Automata, Logics and Formal Verification, GandALF*, volume 119 of *Electronic Proceedings in Theoretical Computer Science EPTCS*, pages 122–135. Open Publishing Association, 2013.
124. J.-L. Lassez, M. J. Maher, and K. Marriott. Unification revisited. In M. Boscarol, L. C. Aiello, and G. Levi, editors, *Foundations of Logic and Functional Programming 1986*, volume 306 of *LNCS*, pages 67–113. Springer, 1988.
125. J. A. Laxminarayana, J. M. Sempere, and G. Nagaraja. Learning distinguishable linear grammars from positive data. In G. Paliouras and Y. Sakakibara, editors, *Grammatical Inference: Algorithms and Applications; 7th International Colloquium ICGI*, volume 3264 of *LNCS/LNAI*, pages 279–280. Springer, 2004.
126. S.-Y. Le, R. Nussinov, and J. V. Maizel. Tree graphs of RNA secondary structures and their comparisons. *Computers and Biomedical Research*, 22:461–473, 1989.
127. A. Lemay, J. Niehren, and R. Gilleron. Learning n-ary node selecting tree transducers from completely annotated examples. In Y. Sakakibara, S. Kobayashi, K. Sato, T. Nishino, and E. Tomita, editors, *Grammatical Inference: Algorithms and Applications, ICGI*, volume 4201 of *LNCS*, pages 253–267. Springer, 2006.
128. A. Lemay, S. Maneth, and J. Niehren. A learning algorithm for top-down XML transformations. In *Proceedings of the Twenty-Ninth ACM SIGMOD-SIGACT-SIGART Symposium on Principles of Database Systems, PODS*, pages 285–296, 2010.
129. B. Levine. Derivatives of tree sets with application to grammatical inference. *IEEE Transactions on Pattern Analysis and Machine Intelligence*, 3:285–293, 1981.

130. B. Levine. The use of tree derivatives and a sample support parameter for inferring tree systems. *IEEE Transactions on Pattern Analysis and Machine Intelligence*, 4:25–34, 1982.
131. L. S. Levy and A. K. Joshi. Skeletal structural descriptions. *Information and Control (now Information and Computation)*, 39:192–211, 1978.
132. D. D. Lewis and W. A. Gale. A sequential algorithm for training text classifiers. In W. B. Croft and C. J. van Rijsbergen, editors, *Proceedings of the 17th Annual International ACM-SIGIR Conference on Research and Development in Information Retrieval (Special Issue of the SIGIR Forum)*, pages 3–12. ACM/Springer, 1994.
133. Y. Liu, Y. Huang, Q. Liu, and S. Lin. Forest-to-string statistical translation rules. In *Proceedings of the 45th Annual Meeting of the Association of Computational Linguistics*, pages 704–711. Association for Computational Linguistics, 2007.
134. C. Löding. Basics on tree automata. In D. D'Souza and P. Shankar, editors, *Modern Applications of Automata Theory*, volume 2 of *IISc Research Monographs Series*, pages 80–109. World Scientific, 2012.
135. A. Lopez. Statistical machine translation. *ACM Comp. Surv.*, 40(3):8:1–8:49, 2006.
136. D. López and S. España Boquera. Error-correcting tree language inference. *Pattern Recognition Letters*, 23(1-3):1–12, 2002.
137. D. López and I. Piñaga. Syntactic pattern recognition by error correcting analysis on tree automata. In F. J. Ferri et al., editors, *Advances in Pattern Recognition, Joint IAPR International Workshops SSPR+SPR'2000*, volume 1876 of *LNCS*, pages 133–142, 2000.
138. D. López, J. M. Sempere, and P. García. Error correcting analysis for tree languages. *International Journal of Pattern Recognition and Artificial Intelligence*, 14(3):357–368, 2000.
139. D. López, J. Ruiz, and P. García. Inference of k-piecewise testable tree languages. In D. Chen and X. Cheng, editors, *Pattern Recognition and String Matching*, pages 341–352. Kluwer Academic Publishers, 2002.
140. D. López, J. M. Sempere, and P. García. Inference of reversible tree languages. *IEEE Transactions on Systems, Man and Cybernetics*, 34(4):1658–1665, August 2004.
141. D. López, J. Calera-Rubio, and A.-J. Gallego-Sánchez. Inference of k-testable directed acyclic graph languages. In J. Heinz, C. de la Higuera, and T. Oates, editors, *Proceedings of the Eleventh International Conference on Grammatical Inference, ICGI*, volume 21 of *JMLR Workshop and Conference Proceedings*, pages 149–163. Journal of Machine Learning Research, 2012.
142. S. Lucas, E. Vidal, A. Amiri, S. Hanlon, and J. C. Amengual. A comparison of syntactic and stastistical techniques for off-line OCR. In R. C. Carrasco and J. Oncina, editors, *Proceedings of the Second International Colloquium on Grammatical Inference (ICGI-94): Grammatical Inference and Applications*, volume 862 of *LNCS/LNAI*, pages 153–167, Berlin, 1994. Springer.
143. B. Luo, A. Robles-Kelly, A. Torsello, R. C.Wilson, and E.R. Hancock. Clustering shock trees. In *3rd IAPR-TC15 Workshop on Graph-based Representations in Pattern Recognition*, pages 217–226, 2001.
144. B. Luo, A. Robles-Kelly, A. Torsello, R. C. Wilson, and E. R. Hancock. Discovering shape categories by clustering shock trees. In W. Skarbek, editor, *Computer Analysis of Images and Patterns, 9th International Conference, CAIP 2001*, volume 2124 of *LNCS*, pages 152–160. Springer, 2001.
145. F. M. Luque, A. Quattoni, B. Balle, and X. Carreras. Spectral learning for non-deterministic dependency parsing. In W. Daelemans, M. Lapata, and L. Màrquez, editors, *EACL 2012, 13th Conference of the European Chapter of the Association for Computational Linguistics*, pages 409–419. The Association for Computer Linguistics, 2012.
146. E. Mäkinen. On inferring linear single-tree languages. *Information Processing Letters*, 73:1–3, 2000.
147. A. Maletti. Learning deterministically recognizable tree series — revisited. In S. Bozapalidis and Rahonis G, editors, *Proc. 2nd Int. Conf. Algebraic Informatics*, volume 4728 of *LNCS*, pages 218–235. Springer, 2007.

148. A. Maletti. Survey: Tree transducers in machine translation. In H. Bordihn, R. Freund, M. Holzer, T. Hinze, M. Kutrib, and F. Otto, editors, *Second Workshop on Non-Classical Models for Automata and Applications, NCMA*, volume 263 of *books@ocg.at*, pages 11–32. Austrian Computer Society, 2010.
149. W. Martens, F. Neven, T. Schwentick, and G. J. Bex. Expressiveness and complexity of XML schema. *ACM Transactions on Database Systems*, 31(3):770–813, 2006.
150. O. Maruyama and S. Miyano. Inferring a tree from walks. *Theoretical Computer Science*, 161:289–300, 1996.
151. H. Matsui, K. Sato, and Y. Sakakibara. Pair stochastic tree adjoining grammars for aligning and predicting pseudoknot RNA structures. *Bioinformatics*, 21(11):2611–2617, 2005.
152. S. Matsumoto and T. Shoudai. Learning of ordered tree languages with height-bounded variables using queries. In S. Ben-David, J. Case, and A. Maruoka, editors, *Algorithmic Learning Theory, 15th International Conference, ALT*, volume 3244 of *LNCS/LNAI*, pages 425–439. Springer, 2004.
153. S. Matsumoto, Y. Hayashi, and T. Shoudai. Polynomial time inductive inference of regular term tree languages from positive data. In *Algorithmic Learning Theory, ALT*, LNCS/LNAI, pages 212–227, 1997.
154. S. Matsumoto, T. Shoudai, T. Uchida, T. Miyahara, and Y. Suzuki. Learning of finite unions of tree patterns with internal structured variables from queries. *IEICE Transactions*, 91-D(2):222–230, 2008.
155. H. Mi, L. Huang, and Q. Liu. Forest-based translation. In *Proceedings of the 46th Annual Meeting of the Association for Computational Linguistics ACL-08: Human Language Technologies HLT*, pages 192–199. Association for Computational Linguistics, 2008.
156. S. Miyano. Learning theory toward genome informatics. *IEICE Transactions*, 78-D(5):560–567, 1995.
157. S. Miyano, A. Shinohara, and T. Shinohara. Polynomial-time learning of elementary formal systems. *New Generation Comput.*, 18(3):217–242, 2000.
158. A. Moschitti. Making tree kernels practical for natural language learning. In D. McCarthy and S. Wintner, editors, *EACL 2006, 11st Conference of the European Chapter of the Association for Computational Linguistics*, pages 113–120. The Association for Computer Linguistics, 2006.
159. A. Moschitti and F. M. Zanzotto. Fast and effective kernels for relational learning from texts. In Z. Ghahramani, editor, *Machine Learning, Proceedings of the Twenty-Fourth International Conference, ICML*, volume 227 of *ACM International Conference Proceeding Series*, pages 649–656. ACM, 2007.
160. T. Motoki, T. Shinohara, and K. Wright. The correct definition of finite elasticity: Corrigendum to identification of unions. In *COLT'91*, page 375. Morgan Kaufmann, 1991.
161. D. Muller and P. E. Schupp. Alternating automata on infinite objects, determinacy and Rabin's theorem. In M. Nivat and D. Perrin, editors, *Automata on infinite words*, volume 192 of *LNCS*, pages 100–107. Springer, 1984.
162. R. Nesson, S. Shieber, and A. Rush. Induction of probabilistic synchronous tree-insertion grammars for machine translation. In *Proceedings of the 7th Conference of the Association for Machine Translation in the Americas*, pages 128–137, Boston, Massachusetts, 2006. AMTA.
163. F. Neven. Automata, logic, and XML. In J. Bradfield, editor, *Computer Science Logic; 16th International Workshop, CSL 2002*, volume 2471 of *LNCS*, pages 2–26. Springer, 2002.
164. F. Neven and T. Schwentick. Automata-and logic-based pattern languages for tree-structured data. In L. E. Bertossi, G. O. H. Katona, K.-D. Schewe, and B. Thalheim, editors, *Semantics in Databases, Second International Workshop, Dagstuhl 2001*, volume 2582 of *LNCS*, pages 160–178. Springer, 2003.
165. T. P. Nguyen, A. Shimazu, Tu-B. Ho, M. Le Nguyen, and V. V. Nguyen. A tree-to-string phrase-based model for statistical machine translation. In *CoNLL 2008: Proceedings of the 12th Conference on Computational Natural Language Learning*, pages 143–150, 2008.
166. J. Niehren, J. Champavère, R. Gilleron, and A. Lemay. Query Induction with Schema-Guided Pruning Strategies. *Journal of Machine Learning Research*, 14:927–964, 2013.

167. J. Ruiz Ochando. *Familias de lenguajes explorables: inferencia inductiva y caracterización algebraica*. PhD thesis, Departamento de Sistemas Informáticos y Computación, Universidad Politécnica de Valencia, 1999.
168. J. Oncina and P. García. Identifying regular languages in polynomial time. In H. Bunke, editor, *Advances in Structural and Syntactic Pattern Recognition*, pages 99–108. World Scientific, 1992.
169. M. Pelillo, K. Siddiqi, and S. Zucker. Matching hierarchical structures using association graphs. *IEEE Transactions on Pattern Analysis and Machine Intelligence*, 21(11):1105–1120, 1999.
170. S. Petrov, L. Barrett, R. Thibaux, and D. Klein. Learning accurate, compact, and interpretable tree annotation. In N. Calzolari, C. Cardie, and P. Isabelle, editors, *ACL 2006, 21st International Conference on Computational Linguistics and 44th Annual Meeting of the Association for Computational Linguistics*. The Association for Computer Linguistics, 2006.
171. G. D. Plotkin. A note on inductive generalization. In B. Meltzer and D. Mitchie, editors, *Machine Intelligence*, volume 5, pages 153–163. Edinburgh University Press, 1970.
172. C. Quirk, A. Menezes, and C. Cherry. Dependency treelet translation: Syntactically informed phrasal SMT. In *Proceedings of the 43rd Annual Meeting of the ACL*, pages 271–279. Association for Computational Linguistics, 2005.
173. S. Raeymaekers, M. Bruynooghe, and J. Van den Bussche. Learning (k, l)-contextual tree languages for information extraction from web pages. *Machine Learning*, 71(2-3):155–183, 2008.
174. J. R. Rico-Juan, J. Calera-Rubio, and R. C. Carrasco. Probabilistic k-testable tree languages. In A. L. Oliveira, editor, *Grammatical Inference: Algorithms and Applications, 5th International Colloquium (ICGI 2000)*, volume 1891 of *LNCS/LNAI*, pages 221–228. Springer, 2000.
175. J. Rogers. Strict LT2 : Regular :: Local : Recognizable. In *LACL*, pages 366–385, 1996.
176. G. Rozenberg and A. Salomaa, editors. *Handbook of Formal Languages (3 volumes)*. Springer, 1997.
177. G. Rozenberg and A. Salomaa, editors. *Handbook of Formal Languages, Volume III*. Berlin: Springer, 1997.
178. J. Ruiz and P. García. Learning k-piecewise testable languages from positive data. In L. Miclet and C. de la Higuera, editors, *Proceedings of the Third International Colloquium on Grammatical Inference (ICGI-96): Learning Syntax from Sentences*, volume 1147 of *LNCS/LNAI*, pages 203–210. Springer, 1996.
179. H. Rulot and E. Vidal. An efficient algorithm for the inference of circuit-free automata. In G. Ferraté et al., editors, *Syntactic and Structural Pattern Recognition (Proc. of the NATO Advanced Research Workshop, held 1986 in Barcelona)*, volume 45 of *ASI Series F*, pages 173–184. Springer, 1988.
180. H. M. Rulot. *ECGI. Un algoritmo de inferencia gramatical mediante corrección de errores*. PhD thesis, Departamento de Sistemas Informáticos y Computación, Universidad Politécnica de Valencia, Spain, 1992.
181. Y. Sakakibara. Learning context-free grammars from structural data in polynomial time. *Theoretical Computer Science*, 76:223–242, 1990.
182. Y. Sakakibara. Efficient learning of context-free grammars from positive structural examples. *Information and Computation*, 97(1):23–60, March 1992.
183. Y. Sakakibara. Pair hidden Markov models on tree structures. In *Proceedings of the Eleventh International Conference on Intelligent Systems for Molecular Biology, ISMB (Supplement of Bioinformatics)*, pages 232–240, 2003.
184. Y. Sakakibara. Grammatical inference in bioinformatics. *IEEE Transactions on Pattern Analysis and Machine Intelligence*, 27:1051–1062, 2005.
185. Y. Sakakibara, M. Brown, R. Hughey, I. S. Mian, K. Sjölander, R. C. Underwood, and D. Haussler. Stochastic context-free grammars for tRNA modeling. *Nucleic Acids Research*, 22:5112–5120, 1994.
186. T. Schwentick. Automata for XML – a survey. *Journal of Computer and System Sciences*, 73(3):289–315, 2007.

187. J. M. Sempere and D. López. Learning decision trees and tree automata for a syntactic pattern recognition task. In F. J. Perales López, A. C. Campilho, N. Pérez de la Blanca, and A. Sanfeliu, editors, *Pattern Recognition and Image Analysis, First Iberian Conference, IbPRIA*, volume 2652 of *LNCS*, pages 943–950. Springer, 2003.
188. J. M. Sempere and D. López. Identifying P rules from membrane structures with an error-correcting approach. In H. J. Hoogeboom, G. Paun, G. Rozenberg, and A. Salomaa, editors, *Workshop on Membrane Computing*, volume 4361 of *LNCS*, pages 507–520. Springer, 2006.
189. J. M. Sempere and D. López. Characterizing membrane structures through multiset tree automata. In G. Eleftherakis, P. Kefalas, G. Paun, G. Rozenberg, and A. Salomaa, editors, *Workshop on Membrane Computing*, volume 4860 of *LNCS*, pages 428–437. Springer, 2007.
190. C. E. Shannon. A mathematical theory of communication. *The Bell System Technical Journal*, 27:379–423 & 623–656, 1948.
191. B. A. Shapiro and K. Zhang. Comparing multiple RNA secondary structures using tree comparisons. *Computer Applications in the Biosciences*, 6(4):309–318, 1990.
192. T. Shoudai, T. Uchida, and T. Miyahara. Polynomial time algorithms for finding unordered tree patterns with internal variables. In R. Freivalds, editor, *Fundamentals of Computation Theory, FCT*, volume 2138 of *LNCS*, pages 335–346. Springer, 2001.
193. K. Siddiqi, A. Shakoufandeh, S. Dickinson, and S. Zucker. Shock graphs and shape matching. *International Journal of Computer Vision*, 35:13–22, 1999.
194. K. Siddiqi, S. Bouix, A. Tannebaum, and S. Zucker. Hamilton-Jacobi skeletons. *International Journal of Computer Vision*, 48:215–231, 2002.
195. R. Siromoney, L. Mathew, K. G. Subramanian, and V. R. Dare. Learning of recognizable picture languages. In A. Nakamura et al., editors, *Parallel Image Analysis, ICPIA*, volume 654 of *LNCS*, pages 247–259, 1992.
196. K. G. Subramanian, A. R. Sagaya Mary, and P. Helen Chandra. On the inference of linear single tree grammars from positive structural information. In B. Prasad, editor, *3rd Indian International Conference on Artificial Intelligence, IICAI*, pages 104–111, 2007.
197. Y. Suzuki, T. Shoudai, S. Matsumoto, and T. Miyahara. Polynomial time inductive inference of ordered tree languages with height-constrained variables from positive data. In C. Zhang, H. W. Guesgen, and W.-K. Yeap, editors, *Trends in Artificial Intelligence, 8th Pacific Rim International Conference on Artificial Intelligence, PRICAI*, volume 3157 of *LNCS*, pages 211–220. Springer, 2004.
198. Y. Suzuki, T. Miyahara, T. Shoudai, T. Uchida, and Y. Nakamura. Discovery of maximally frequent tag tree patterns with height-constrained variables from semistructured web documents. In *International Workshop on Challenges in Web Information Retrieval and Integration, WIRI*, pages 104–112. IEEE Computer Society, 2005.
199. Y. Suzuki, T. Shoudai, T. Uchida, and T. Miyahara. Ordered term tree languages which are polynomial time inductively inferable from positive data. *Theoretical Computer Science*, 350(1):63–90, 2006.
200. J. W. Thatcher. Characterizing derivation trees of context-free grammars through a generalization of finite automata theory. *Journal of Computer and System Sciences*, 1:317–322, 1967.
201. C. I. Tîrnaucă and C. Tîrnaucă. Learning regular tree languages from correction and equivalence queries. *Journal of Automata, Languages and Combinatorics*, 12(4):501–524, 2007.
202. C. Tirnăucă. Correction queries in active learning. In C. Martín-Vide, editor, *Scientific Applications of Language Methods*, volume 2 of *Mathematics, Computing, Language, and Life: Frontiers in Mathematical Linguistics and Language Theory*, pages 387–419. Imperial College Press, 2011.
203. S. Tirthapura, D. Sharvit, P. Klein, and B. B. Kimia. Indexing based on edit-distance matching of shape graphs. In *SPIE International Symposium on Voice, Video, and Data Communication*, pages 25–36, 1998.
204. A. Torsello and E. R. Hancock. Learning structural variations in shock trees. In T. Caelli, A. Amin, R. P. W. Duin, M. Kamel, and D. de Ridder, editors, *Structural, Syntactic, and Statistical Pattern Recognition, SSPR and SPR 2002*, volume 2396 of *LNCS*, pages 113–122. Springer, 2002.

205. A. Torsello and E. R. Hancock. Computing approximate tree edit distance using relaxation labelling. *Pattern Recognition Letters*, 24:1089–1097, 2003.
206. A. Torsello and E. R. Hancock. Learning shape-classes using a mixture of tree-unions. *IEEE Transactions on Pattern Analysis and Machine Intelligence*, 28(6):954–967, 2006.
207. O. Unold. Grammar-based classifier system for recognition of promoter regions. In B. Beliczynski, A. Dzielinski, M. Iwanowski, and B. Ribeiro, editors, *Adaptive and Natural Computing Algorithms, 8th International Conference, ICANNGA (1)*, volume 4431 of *LNCS*, pages 798–805. Springer, 2007.
208. O. Unold and L. Cielecki. How to use crowding selection in grammar-based classifier system. In *Proceedings of the Fifth International Conference on Intelligent Systems Design and Applications, ISDA*, pages 124–129. IEEE Computer Society, 2005.
209. O. Unold and G. Dabrowski. Use of learning classifier system for inferring natural language grammar. In T. Kovacs, X. Llorà, K. Takadama, P. L. Lanzi, W. Stolzmann, and S. W. Wilson, editors, *Learning Classifier Systems, International Workshops, IWLCS*, volume 4399 of *LNCS*, pages 17–24. Springer, 2007.
210. L. G. Valiant. A theory of the learnable. *Communications of the ACM*, 27:1134–1142, 1984.
211. A. Vaswani, H. Mi, L. Huang, and D. Chiang. Rule Markov models for fast tree-to-string translation. In D. Lin, Y. Matsumoto, and R. Mihalcea, editors, *The 49th Annual Meeting of the Association for Computational Linguistics: Human Language Technologies, Proceedings of the Conference, ACL*, pages 856–864. The Association for Computer Linguistics, 2011.
212. J. T.-Li Wang, B. A. Shapiro, D. Shasha, K. Zhang, and K. M. Currey. An algorithm for finding the largest approximately common substructures of two trees. *IEEE Transactions on Pattern Analysis and Machine Intelligence*, 20(8):889–895, 1998.
213. D. Wood. Standard generalized markup language: Mathematical and philosophical issues. In J. van Leeuwen, editor, *Computer Science Today: Recent Trends and Developments*, volume 1000 of *LNCS*, pages 344–365. Springer, 1995.
214. K. Wright. Identification of unions and languages drawn from an identifiable class. In *Conference on Learning Theory, COLT*, pages 328–333. Morgan Kaufmann, 1989.
215. H. Yamasaki and T. Shoudai. A polynomial time algorithm for finding a minimally generalized linear interval graph pattern. *IEICE Transactions*, 92-D(2):120–129, 2009.
216. T. Yokomori, N. Ishida, and S. Kobayashi. Learning local languages and its application to protein α-chain identification. In *Proc. 27th Hawaii International Conference on System Sciences*, pages 113–122, 1994.
217. Y. Yoshimura, T. Shoudai, Y. Suzuki, T. Uchida, and T. Miyahara. Polynomial time inductive inference of cograph pattern languages from positive data. In S. Muggleton, A. Tamaddoni-Nezhad, and F. A. Lisi, editors, *Inductive Logic Programming – 21st International Conference, ILP 2011*, volume 7207 of *LNCS*, pages 389–404. Springer, 2012.
218. R. Yoshinaka and M. Kanazawa. Distributional learning of abstract categorial grammars. In S. Pogodalla and J.-P. Prost, editors, *Logical Aspects of Computational Linguistics — 6th International Conference, LACL*, volume 6736 of *LNCS*, pages 251–266. Springer, 2011.
219. F. Zhai, J. Zhang, Yu Zhou, and C. Zong. Unsupervised tree induction for tree-based translation. *Transactions of the Association for Computational Linguistics*, 1:243–254, 2013.

Chapter 8
Learning the Language of Biological Sequences

François Coste

Abstract The application to biological sequences is an appealing challenge for Grammatical Inference. While some first successes have already been recorded, such as the inference of profile Hidden Markov Models or stochastic Context-Free Grammars which are now part of the classical Bioinformatics toolbox, it is still a nice and open source of problems or inspiration for our research, with the possibility to apply our ideas to real fundamental applications. In this chapter, we survey biological sequences' main specificities and how they are handled in Pattern/Motif Discovery in order to introduce the important concepts and techniques used and present the latest successful approaches in that field by Grammatical Inference.

8.1 Linguistic Metaphor

New sequencing technologies are giving access to an ever increasing amount of DNA, RNA or protein sequences for more and more species. One major challenge in the post-genomic era is now to decipher this set of genetic sequences composing what has been popularly named "the language of life" [1].

As witnessed by this expression, the linguistic metaphor has been used for a long time in genetics. Indeed, the discovery of the double helix structure of DNA in 1953 showed that the genetic information contained in this biological macromolecule can be represented by two (long) complementary sequences over a four-letter alphabet $\{A, C, G, T\}$ symbolizing the *nucleotides*, the complementary letters (called Watson–Crick *base pairs*) being A–T and C–G. This genetic information is used to construct and operate a living organism by the *transcription* when needed of pieces of DNA sequences, named *genes*, into RNA single strand macromolecules which can also be represented by a sequence on almost the same four-letter alphabet $\{A, C, G, U\}$, where T has been replaced by its unmethylated form U. Sequences of RNAs coding for proteins are in turn

F. Coste (✉)
Inria Rennes—Bretagne Atlantique, Campus de Beaulieu, 35042
Rennes Cedex, France
e-mail: francois.coste@inria.fr

J. Heinz and J.M. Sempere (eds.), *Topics in Grammatical Inference*,
DOI 10.1007/978-3-662-48395-4_8

translated into protein sequences of *amino acid residues*, over the 20 amino acid's alphabet $\{A, C, D, E, F, G, H, I, K, L, M, N, P, Q, R, S, T, V, W, Y\}$, that determine their three-dimensional conformations and functions in the cells (see for instance [2] for a more detailed introduction to the production of RNAs and proteins encoded in DNA). Sequences are thus at the core of storage of heredity information and its expression into the functional units of the cells: the natural language metaphor arises then quickly. This metaphor may be convenient for vulgarization but can also be a source of inspiration for scientists trying to discover the functional units of the genome and how this "text" is structured.

Applying computational linguistics tools to represent, understand and handle biological sequences is a natural continuation of the linguistic metaphor. Using formal grammars, such as the ones introduced in 1957 by Noam Chomsky [3] to describe natural languages and study syntax acquisition by children, has been advocated in particular by Searls: his articles provide a good introduction to the different levels of expressiveness required to model biological macromolecules by grammatical formalisms [4–8]. Basically, copies and long-distance correlations are common in genomic sequences, calling rapidly for context-sensitive grammars in Chomsky's hierarchy to model them, which makes parsing unworkable. As in linguistics, a solution to getting polynomial-time parsing and still representing many of the non-local constraints from genomic sequences is to use mildly context-sensitive languages [9]. Along this way, Searls introduced String Variable Grammars as an expressive formalism for describing the language of DNA that has led to several generic practical parsers: the precursor Genlang [10] and its successors Stan [11], Patscan [12], Patsearch [13] and Logol [14]. Many specialized parsers have also been devised, as for instance RNAMotif [15], RNAbob [16], Hypasearch [17, 18], Palingol [19] and Structator [20], tailored to handle efficiently RNA stem-loop secondary structures.

But one has still to design the grammar. In contrast with all the expertise available on natural languages, little is known about the syntax of DNA and the functional/semantic role of its parts. For instance, how are the equivalents of "words", "sentences" and even "punctuation marks" defined? In some specific cases, expert knowledge can be used to build a grammar, eventually by successive trial-and-error refinements with respect to the sequences retrieved by the model. In the other cases, expert knowledge is missing or is insufficient.

On the other hand, a huge number of genomic sequences are available, opening the door to grammar inference from these sequences. In this chapter, we will present advances made towards the big challenge of learning automatically the language of genomic sequences. The first step we consider is to discover what the genomic "words" are: this is mainly the domain of Motif Discovery, and related work is presented in Sect. 8.2. The second step is then to learn the "syntax" governing the admissible chaining of "words" in macromolecules: this is the classical goal of Grammatical Inference and we present the first successes obtained at the intersection of this field and Bioinformatics in Sect. 8.3.

8.2 Discovering and Modeling Biological Words

"Words" can be looked at different levels in DNA, requiring different levels of modelization. We investigate in this section how this has been classically handled in Bioinformatics from the simplest historical first steps, introducing and illustrating some specificities of biological sequences, to the more elaborate techniques from today's state of the art.

8.2.1 Short DNA Words

Simple Words A classical example of an identified DNA substring is AAGCTT (on the upper strand and its complement TTCGAA on the lower strand), that is specifically recognized in *H. influenzae* bacteria by one of its enzymatic proteins named HindIII that cleaves the double strand DNA of invading viruses at the sites where this substring occurs, while the bacteria's occurrence sites of the substring in its DNA are protected from cleavage by a prior methylation. The HindIII protein is said to be a restriction enzyme. More than 800 different restriction enzymes and more than 100 corresponding recognition sequences have been identified in bacterial species, with important applications in genetic engineering. These recognition sequences show a great variability among species, many of them being palindromic on complementary strands (meaning the sequence reads the same backwards and forwards in complementary DNA strands, like in AAGCTT and TTCGAA), reflecting that both strands of DNA have to be cut, often by a complex of two identical proteins operating on each strand. The main characteristic of these substrings is their short length (about four to eight base pairs), that makes them likely to appear frequently in any genome, providing them an efficient defense against unknown invading viruses. These sequences are thus rather ubiquitous and do not support information by themselves (they are only substrings recognized by the restriction enzymes), and it could be discussed whether they are "words" in a linguistic sense.

Conserved Words Another example of a well-known short sequence is the Pribnow box, early identified in the DNA of *E. coli* bacteria. It was discovered by Pribnow [21] by looking at the DNA sequences around six, experimentally determined, starting points of the transcription of genes into RNAs by a molecule named RNA-polymerase. Would you find in these sequences, shown hereafter and aligned on the known transcription start site formatted in bold, the protein binding site initiating the transcription by the RNA-polymerase?

```
Site1:   ...AAGTAAACACGGTACGATGTACCAC A TGAAACGACAGTGAGTCA...
Site2:      ...TGCTTCTGACTATAATAGACAGG G TAAAGACCTGATTTTTGA...
Site3:     ...TTTATTGCAGCTTATAATGGTTAC A AATAAAGCAATAGCA...
Site4:      ...CCACTGGCGGTGATACTGAGCAC A TCAGCAGGACGCACTGAC...
Site5:      ...CGTCATTTGATATGATGCGCCCC G CTTCCCGATAAGGGAGCA...
Site6:      ...CTTCCGGCTCGTATGTTGTGTGG A ATTGTGAGCGGATAACAA...
```

By looking carefully at the sequences, one can find a conserved region (underlined below), located about 10 positions before the transcription start site, that may have been conserved despite mutations for its function through natural selection:

```
Site1:    ...AAGTAAACACGG TACGATG TACCAC A TGAAACGACAGTGAGTCA...
Site2:      ...TGCTTCTGAC TATAATA GACAGG G TAAAGACCTGATTTTTGA...
Site3:    ...TTTATTGCAGCT TATAATG GTTAC A AATAAAGCAATAGCA...
Site4:     ...CCACTGGCGGT GATACTG AGCAC A TCAGCAGGACGCACTGAC...
Site5:      ...CGTCATTTGA TATGATG CGCCCC G CTTCCCGATAAGGGAGCA...
Site6:     ...CTTCCGGCTCG TATGTTG TGTGG A ATTGTGAGCGGATAACAA...
```

Consensus Sequences and Motifs Looking at the underlined alignment of this conserved region, only two positions (the second and the sixth) are strictly conserved out of seven and the farthest sequences share only three identical positions for four mismatches. But the *consensus sequence* TATAATG of the alignment, built by keeping only the most abundant letter at each position, appears with no more than two mismatches, and one may consider it as an archetypal (eventually ancestral) sequence for the region and the other sequences as its variants by meaningless mutations. Searching for this consensus sequence TATAATG without mismatch, we would retrieve only one of the six conserved sites and we would expect one match per $4^7 \simeq 16{,}000$ bp in whole DNA. Allowing one mismatch, we would retrieve three of the six sites and we would expect one match per 700 bp. Allowing two mismatches, we would retrieve all the sites but we would expect one match per 70 bp, which is likely to be too much.

We can remark that the nucleotides A and G are evenly distributed at the fourth underlined position and TATGATG would also have been a good candidate consensus sequence. Actually, it would be more informative to know that the fourth position has to be a purine (*A* or *G* bases) and, as done by Pribnow, we can use the *consensus motif* TAT[AG]ATG (where brackets specify a set of alternative bases at the position) to designate the sequences probably engaged by RNA polymerase. This motif retrieves two sites for one expected match per 8,000 bp, and five sites if one error is allowed for one expected match per 400 bp. If we assume that the base at the fifth position is not important, we can also relax the consensus to TAT[AG]xTG where x is a wildcard for any base. This consensus retrieves four of the sites with about one match per 2,000 bp and all the sites if one error is allowed with a match per 100 bp. And choosing the full *consensus* [GT]A[CT][AG][ACT]T[AG] would recognize all the sites and would expect a match per 350 bp. As shown in this example, consensus offers many ways to model a word and its possible variants in DNA, ranging from consensus sequences allowing a limited number of errors to full consensus motifs, with all the intermediate ambiguity/sensitivity trade-offs.

"De novo" discovery of such words can be done by enumerating them and returning those over-represented in a collection of genome sequences, i.e. occurring more frequently than expected by chance. This approach has been successful in Motif Discovery, particularly for the discovery of short words and rather simple motifs (to enable a practical enumeration, even if efficient datastructures can be used and enumerating only the motifs that have sufficient support in the sequences can help); see [22, 23] for details.

Position Specific Matrices Yet, consensus sequences or motifs are not completely satisfactory for representing and discovering biological words. Taking again the example of the full consensus motif [TG]A[TC][GA][ACT]T[GA], do we really want GACACTA to be recognized like TATAATG? Or if a more specific consensus, such as the consensus sequence TATAATG, is chosen, allowing a limited number of errors, how is it possible to express that some positions can mutate more easily than others and that some base mutations occur more likely at some positions? Moreover, while conserved on average, the binding sites involved in the initiation of transcription occur rarely as exact matches of their specific consensus sequence. On average in bacteria, only half of the positions in each site match with the consensus sequence. A first explanation is that bindings have to be reversible. Different affinities with binding proteins enable as also to tune at a fine level the concentration of the RNA (and eventually protein) genes expressed in the cell, which would be interesting to estimate from the motif.

Weighting the consensus motifs addresses these issues. This is usually done on the basis of a summary of the sites by their base count at each position in a *position-specific count matrix* (PSCM). For the Pribnow sites example, the PSCM for the aligned conserved region would be:

	1	2	3	4	5	6	7
A	0	6	0	3	4	0	1
C	0	0	1	0	1	0	0
G	1	0	0	3	0	0	5
T	5	0	5	0	1	6	0

If we denote by $o_i(a)$ the observed count of base a at position i of the sites, estimation of probability of a at i in the site is given by:

$$\hat{p}_i(a) = \frac{o_i(a)}{\sum_{a' \in \{A,C,G,T\}} o_i(a')}.$$

Under the strong assumption that the probability of a base at a position depends only on the position, the probability of a sequence on $a_1 a_2 \ldots a_k$ given a *position-specific probability matrix* (PSPM) $\mathbf{P} = [\mathbf{p_1}, \mathbf{p_2} \ldots, \mathbf{p_k}]$ is $\Pi_{i=1}^{k} p_i(a_i)$. For instance, for the example above, the probability of TATAATG would be $\frac{5}{6} \times \frac{6}{6} \times \frac{5}{6} \times \frac{3}{6} \times \frac{4}{6} \times \frac{6}{6} \times \frac{5}{6} \simeq 1.2 \times 10^{-4}$ while for GACACTA it would only be $\frac{1}{6} \times \frac{6}{6} \times \frac{1}{6} \times \frac{3}{6} \times \frac{1}{6} \times \frac{6}{6} \times \frac{1}{6} \simeq 6.8 \times 10^{-5}$. By way of comparison, both sequences would have a probability of $(\frac{1}{4})^7 \simeq 6.1 \times 10^{-5}$ of being generated randomly by an equiprobable choice of the bases.

In the genome of *S. cerevisiae* which contains 64 % of A and T, the probability of TATAATG and GACACTA would respectively be 2×10^{-4} and 6×10^{-6}, making the second word more exceptional and thus more interesting than the first word with respect to this background model. When positions are assumed to be independent, the odd-score of the probability of a sequence $a_1 a_2 \ldots a_k$ by $[\mathbf{p_1}, \mathbf{p_2} \ldots, \mathbf{p_k}]$ with respect to its probability in a background model where each base a has a probability $p(a)$

can directly be computed by $\Pi_{i=1}^{k} \frac{p_i(a_i)}{p(a_i)}$, and the comparison with respect to expected background probability can directly be embedded in a *Position Weight Matrix* (PWM) [24], also called Position-Specific Weight Matrix (PSWM) or *Position-Specific Scoring Matrix* (PSSM), in logarithm form to facilitate computation (sum instead of product and better precision for rounded computation). In a PWM, the score of base a at position i is usually defined by

$$s_i(a) = \log_2 \frac{p_i(a)}{p(a)}$$

and the score of a sequence $a_1 a_2 \ldots a_k$ is given by

$$S(a_1 a_2 \ldots a_k) = \sum_{i=1}^{k} s_i(a_i).$$

The PWM computed from the PSCM above, assuming that the bases are equiprobable in the background model ($p(A) = p(C) = p(G) = p(T)$), would be:

	1	2	3	4	5	6	7
A	$-\infty$	2	$-\infty$	1	1.42	$-\infty$	-0.58
C	$-\infty$	$-\infty$	-0.58	$-\infty$	-0.58	$-\infty$	$-\infty$
G	-0.58	$-\infty$	$-\infty$	1	$-\infty$	$-\infty$	1.74
T	1.74	$-\infty$	1.74	$-\infty$	-0.58	2	$-\infty$

Bases over-represented with respect to background probability have positive scores, while under-represented bases have negative scores. Using a sliding window of width k, PWM can assign a score at each site of a genome reflecting its likelihood of being part of the motifs. The highest score for a sequence with the PWM above is 11.64, obtained for TATAATG, while the lowest score (except $-\infty$) is 2.68, obtained for GACACTA.

First Pseudocounts Let us remark that a mutation from A to G at the fifth position of the best sequence TATAATG will directly result in $-\infty$ score. Nucleotides that occur rarely in the motif at a specific position may not be seen in a small sample by chance but will force the probability of any sequence containing one of these missing nucleotides to 0. Pseudocounts are thus usually added to compensate for small samples counts. This can be done by adding systematically 1 to the observed counts, and the estimate of probability of a at i will then be:

$$\hat{p}_i(a) = \frac{o_i(a) + 1}{\sum_{a'} (o_i(a') + 1)}.$$

More elaborate pseudocounts can be used; for instance, in

$$\hat{p}_i(a) = \frac{o_i(a) + A\, p(a)}{\sum_{a'} (o_i(a') + A\, p(a'))}$$

the pseudocount added is proportional to background probability $p(a)$ and the weight A given to the prior. Choosing $A = 2$ and keeping the equiprobability of the bases, the PWM on the Pribnow example would be

	1	2	3	4	5	6	7
A	−2.00	1.70	−2.00	0.81	1.17	−2.00	−0.42
C	−2.00	−2.00	−0.42	−2.00	−0.42	−2.00	−2.00
G	−0.42	−2.00	−2.00	0.81	−2.00	−2.00	1.46
T	1.46	−2.00	1.46	−2.00	−0.42	1.70	−2.00

and we would have $S(\text{TATAATG}) = 9.76$, $S(\text{GACACTA}) = 2.53$ and $S(\text{TATAGTG}) = 6.59$, the minimal score being -14. By adding pseudocounts, all the sequences have a score strictly greater than $-\infty$. One can still discriminate a set of sequences by choosing a cut-off value, chosen as a compromise between desired recall and precision, with the advantage over sequence consensus or motifs of being better suited for the representation of similar sequences without a strict conservation per position.

Measuring Conservation The conservation of a site can be evaluated according to a measure named information content [25] that measures the information gain on the site provided by the PSPM with respect to a uniform random choice of the bases. The information content IC_i at position i is given by the formula:

$$IC_i = 2 + \sum_a p_i(a) \log_2 p_i(a).$$

Assuming positional independence, information content of the complete site is simply the sum of the information contents:

$$IC = \sum_{i=1}^{k} IC_i.$$

Information content is the basis of a convenient visualization of PSPM named sequence logos [26] that displays simultaneously conservation of each position, and their proportional base composition (see Fig. 8.1).

Information content can be generalized to account for the background model with biased base probability distribution $\mathbf{p}$:

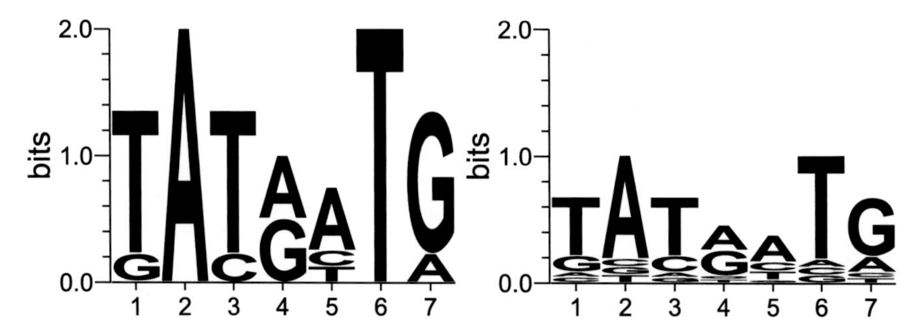

Fig. 8.1 Sequence logos for the Pribnow example (*left* without pseudocounts, *right* with pseudocounts). Height of stacks of symbols shows the information content of the position and the relative heights of the bases indicates their probability at the position (logo generated with WebLogo 3.3 [27])

$$IC_{i||\mathbf{p}} = \sum_a p_i(a) \log_2 \frac{p_i(a)}{p(a)}$$

$IC_{i||\mathbf{p}}$ is known as the relative entropy (a.k.a. Kullback-Liebler divergence [28]) and measures how much the $p_i(x)$ diverge from the background distribution $\mathbf{p}$ at the position. Let us note that when $\forall a \in \{A, C, G, T\}$, $p(a) = \frac{1}{4}$, the formula can be rewritten into IC_i. The generalized information content of the site is once again the sum over the positions $IC_{||\mathbf{p}} = \sum_{i=1}^{k} IC_{i||\mathbf{p}}(i)$: it measures how much the distribution defined by the PSPM contrasts with the distribution obtained by a Bernoulli-like process.

Information content is thus related to how exceptionally conserved is the set of underlying words with respect to such background models. It is thus a good objective function for PWM motif discovery programs that aim at identifying such sets of words in a set of sequences (for instance, to find binding sites near the transcription sites as in the Pribnow example). In its simplest setting, the problem can be stated as looking for a word of length k per sequence such that the corresponding information content, or a related score, is maximized. Many strategies for the exploration of the search space have been proposed. This includes the greedy algorithm consensus [29–31], expectation maximization algorithms like MEME [32] and several algorithms based on a Gibbs sampling strategy: Gibbs [33–35], AlignACE [36], MotifSampler [37] or BioProspector [38]. The scores used are information content (IC) (consensus, MotifSampler), log-likelihood ratio (LLR) (MotifSampler, Gibbs), E-value of the log-likelihood (MEME) or E-value of the IC (consensus).

Usage PWM/PSSM are widely used in popular databases such as TRANSFAC [39] and JASPAR [40] to model binding sites, identified experimentally by techniques such as SELEX or now ChIP-Seq, with the help of motif discovery programs to refine the site localization, and are then available to scan new genomes for the prediction of putative binding sites. There is still a large number of false positives, and regulation in more complex organisms than bacteria is still incompletely understood. Whether those sites are actually bound by a protein and play a functional role in transcription, and under what conditions, must still be determined experimentally by traditional molecular techniques like promoter bashing, reporter gene assays, ChIP experiments, etc.

8.2.2 Longer Words

Binding sites, involved in the regulation of the transcription of DNA genes into RNA and the production of proteins, are examples of short words in DNA. Gene coding for RNA or proteins, that are the functional products of DNA in the cell, can also easily be considered as (longer) DNA words.

In the context of natural evolution, genes as well as other DNA sequences, are subject to genomic mutations (substitutions, insertions, deletions or recombinations) under natural selection pressure. Most of these mutations are lethal or harmful, but about a third of them are either neutral or weakly beneficial. There is thus a sequence conservation of the genes transmitted among the individuals or the species, but with substitutions of bases and insertions or deletions of (eventually stretches of) bases. Biologists use the term of "homologs" to designate sequences inherited in two species by a common ancestor. Homology is the base of comparative genomics to annotate the sequences that can be considered as variants of the same word. But homology does not imply necessarily that function is preserved. The TIGRFAM protein database introduced the term "equivalogs" to designate homologs that are conserved with respect to function since their last common ancestor. This later concept matches more closely the linguistic closely of a "word" (with literal or practical meaning) but is more difficult to establish, especially *in silico*.

Similarity of Two Proteins Homology of two proteins can be estimated by aligning their sequences so as to optimize the number of exact matches between aligned amino acids and by reporting the percentage of identity between the two aligned sequences. To better evaluate their functional kinship, it is better to take into account the different physico-chemical properties of the amino acids (see Fig. 8.2). For instance, if the electric charge of an amino acid is important for the function of the protein, the function is more likely to be conserved by mutations preserving this charge. In some other cases, the hydrophoby of the amino acid will be its important feature.

Substitution matrices such as Blosum62 [42] score the similarity of amino acids according to their propensity to be exchanged with each other in blocks of conserved regions (Table 8.1). Such matrices reflects the mean physico-chemical similarity between amino acids under natural selection pressure, as well as some similarity or redundancy of the genetic code.

Substitution matrices provide a way to score the similarity (instead of their percentage of identity) of two proteins by aligning their sequence of amino acids so as to maximize the sum of the amino acid substitution scores. This can be computed in quadratic time by a dynamic programming *global alignment* algorithm known as the

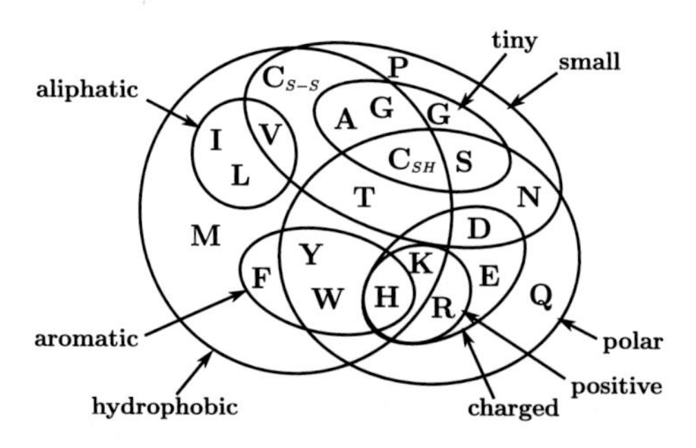

Fig. 8.2 Venn diagram of amino acid properties (adapted from: [41])

Table 8.1 BLOSUM62 substitution matrix

	C	S	T	P	A	G	N	D	E	Q	H	R	K	M	I	L	V	F	Y	W	
C	9																				C
S	-1	4																			S
T	-1	1	5																		T
P	-3	-1	-1	7																	P
A	0	1	0	-1	4																A
G	-3	0	-2	-2	0	6															G
N	-3	1	0	-2	-2	0	6														N
D	-3	0	-1	-1	-2	-1	1	6													D
E	-4	0	-1	-1	-1	-2	0	2	5												E
Q	-3	0	-1	-1	-1	-2	0	0	2	5											Q
H	-3	-1	-2	-2	-2	-2	1	-1	0	0	8										H
R	-3	-1	-1	-2	-1	-2	0	-2	0	1	0	5									R
K	-3	0	-1	-1	-1	-2	0	-1	1	1	-1	2	5								K
M	-1	-1	-1	-2	-1	-3	-2	-3	-2	0	-2	-1	-1	5							M
I	-1	-2	-1	-3	-1	-4	-3	-3	-3	-3	-3	-3	-3	1	4						I
L	-1	-2	-1	-3	-1	-4	-3	-4	-3	-2	-3	-2	-2	2	2	4					L
V	-1	-2	0	-2	0	-3	-3	-3	-2	-2	-3	-3	-2	1	3	1	4				V
F	-2	-2	-2	-4	-2	-3	-3	-3	-3	-3	-1	-3	-3	0	0	0	-1	6			F
Y	-2	-2	-2	-3	-2	-3	-2	-3	-2	-1	2	-2	-2	-1	-1	-1	-1	3	7		Y
W	-2	-3	-2	-4	-3	-2	-4	-4	-3	-2	-2	-3	-3	-1	-3	-2	-3	1	2	11	W
	C	S	T	P	A	G	N	D	E	Q	H	R	K	M	I	L	V	F	Y	W	

Frequently observed substitutions receive positive scores and seldom observed substitutions are given negative scores (log odds ratio)

Needleman–Wunsch algorithm [43], that copes also with *insertions and deletions* of subsequences that are common in DNA sequences by the addition of affine penalty scores for ‘gaps’.

Global alignment enables one to compare two protein sequences over their whole length, but many proteins are composed of several domains that are stable units of protein spatial structures able to fold autonomously. Domains may have existed, or may still exist, as independent proteins: they constitute the protein building blocks selected by evolution and recombined in different arrangements to create proteins with different functions. Comparing proteins at this level requires local rather than global alignments. The best *local alignment* of two sequences can be computed by the Smith–Waterman algorithm [44], a variation of the global alignment dynamic programming algorithm not penalizing gaps at both ends of the sequences. To search an entire database for homologous (sub-)sequences of a given protein sequence in reasonable time, heuristic and approximate local alignment algorithms have been developed, such as FASTA [45] or BLAST [46], one of the most widely used bioinformatics programs.

Modeling Conserved Protein Sequences When getting more than two related protein sequences, switching from pairwise sequence alignment to *multiple sequence alignment* enables one to identify evolutionarily or structurally conserved regions and key positions in all the sequences. Most formulations of multiple sequence alignment lead to NP-complete problems; therefore, classical multiple sequence alignment programs rely on heuristics. Most of them perform global multiple sequence alignment such as ClustalW [47], T-Coffee [48], Probcons [49], MUSCLE [50] or MAFFT [51]. Local multiple sequence alignment can be found by the methods cited above to build PWM, the set of conserved k-words being a specific case of local alignment without gaps. In between global and local alignment, DIALIGN [52, 53] proposes an original approach based on significant local pairwise alignment of segments that enables it to identify a set of multiple sequence local alignments shared by all the sequences without any gap penalty.

Profile HMM Modeling locally conserved regions identified by multiple sequence alignment can be done once again with PWM. To handle larger regions with insertions and deletions, PWMs have been generalized to so-called *profile* models by the addition of insertion/deletion penalties at each position [55] and furthermore to *profile Hidden Markov Models* (pHMM) by adding also probabilities for entering into insertion, deletion or matching mode at next position given the current position and mode [56, 57]. Namely, pHMMs are hidden Markov models with a predefined specific k-position left-to-right architecture, with three (hidden) states per position (see Fig. 8.3): a *match* state generating amino acids according to the conserved position distribution (the equivalent of a PWM column), an *insert* state generating amino acids with respect to their distribution in gaps (by default, their background probability) and a *delete* silent state enabling passing a match state without emitting any amino acid.

Transitions are only allowed between states from one position to the next one and are probabilized, enabling one to tune the likelihood of inserting or deleting amino acids at each position and the likelihood of continuing insertions or deletions after entering one of these modes, as seen in protein sequence families.

If the topology of a pHMM is set, its probabilistic parameters can be estimated from available sequences of the family by a classical Expectation-Maximization scheme such as the Baum–Welch algorithm [58]. Nevertheless, the classical workflow in Bioinformatics is rather to start from a multiple sequence alignment of the sequences, assign for each column of the alignment involving enough sequences (say more than half of the family) a match state (and its insertion and deletion companion states) and convert observed counts of symbol emissions and state transitions into probabilities from the alignment.

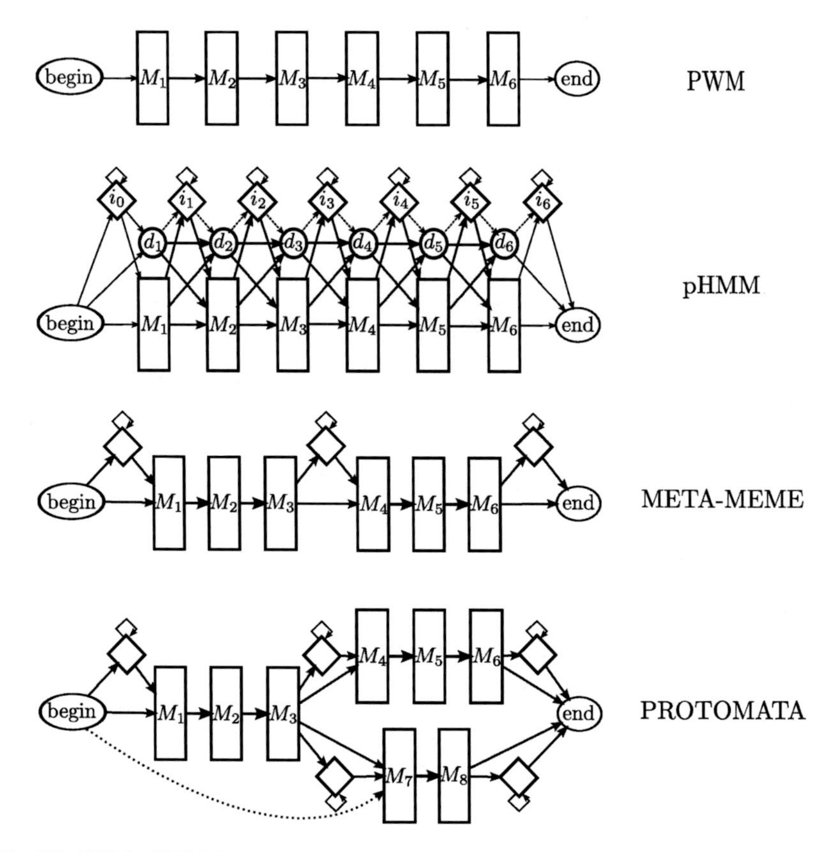

Fig. 8.3 PWM, pHMM, Meta-MEME and Protomata types of architecture (inspired from [54])

Elaborate Pseudocounts Even if the topology of pHMM is simple, the number of free parameters to estimate is still big compared to the number of sequences usually available. Much work has thus been done to avoid overspecialization and compensate for the lack of data or its biases by the development of transition regularizers, sequence weighting schemes and, especially, sophisticated pseudocount schemes based on the usage and elaboration of a priori knowledge on amino acid substitutability. As a matter of fact, the alphabet size of amino acids is greater than that of nucleotides, and targeted characterizations with pHMMs tend to be longer than with PWMs: pseudocounts are thus even more important here.

Classical pseudocounts presented above for nucleotides can be used, but taking into account the substitutability preferences of amino acids arising from their shared physico-chemical properties leads to better performances. A first way is to use available substitution matrices such as BLOSUM62: if we denote by $m(a|b)$ the probability of having a mutation to a from b, derived from the corresponding score in the chosen substitution matrix (see [42]), an intuitive scheme introduced for PWM with many variants [59] is to make each amino acid b contribute to pseudocounts of amino acid a in proportion to its abundance at the position and its probability $m(a|b)$ of mutating into a. If we denote by $m_i(a) = \sum_b \frac{o_i(b)}{\sum_{b'} o_i(b')} m(a|b)$ the probability of getting a by mutation of residues at position i, an estimate for the probability of a in i can be:

$$\hat{p}_i(a) = \frac{o_i(a) + A\, m_i(a)}{\sum_{a'} (o_i(a') + A\, m_i(a'))}.$$

This pseudocount scheme has the advantage of interpolating between the score of pairwise alignment, such as with BLAST, when a small number of sequences is available (consider for instance the case of only one sequence and $A \gg 1$) and the maximum likelihood approach when more sequences are available (when $\sum_a o_i(a) \gg A$). In practice, A has to be chosen to tune the importance of pseudocounts with respect to observed counts, classical proposed policies being to choose $\min(20, \sum_a o_i(a))$ [60] or $5R$ [59] where R is the number of different amino acids observed in the column, a simple measure of its diversity.

This pseudocount scheme performs well but does not take full advantage of the column composition knowledge. Instead of distributing pseudocounts from each amino acid count independently, one may wish to distribute them according to the whole column distribution. For instance, if the column is biased towards small hydrophobic amino acids, one would like to bias the pseudocounts towards this combination of physico-chemical properties. To this end, [62] proposed using Dirichlet mixture densities as a means of representing prior information about typical amino acid column distributions in multiple sequence alignments and derived the formulas to compute the corresponding posterior distributions given observed counts in the Bayesian framework.

Dirichlet mixtures can be thought of as mixtures of M pseudocount vectors $\boldsymbol{\alpha}_1, \ldots, \boldsymbol{\alpha}_M$ corresponding to M different typical distributions of amino acids having each a prior probability of $q_j, 1 \leq j \leq M$, where each Dirichlet density $\boldsymbol{\alpha}_j = (\alpha_j(A), \alpha_j(C), \ldots, \alpha_j(W))$ contains the appropriate amino acid pseudocounts (the equivalent of $A\, p(a)$ or $A\, m_i(a)$ in the pseudocount formulas above) for the typical distribution j.

An example of a Dirichlet mixture from [61] is given in Table 8.2. This Dirichlet mixture and more recent ones can be found on the site of the Bioinformatics and Computational Biology group at UCSC at http://compbio.soe.ucsc.edu/dirichlets/. These mixtures were estimated by maximum likelihood inference from the columns of available large "gold standard" datasets of protein multiple alignments that are assumed to be accurate and representative.

According to the authors, the mixture shown here was one of their first really good Dirichlet mixtures. It is composed of nine components that favor each a different distribution of amino acids biased towards one or several physico-chemical properties from Fig. 8.2: for instance, Dirichlet density α_2 favors aromatic amino acids (Y, F, W, H) by assigning them higher pseudocounts (relatively to what would be expected from their background frequency; see [61] for details) while α_5 favors aliphatic or large and non-polar amino acids. The last component is specific: it favors columns with few different amino acids, with a preference for P, G, W or C, by

Table 8.2 Parameters of Blocks9, a nine components Dirichlet mixture prior [61]

j	1	2	3	4	5	6	7	8	9
$\alpha_j(A)$	0.271	0.021	0.561	0.070	0.041	0.116	0.093	0.452	0.005
$\alpha_j(C)$	0.040	0.010	0.045	0.011	0.015	0.037	0.005	0.115	0.004
$\alpha_j(D)$	0.018	0.012	0.438	0.019	0.006	0.012	0.387	0.062	0.007
$\alpha_j(E)$	0.016	0.011	0.764	0.095	0.010	0.018	0.348	0.116	0.006
$\alpha_j(F)$	0.014	0.386	0.087	0.013	0.154	0.052	0.011	0.284	0.003
$\alpha_j(G)$	0.132	0.016	0.259	0.048	0.008	0.017	0.106	0.140	0.017
$\alpha_j(H)$	0.012	0.076	0.215	0.077	0.007	0.005	0.050	0.100	0.004
$\alpha_j(I)$	0.023	0.035	0.146	0.033	0.300	0.797	0.015	0.550	0.002
$\alpha_j(K)$	0.020	0.014	0.762	0.577	0.011	0.017	0.094	0.144	0.005
$\alpha_j(L)$	0.031	0.094	0.247	0.072	0.999	0.286	0.028	0.701	0.006
$\alpha_j(M)$	0.015	0.022	0.119	0.028	0.210	0.076	0.010	0.277	0.001
$\alpha_j(N)$	0.048	0.029	0.442	0.080	0.006	0.015	0.188	0.119	0.004
$\alpha_j(P)$	0.054	0.013	0.175	0.038	0.013	0.015	0.050	0.097	0.009
$\alpha_j(Q)$	0.021	0.023	0.531	0.185	0.020	0.011	0.110	0.127	0.004
$\alpha_j(R)$	0.024	0.019	0.466	0.507	0.015	0.013	0.039	0.144	0.007
$\alpha_j(S)$	0.216	0.029	0.583	0.074	0.012	0.028	0.119	0.279	0.003
$\alpha_j(T)$	0.147	0.018	0.446	0.072	0.036	0.088	0.066	0.358	0.004
$\alpha_j(V)$	0.065	0.036	0.227	0.043	0.180	0.944	0.025	0.662	0.003
$\alpha_j(W)$	0.004	0.072	0.030	0.011	0.013	0.004	0.003	0.062	0.003
$\alpha_j(Y)$	0.010	0.420	0.121	0.029	0.026	0.017	0.019	0.199	0.003
$\sum_a \alpha_j(a)$	1.181	1.356	6.664	2.081	2.081	2.568	1.766	4.988	0.100
q_j	0.183	0.058	0.090	0.079	0.083	0.091	0.116	0.066	0.234

assigning tiny pseudocounts to all amino acids so that the observed count will dominate. This component has the highest prior probability ($q_9 = 0.234$) since many positions in alignments exhibit a unique conserved amino acid, followed by the first component ($q_1 = 0.183$) that favors small neutral amino acids that appear to be often mixed together in alignment columns, while the more specific density of the second component has the lowest prior probability of the mixture ($q_2 = 0.058$).

Basically, the Dirichlet density $\boldsymbol{\alpha_j}$ of a Dirichlet mixture component embeds a prior in the form of a pseudocount that enables one to compute the posterior probability $\hat{p}_i(a|\boldsymbol{\alpha_j})$ of each amino acid a from observed counts at position i with respect to this prior by:

$$\hat{p}_i(a|\alpha_j) = \frac{o_i(a) + \alpha_j(a)}{\sum_{a'} \left(o_i(a') + \alpha_j(a')\right)}.$$

This formula can be extended to a mixture of M Dirichlet densities $\Theta = (\boldsymbol{\alpha_1}, \ldots, \boldsymbol{\alpha_M}, q_1, \ldots, q_M)$ by distributing these probabilities proportionally to the likelihood $p_i(j)$ of each component for the observed count distribution:

$$\hat{p}_i(a|\Theta) = \sum_{j=1}^{M} p_i(j) \frac{o_i(a) + \alpha_j(a)}{\sum_{a'} \left(o_i(a') + \alpha_j(a')\right)}.$$

$p_i(j)$ is named the *posterior mixture coefficient* of component j and can be estimated by application of Bayes rule from the *prior Dirichlet mixture coefficient* q_j and the likelihood of the observed counts for component j determined by density $\boldsymbol{\alpha_j}$:

$$\hat{p}_i(j) = \frac{q_j \; p(\mathbf{o_i}|\boldsymbol{\alpha_j})}{\sum_{j'=1}^{M} q_{j'} \; p(\mathbf{o_i}|\boldsymbol{\alpha_{j'}})}$$

where $p(\mathbf{o_i}|\boldsymbol{\alpha_j})$, the likelihood of the observed counts according to Dirichlet density $\boldsymbol{\alpha_j}$, is given by the complicated but simple to calculate formula

$$p(\mathbf{o_i}|\boldsymbol{\alpha_j}) = \frac{(\sum_a o_i(a))!}{\prod_a o_i(a)!} \cdot \frac{\prod_a \Gamma(o_i(a) + \alpha_j(a))}{\Gamma(\sum_a o_i(a) + \alpha_j(a))} \cdot \frac{\Gamma(\sum_a \alpha_j(a))}{\prod_a \Gamma(\alpha_j(a))}$$

where $\Gamma(x)$, the gamma function, is the standard continuous generalization of the integer factorial function.

These formulas obtained by Bayesian inference provide a powerful pseudocount scheme to estimate the distribution at a position from a small number of observation counts and priors on different typical column amino acid distributions. From more than hundred sequences required to build a good characterization of a family of homologous sequences, one comes down to fifty sequences, or even as few as ten or twenty examples with the latest pseudocount schemes.

Usage Profile HMMs have thus become a method of choice for the classification and the annotations of homologous protein sequences. Instead of using BLAST to search in a database of annotated sequences for one homolog to the sequence to annotate, the idea is to build first a pHMM for each family of homologous sequences and then to predict to which family the sequence belongs by testing which pHMM recognize it. This way, information from the whole family, rather than from only one sequence, can be used for more sensitive annotation. The most popular pHMM packages are HMMER (pronounced hammer) [54] and SAM [63]. The HMMER package is used in particular in the PFAM [64, 65] and TIGRFAM [66] databases gathering alignments and pHMM signatures for domains and proteins that are widely used by biologists for the annotation of new sequenced genomes. The SAM package is more directed towards the recognition of a remote homolog sharing a common structural fold: it was applied to search for protein structure templates in several structure prediction competitions CASP [67] and it is used by the SUPERFAMILY [68] library of profile hidden Markov models that represent all proteins of known 3D structure.

Thanks to the work done to require fewer and fewer examples by the incorporation of a priori knowledge on the similarity of homologous sequences, the recent trend has been to build a pHMM starting from only one proteic sequence as initiated by PSI-BLAST with PWM [69] to provide a more sensitive alternative to BLAST. Starting from a unique query sequence, the strategy is to bootstrap the search with close homologs: a pHMM is built from the query sequence and then progressively refined by searching and including iteratively the most significant sequence matches in comprehensive sequence databases such as UniProt [70] or the non-redundant (nr) database from NCBI [71]. The result of this procedure is a sensitive pHMM and the retrieved homologous sequences to the query. This strategy was used by SAM-T98 and its successor SAM-T2K for the CASP competitions [72–74]. pHMM packages implementing this strategy with fast heuristic prefilters, such as in the new HMMER3 [75], are now as fast as BLAST. The idea has been pushed one step further by HHSearch [76] and its filtered speeded-up version HHBlits [77] that preprocess the sequences from the databases to group them in sets of close homologs represented by a pHMM and perform then iterative pHMM-pHMM alignments to obtain more sensitive results for the search of remote homologs sharing the same structural fold, helped by sequence context-specific pseudocounts.

Modeling Conserved RNA Sequences Profile HMMs have been especially successful for modeling protein homologs and they are also starting to be used for modeling DNA homologs [78, 79]. However, they are not adapted so well for modeling RNA not translated into proteins. These so-called non-coding RNA (ncRNA) molecules play vital roles in many cellular processes. One of the best known examples of functional ncRNA is the family of transfer RNAs (tRNA) that is central for the synthesis of proteins. A tRNA molecule is shown in Fig. 8.4: one can see from this example that, like proteins, RNAs are single-strand molecules that fold into a three-dimensional structure ("tertiary structure") that determines the function, and, as in DNA, the complementarity between the bases (A–U and C–G) is a key determinant of RNA structure that is typically composed of short helices packed together and is often simply represented by the base pairing on the sequence ("secondary structure").

The contiguous paired bases that form the helices, named stems, predominantly occur in a nested fashion in the RNA sequences as complementary palindromic subsequences. These kinds of long-distance correlations in the sequence that are crucial for RNA structure are typically context-free and lie beyond the expressiveness of pHMMs that are restricted to position-based characterizations.

RNA and Context-Free Grammars In Fig. 8.5, an example is given of how a context-free Grammar can be designed in a straightforward way to capture the non crossing base pairing. The idea is to have a pair matching rule $S_i \rightarrow aS_{i+1}b$ for each paired base (a, b) and a base matching rule of the form $S_i \rightarrow aS_{i+1}$ or $S_i \rightarrow S_{i+1}a$ for each unpaired base a. By ordering this rule with respect to sequence order and introducing a branching rule $B_i \rightarrow S_iS_j$ to chain successive nested structures, one gets a grammar recognizing the RNA sequence with a derivation tree mirroring its secondary structure.

The secondary structure is often more conserved than the sequence of non-coding RNAs: mutations in one strand of a stem are often compensated for by a mutation in the complementary strand. These compensatory mutations restore base pairing at

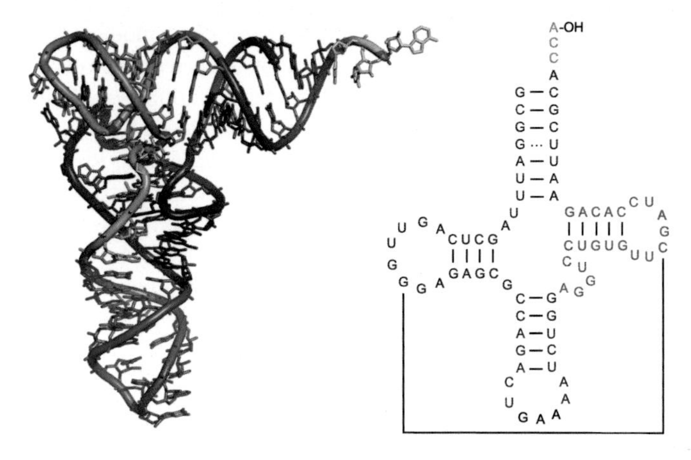

Fig. 8.4 Tertiary (*left*) and secondary (*right*) structure of yeast tRNA-Phe

Grammar $G = \langle \Sigma = \{A, C, G, U\}, N = \{S_1 \dots S_{17}, B_1, \}, S_1, P \rangle$, s.t. the production rules in P are:

$S_1 \rightarrow \mathrm{A}S_2$,	$S_3 \rightarrow S_4\mathrm{U}$,	$S_6 \rightarrow \mathrm{C}S_7\mathrm{G}$,	$S_9 \rightarrow \mathrm{U}S_{10}$,	$S_{12} \rightarrow \mathrm{G}S_{13}\mathrm{C}$,	$S_{15} \rightarrow S_{16}\mathrm{A}$,
$S_2 \rightarrow \mathrm{A}B_1$,	$S_4 \rightarrow \mathrm{G}S_5\mathrm{C}$,	$S_7 \rightarrow \mathrm{U}S_8$,	$S_{10} \rightarrow \mathrm{C}$,	$S_{13} \rightarrow \mathrm{C}S_{14}$,	$S_{16} \rightarrow \mathrm{A}S_{17}$,
$B_1 \rightarrow S_3S_{11}$,	$S_5 \rightarrow \mathrm{A}S_6\mathrm{U}$,	$S_8 \rightarrow S_9\mathrm{G}$,	$S_{11} \rightarrow \mathrm{G}S_{12}\mathrm{C}$,	$S_{14} \rightarrow \mathrm{G}S_{15}\mathrm{C}$,	$S_{17} \rightarrow \mathrm{C}$

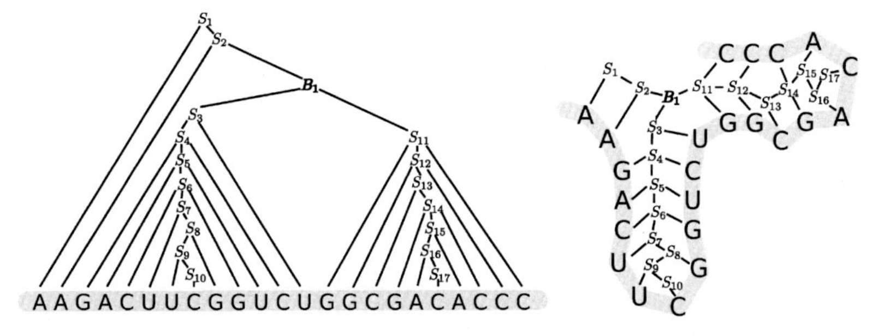

Fig. 8.5 Example of context-free grammar and derivation tree mirroring the secondary structure of an RNA sequence

a position and contribute to the conservation of the RNA secondary structure and therefore its function. Let us remark here that single mutations can also occur on non-paired bases without changing the secondary structures. The grammar above can easily be generalized to cope with these kinds of mutations, preserving the structure that the sequence can undergo. To do so, each pair matching rule can be complemented to match the other complementary pairs of bases and each single base matching rule can also be complemented to match the other bases.

For instance, in the example of Fig. 8.5, $S_4 \rightarrow \mathrm{G}S_5\mathrm{C}$ would be complemented to get a rule $S_4 \rightarrow aS_5\tilde{a}$ for each pair of bases $(a, \tilde{a})$, where $\tilde{a}$ denotes the complementary base to a, and $S_1 \rightarrow \mathrm{A}S_2$ would be complemented to get a rule $S_1 \rightarrow aS_2$ for each base a; and so on for the other matching rules.

Profile SCFG By doing so, the resulting grammar would model the secondary structure and would lose the information of the initial RNA sequence even if this can be important for homology search or functional characterization. A trade-off between sequence and secondary structure conservation can be achieved by weighting differently each base or pair of bases matched by each rule according to its probability of occurring at the position. At this point, the obtained grammar could be seen as a stochastic context-free counterpart of the (regular) PWMs seen above, allowing us to match a base a at one position i with weight $w_{i,a}$ as with a PWM by a base matching rule $S_i \rightarrow aS_{i+1}/w_{i,a}$, but allowing us also to match paired bases $(a, \tilde{a})$ at paired positions (i, j) with a weight $w_{i,(a,\tilde{a})}$ by a pair matching rule $S_i \rightarrow aS_{i+1}\tilde{a}/w_{i,(a,\tilde{a})}$. To obtain the context-free counterpart of pHMM, named *profile stochastic context-free grammars* (pSCFG) [81] or *covariance models* (CM) [82], each matching rule S_i is completed with position-based deletion rules (of the form $S_i \rightarrow S_{i+1}/w_i^{del}$) and insertion rules (of the form $I_i \rightarrow aI_i/w_{i,a}^{ins}$ or $I_i \rightarrow I_ia/w_{i,a}^{ins}$). For positions matching one base, this is done as for pHMM. For positions matching paired bases, deletion and insertion rules are added in a similar way but taking care to enable insertion or deletion on each side (left or right) of the nested sequence, which requires the equivalent of six states instead of three by position.

As with pHMMs, pSCFG's parameters can be trained by likelihood maximization approaches from a set of aligned sequences, but this requires additionally an RNA consensus (nested) secondary structure indicating the paired bases and the unpaired bases to set up the topology. This secondary structure can be known for one of the aligned sequences, be predicted by free energy minimization on a sequence or be the inferred common secondary structure from a set of multiple, homologous sequences. In Fig. 8.6, an example of three aligned RNA sequences with such a secondary structure is given with nested '>' and '<' indicating the paired positions, 'x' the unpaired positions and '.' the insertions with respect to the structure. From this information, one can automatically only keep the matching positions sufficiently shared among the sequence to get the paired ('>', '<') and unpaired ('x') matching positions of the pSCFG corresponding to the template secondary structure displayed on the left of Fig. 8.6. Each matching position is systematically completed with its companion insertion/deletion rules to get the complete pSCFG topology and parameters can

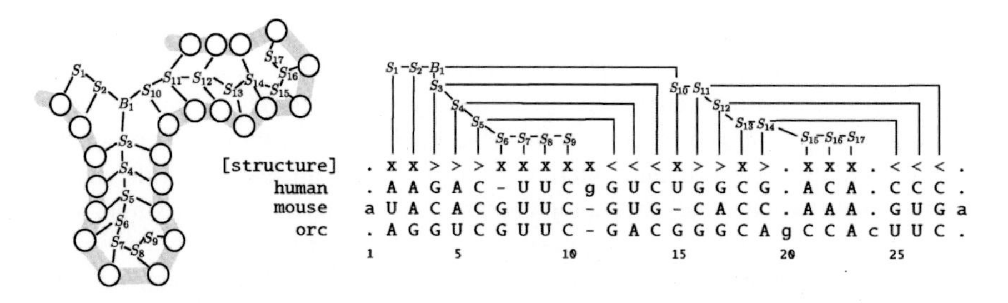

Fig. 8.6 Setting pSCFG's topology from multiple sequence alignment annotated by a secondary structure (example adapted from [80])

then be trained to maximize the likelihood of the alignment, eventually completed by pseudocounts.

Usage By using a context-free representation, PCFGs and CM extend pHMMs nicely to handle not only the base distribution at each position but also the pairs of base distribution at each (nested) paired position, capturing this way an important structural feature of ncRNA sequences that make it suitable to retrieve successfully RNA homologs. The Rfam database [83] that is an authoritative collection of non-coding RNA families represents each family by a multiple sequence alignment, predicted secondary structure and CM, and is powered by Infernal [84], the kinship software package to HMMER dedicated to modeling RNA with CM.

To get finer results on the characterization of ncRNA, one would need to be able to represent also cross-correlations such as pseudoknots (typical RNA structures with two stems in which half of one stem is intercalated between the two halves of another stem), which with all the computational hardness that they involve, is beyond the generative power of context-free grammars. Even if some proposals have been made to represent this kind of struture by grammatical models [85–88], learning such models will be extremely difficult. Finding good representations with practical computational time for learning that kind of correlation on genomic sequences is still an open and challenging research area.

Towards Sentences So far, we have seen approaches modeling homologous proteins or RNA genes in their maximal alignable length. To find more distant homologs or to focus on functionally important parts of the sequences, other approaches prefer to target the identification and the characterization of the most conserved parts shared by a set of sequences.

For instance, Meta-MEME [89] is based on an iterative search by MEME of a set of significant local alignments on a set of DNA or proteic sequences [32] that are used to build a simplified profile HMM where all the delete states are removed and only the insertion states between each block modeling a local alignment found are kept (see Fig. 8.3).

Pratt [90] searches for even more strict conservation: instead of local alignments on all the sequences, it searches by enumeration for interspaced strictly conserved amino acid or nucleotide symbols occurring in a sufficiently large subset of the sequences and then refines heuristically these patterns with new matching components offering a choice between sets of symbols. The patterns potentially returned by Pratt are composed of a suite of symbols or choice of symbols separated by wildcards indicating an insertion of a stretch of symbols bounded by a minimal and a maximal length. To remain feasible, the search has to be constrained by many user-defined parameters limiting the size of the pattern and the number of insertions, the program returning then the best patterns in this search space with respect to an information base or a minimum description length score.

An example of a well-known pattern is the C2H2 signature of 'zinc finger' in proteins: C-x(2,4)-C-x(3)-[LIVMFYWC]-x(8)-H-x(3,5)-H, read as a C followed by two to four amino acids, then a C followed by three amino acids and then one of the amino acid chosen in [LIVMFYWC] followed by eight amino acids, an H, three to five amino acids and finally an H. These patterns are among the most expressive patterns used in Bioinformatics and can be seen as the deterministic counterpart of the Meta-MEME models, with blocks arising from exact conservation rather than from local similarity. They are known as Prosite's patterns from the name of the database [91] that popularized them as exact signatures of many domains, families and functional sites on proteins. While the patterns in Prosite were initially mostly built semi-automatically from multiple sequence alignments, Pratt is now the default pattern discovery software proposed to users on Prosite's website to find patterns without the need for a sequence alignment.

These later methods enable one to discover shorter functional or structural conserved units than genes or domains—the highly conserved blocks of Meta-MEME in all sequences or the adjacent groups of conserved positions identified by Pratt in a sequence subset—introducing each unit as a new potential genomic word or the succession of these units as a more complex, interspaced in the sequence (but eventually close in space), word.

8.3 Learning Syntax

So far, we have seen the state-of-the-art methods actually used in practice by biologists to discover and model (conserved) words in genomic sequences. The achievements in Bioinformatics for expressive characterizations are strongly linked with multiple sequence alignments, resulting in position-specific signatures that represent a suite of independent, uncorrelated conserved positions (or pairs of positions for RNA), eventually augmented with the ability to insert symbols between these positions or to skip some of them. Learning is then based on (1) the choice by the expert of the most adequate simple topology, (2) the identification and alignment of the conserved positions among the sequences and, for stochastic models, (3) training the parameters to maximize the likelihood of the sample with respect to priors.

In this section, we are interested in overtaking the position-specific characterization of (conserved) words. In particular, we would like to learn models with dependencies between the symbols of the sequences. In other words, this would allow us to make progress towards the goal of learning not only the words but also the syntax (the grammar) of genomic sequences. The difficulty is that, with dependencies being unknown, one cannot then cannot anymore rely on predefined topologies such as the pHMMs or pSCFGs: the structure of the grammar has to be learnt from the sample, which constitutes a complete Grammatical Inference task and a challenging application for that field.

Learning k-Testable Languages A first step towards learning grammars on genomic sequences is the early work of Yokomori et al. [92, 93] on learning automata representing locally k-testable languages applied to the identification of hemoglobin α-chains. The class of locally k-testable languages, very similar to the class of k-testable languages in the strict sense [94, 95], is linked to *n-grams* and, more

Fig. 8.7 Dayhoff's and binary amino acid encodings used in [92, 93, 96, 97]

Dayhoff's coding

AminoAcids	Properties	Symbol
C	Sulfurpoly merization	a
G, S, T, A, P	Small	b
D, E, N, Q	Acid and amide	c
R, H, K	Basic	d
L, V, M, I	Hydrophobic	e
Y, F, W	Aromatic	f

Binary coding

Amino Acids	Hydrophoby index	Symbol
A, C, F, G, I, L, M, N, S, T, V, W, Y	High	0
D, E, H, K, P, Q, R	Low	1

biologically, to (persistent) splicing systems. Languages of this class have the property that it is sufficient to parse the substrings of length k to decide whether a sequence is accepted or not; dependencies are therefore limited to the length k but cover all the length of the sequences in contrast to motifs. Given k, learning such a language can be done by a simple efficient algorithm building an automaton memorizing the subwords of length k appearing in the positive sample and the corresponding one-letter admissible transitions between them. This algorithm ensures identification in the limit of k-testable languages when k is known. In practice, however, the value of k is estimated by cross-validation and is usually small, the inference being then less subject to over-specialization. To apply this simple inference algorithm to proteins, Yokomori et al. reduce the 20 letter alphabet to a six letter alphabet, clustering amino acids according to main substitutability classes following Dayhoff's coding method, or drastically to a binary alphabet according to hydropathy (see Fig. 8.7). Recoding the sequences with these reduced alphabets help greatly the generalization and enables us to bootstrap the inference by some biological knowledge on amino acids similarities.

This first work is the root of recent studies applying similar approaches to learn grammatical models for the prediction of coiled-coil proteins [96] and transmembrane regions in proteins [97], whose performances are close to those of dedicated tools built with human expertise. In these works, the application scope of learning a k-testable language is extended from a sequence classification to a sequence labeling task through preliminary sequence recoding and automata to transducer post transformation. Sequence recoding is done first by reducing the alphabet according to Dayhoff's code as in [92, 93] but the alphabet is hereafter augmented by combining letters of the reduced alphabet with their label in the labeled sequences forming the training sample for the task. For instance, using an example from [97], one protein sequence of the training set,

```
M R V T A P R T L L L L L W G A V A L T E T W A G S H S M R,
```

would be encoded first following Dayhoff's coding into

```
e d e b b b d b e e e e e f b b e b e b c b f b b b d b e d
```

and, from its known transmembrane topology, this sequence could be labeled as follows (see [97] for alternative labeling):

```
e d e b b b d b e e e e e f b b e b e b c b f b b b d b e d
O O O C M M M M M M D I I I I I I I I A M M M M M B O O O O
```

where M labels residues in transmembrane regions, I labels residues in the cell while O labels residues out of the cell and A, B, C, D label the shift from outer/inner regions to/from transmembrane regions. Then, in the augmented alphabet, composed of a symbol xL for each letter x labeled by L, the sequence encoding the labeled example would begin by the following symbols (separated by white spaces):

```
eO dO eO bC bM bM dM bM eM eM eD eI eI fI bI bI eI bI eI bA cM...
```

By encoding the sequences from the positive sample this way, one can learn a k-testable language by a classical algorithm, such as k-TSSI, designed to learn k-testable languages in the strict sense [94, 95], with the advantage that, as in the morphic generator methodology [98], identical letters can be distinguished by their label during the inference. By transforming each transition labeled by a symbol xL from the learned automaton into a transition by letter x and output label L, one gets back a labeling transducer that can then be weighted and used for the task, eventually with the help of error correcting parsing techniques to compensate for the lack of data. These studies show that grammatical inference techniques can be applied with encouraging results to genomic sequences, even with such a limited class of languages when helped by pertinent pre- and post-processing techniques. We will now focus on learning more expressive grammatical representations of languages, and thus more complex dependencies, on these kinds of sequences.

Learning Automata At the first level of Chomsky's hierarchy (regular languages), we have investigated in our team the inference of full automata to model functional or structural families of protein directly from their complete sequence. RPNI [99, 100], EDSM and Blue-Fringe [101] (see Chap. 4, *On the Inference of Finite State Automata from Positive and Negative Data*, López and García) having been shown to be successful in practice on artificial data, testing these methods on this task was appealing. Our preliminary attempts showed also that these methods, even improved by taking into account similarities of amino acids, were performing very badly on leave-one-out experiments. Our analysis of these results yields that protein sequences whose length is about 300 symbols on average on a 20-letter alphabet, and whose functional parts are not necessarily at the beginning or the end of the sequences, are not well suited for these algorithms relying mainly on common sequence heads and tails for the inference. To avoid these pitfalls, we have proposed shifting from a deterministic to non-deterministic automata and adapt consequently the idea of evidence introduced by EDSM to merge common (similar) substrings rather than common tails, obtaining a first successful application of the classical state merging grammatical inference framework to learn automata on protein sequences: Protomata-Learner [102–107].

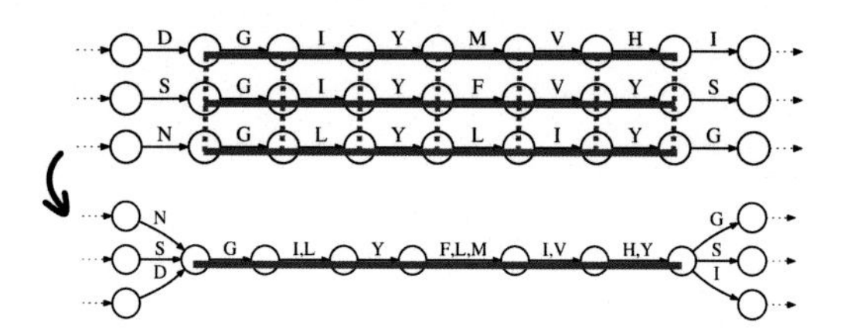

Fig. 8.8 Merging similar substrings

Shifting from a deterministic to a non-deterministic setting in the state merging approach requires simply starting from the Minimal Canonical Automaton (MCA) of the sample set (the non-deterministic automaton that is the union of the canonical automata built on each sample) rather than the Prefix Tree Acceptor (PTA) and proceeding by merging some of its states (without merging for determinisation) [108] or, inspired by EDSM, by merging successively the states on paths labeled by common substrings.

Similar Substring Merging Approach To deal with amino acid similarities, the heuristic has been generalized to look at common *similar* substrings, on the basis of the significantly similar pairs of substrings (named diagonals) precomputed by Dialign to serve as multiple sequence alignment building blocks [53]. A Dialign's diagonal d is a pair of equal-length substrings (d_1, d_2), implicitly aligned from left-to-right and whose similarity $s(d)$ is computed by summing the substitution score (given by a substitution matrix) of the aligned amino acids. Dialign computes also for each diagonal the weight of its similarity $w(d)$ as the negative logarithm of the similarity's p-value, namely of the probability of finding a diagonal of the same length with a greater or equal similarity in random amino acid sequences. The weight measuring how exceptional the similarity of the diagonal is relative to its length enables us to compare diagonal of different lengths and to define similar diagonals: random diagonals ought to have a weight of 0; similar diagonals are thus those whose weight is greater than 0, or greater than a positive weight threshold parameter t if one wants more significant similarity before considering the substrings in the diagonal.

The task is then to distinguish the similar diagonals that are characteristic of the family from those that are similar by chance or for another unrelated reason. This is done in Protomata-Learner by a best-first greedy approach: at each iteration, the best similar substrings are selected by one heuristic (maximizing their support in the training set and also their similarity) and states aligned by these substrings are merged (see Fig. 8.8), discarding from the future choices the remaining similar substrings that are incompatible with the selected ones.

Incompatible daigonals are those with an overlap presenting conflicting alignments inside the diagonals and forcing us to choose at most one of them[1] (see

[1]This corresponds to the preservation constraint from [104] forbidding us to merge together the states resulting from merging a diagonal to prevent identified conserved words from being damaged.

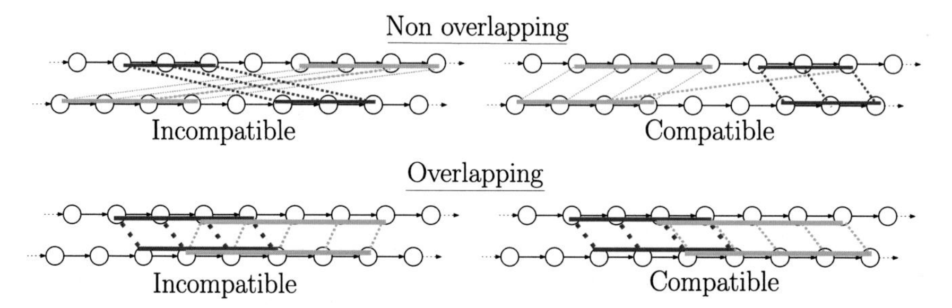

Fig. 8.9 Incompatible and compatible diagonals

Fig. 8.9 top). Another kind of incompatible diagonal can be introduced to help the inference when it is assumed that the protein sequence family does not undergo shuffling mutations (that are unlikely to occur without structure and function change): in that case, the order of the similar substrings in the sequences is preserved and crossing diagonals are incompatible (see Fig. 8.9 bottom). This greedy similar substring merging algorithm halts when no more compatible similar substrings are available for merging, relying so on incompatibilities and on the chosen threshold t to stop the inference. No negative sample is required, the characterization being directed towards maximizing the global unexpected similarity of substrings with respect to random sequences and adopting in this way a Minimum Description Length perspective rather than the discriminative Occam's razor inspiration of RPNI or EDSM.

A New Kind of Alignment The similar substring merging approach of Protomata-Learner under such incompatibility constraints can be linked to the classical Bioinformatics field by considering the sets of similar substrings merged as a new kind of multiple sequence alignment, named *partial local multiple alignment* (PLMA), exhibiting conserved regions that can be local, involving only a contiguous subset of the amino acids in the sequences as defined for classical local alignments, but also partial, involving contiguous amino acids from only a subset of the sequences instead of all the sequences. This later property enables us to represent unrelated conserved regions among subsets of the sequences: instead of being limited to the identification of conserved positions in all the sequences, one can identify alternative conserved words in some sequences, not necessarily aligned, and their chaining, paving thus the way to modeling syntax in addition to conserved words. For the inference of automata, the aligned substrings from conserved regions of the PLMA are merged, weighting eventually amino acid transitions thanks to efficient PWM or pHMM weighting schemes, and insertion states are added to link consecutive conservation regions (see Fig. 8.10), enabling learning topologies that can be seen as a generalization of pHMM or Meta-MEME architectures overtaking these position-specific characterizations by enabling us to model alternative paths (see Fig. 8.3).

Learning Context-Free Grammars Even if automata enable us to take a important step toward more expressive models, they are limited to successive short-term dependencies while it is well known that, from protein folding, residues that are far in the

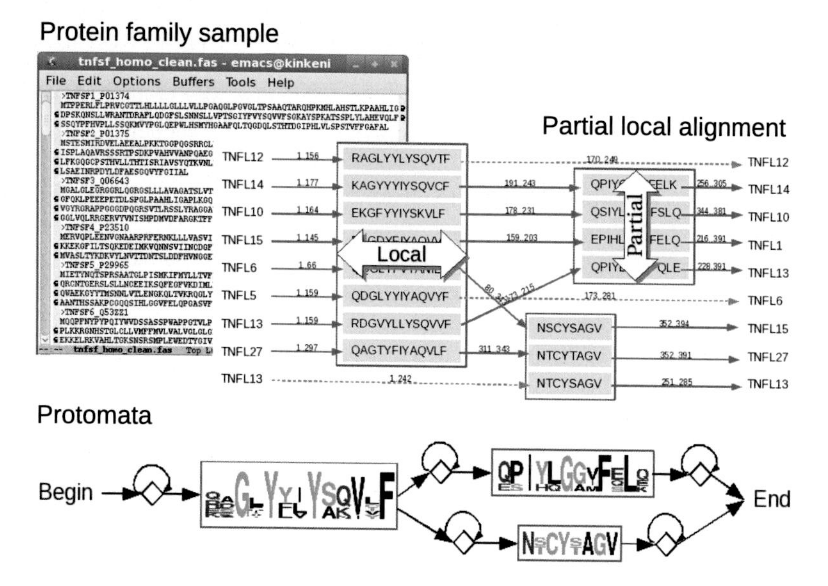

Fig. 8.10 Learning automata by partial local alignment from set of protein sequences

sequence may be close in space and interact together or are simply correlated. To represent this kind of long-distance interaction, one needs to learn more expressive grammatical representations.

From a General Template Grammar A first attempt towards this goal is the framework introduced in [109] based on a genetic algorithm training the weights of a complete stochastic context-free grammars in Chomsky's normal form to maximize the likelihood of the training sample. A complete grammar is such that the rule $A \rightarrow BC$ exists for each non-terminal A, B, C: the number of rules grows thus extremely fast with respect to the chosen number of non-terminals. The framework aims at limiting the number of non-terminals by proposing biasing the topology of the grammar towards nested dependencies and more drastically by an original way of coping with the size of the amino acid alphabet and introducing knowledge on their physico-chemical properties: all the amino acids are generated from only three non-terminals, corresponding to three discretized levels (low, medium or high level) of a chosen property of interest (for instance the van der Waals volume), the probability of generating the amino acid being fixed with respect to these levels (and thus not subject to training). Then a grammar considers amino acids only with respect to one property and if more than one property is of interest, one needs to train several grammars and to combine parsing scores for membership predictions. Experiments restricted to binding site regions of protein sequences and nine non-terminals show a good recognition accuracy on this task and pertinent parse trees illustrating the interest of this kind of context-free model.

By Local Substitutability We have recently proposed a different approach [110] showing the versatility and the efficiency of distributional learning of context-free languages (see Chap. 6, *Distributional Learning of Context-Free and Multiple Context-Free Grammars*, Clark and Yoshinaka) by applying it to protein sequences. PLMAs are used once again, but here as a pre-processing step to deal with amino acid similarity: by using parameters that allows to identify all short highly conserved regions under overlapping and crossing incompatibilities, the sequences are recoded according to these conservation blocks and provided as input for the actual generalization step performed by a grammatical inference algorithm. To be able to parse non-encoded protein sequences, a post-processing of the inferred grammar is performed to replace each terminal corresponding to a conserved region by a new non-terminal generating amino acid from the region (by introducing a succession of new non-terminals for each set of aligned amino acid from the region, in charge of generating indifferently any amino acid from the set) and introduce new non-terminals in charge of generating any amino acid for non-conserved regions. Used this way, PLMAs detect and align similar amino acids but entails almost no generalization when no grammatical inference algorithm is used, as testified by leave-one-out experiments. More surprisingly, when we tried state-of-the-art grammatical inference algorithms learning substitutable [111, 112] or k, l-substitutable [113] context-free languages, based on a formalization of substitutability idea introduced in linguistics by Zellig Harris in the 1950s [114], no additional generalization was performed.

Learning such languages is based on the identification of substrings appearing in a common context, to generalize the language by allowing these substrings to be substituted for each other (a contextual constraint for substitutability being added for k, l-substitutable language): i.e. if xyz and $xy'z$ are both in the training set, then any occurrence of y (or a subset of them for k, l-substitutable language) can be substituted by y', and vice versa, in the language. The problem in the preliminary experiments on protein sequences is that this criterion was never met in the training samples. As a matter of fact, if the sequences are long, observing a double occurrence of the common context (x, z) and a double occurrence of y, given that at least one of these substrings has to be long, has low likelihood in practice. Moreover, these characterizations rely on conserved heads and tails that, as already stated for the inference of automata, are not necessarily informative and conserved in protein sequences.

In [110], we proposed thus a variant of the substitutability generalization criterion that considers *local* rather than global context to define the substitutable substrings: local substitutability criterion states that it is sufficient to have both $xuyvz$ and $x'uy'vz'$ for a common local context (u, v) of sufficient length in the training set to allow us to substitute any occurrence of y (or a subset of them for k, l-substitutable languages) by y'. At the price of adding two additional parameters on the required left and right lengths of the common local context enabling us to define substitutability of the substrings (or only one parameter when right and left contexts are considered symmetrically), one has been able to get a real and pertinent generalization. Thanks to the development and the implementation of a faster algorithm for learning local substitutable context-free grammars, named ReGLiS, combined with the encoded pre- and post-processing scheme, these results have been confirmed on the complete set of protein families used for the testing in [109]: using the entire protein sequences rather than only the short binding site substrings, our leave-one-out experiments show a good recall and a perfect precision [115]. These preliminary results, obtained without any weights on the rules, are really encouraging and should be easily improved. They already show, with other works presented in this section, that the application of grammatical inference can be successful for non-trivial syntactic characterizations of protein families. More generally, learning syntax on genomic sequences is a very nice open playground for grammatical inference, enabling us to apply ideas or techniques from the field but being also a source of inspiration for novel practical and theoretical challenging developments.

8.4 Conclusion

We have presented here the first successful steps towards learning the language of biological sequences. So far, the state of the art is mainly at the word level: the discovery of exceptional words, the alignment of conserved words and their modeling by the parametrization of simple adequate topologies based on biological priors. Some recent advances have also been made on learning non-trivial grammar topologies for proteins but we are only at the beginning of this exciting challenge addressed by Grammatical Inference.

To draw the lines of future research in that field, one can guess that the focus on learning topologies with (long-distance) correlations will continue. In proteins and RNA, it would allow us to capture correlations between positions that are far in the sequences but close in the 3D space. In DNA, the problem seems more complicated since the challenge is then to deal with palindromes and copies, requiring us to use and learn more expressive grammars. For DNA, recent advances have thus rather been on a simpler task: discovering the hierarchical structure of DNA as an instance of the smallest grammar problem, along the lines initiated by Sequitur [116] and its successors [117–123]. These studies have not been presented in this chapter since it is still difficult to assert and compare their biological pertinence, but these approaches based on repeats may help us to better understand what are the important words and where are their occurrences in DNA and to decipher its word structure as a preliminary step to learning grammars. Moreover, the repeats used in these approaches are not that far from the variables used in the current state-of-the-art DNA parsers of the first section. This is an interesting convergence when the goal is to design automatically, or help the expert to design, the grammars for these parsers.

We have proposed in this chapter an overview from a Grammatical Inference point of view of the achievements and open challenges in this research field as well as some keys to enter it. To further investigate this area, we propose a short list of additional reading recommendations.

Further Readings First, Wikipedia (http://www.wikipedia.org/) covers fairly well the related concepts in biology or bioinformatics and these pages are usually well written. Good entry points to Pattern Discovery are [22, 124] while [23] offers a comprehensive algorithmic and theoretical treatment of the subject. For probabilistic models on sequences, an excellent review with a grammatical inference point of view is [125], while the reference books [126, 127] contain non-grammatical machine learning techniques. Finally, on Grammatical Inference, the other chapters of this book should be helpful, as well as the reference book [128].

References

1. Beadle, G.W., Beadle, M.: The language of life: an introduction to the science of genetics. American Institute of Biological Sciences (1966)
2. Clancy, S., Brown, W.: Translation: DNA to mRNA to protein. Nature Education (2008)
3. Chomsky, N.: Syntactic Structures. Mouton (1957)
4. Searls, D.B.: The computational linguistics of biological sequences. In Hunter, L., ed.: Artificial Intelligence and Molecular Biology. AAAI Press (1993) 47–120
5. Searls, D.B.: Linguistic approaches to biological sequences. Computer Applications in the Biosciences **13** (1997) 333–344
6. Searls, D.B.: The language of genes. Nature **420** (2002) 211–217
7. Chiang, D., Joshi, A.K., Searls, D.B.: Grammatical representations of macromolecular structure. Journal of Computational Biology **13** (2006) 1077–1100
8. Searls, D.B.: A primer in macromolecular linguistics. Biopolymers **99** (2013) 203–17
9. Joshi, A.K., Weir, D.J., Vijay-Shanker, K.: The convergence of mildly context-sensitive grammar formalisms. Technical Report MS-CIS-90-01, University of Pennsylvania (1990)
10. Dong, S., Searls, D.B.: Gene structure prediction by linguistic methods. Genomics **23** (1994) 540–551
11. Nicolas, F., Rivals, E.: Hardness results for the center and median string problems under the weighted and unweighted edit distances. J. Discrete Algorithms **3** (2005) 390–415
12. Dsouza, M., Larsen, N., Overbeek, R.: Searching for patterns in genomic data. Trends in Genetics **13** (1997) 497–498
13. Pesole, G., Liuni, S., D'Souza, M.: Patsearch: a pattern matcher software that finds functional elements in nucleotide and protein sequences and assesses their statistical significance. Bioinformatics **16** (2000) 439–450
14. Belleannée, C., Sallou, O., Nicolas, J.: Logol: Expressive Pattern Matching in Sequences. Application to Ribosomal Frameshift Modeling. In Comin, M., Kall, L., Marchiori, E., Ngom, A., Rajapakse, J., eds.: PRIB2014 - Pattern Recognition in Bioinformatics, 9th IAPR International Conference. Volume 8626 of Lecture Notes in Computer Science, Stockholm, Springer (2014) 34–47
15. Macke, T.J., Ecker, D.J., Gutell, R.R., Gautheret, D., Case, D.A., Sampath, R.: Rnamotif, an RNA secondary structure definition and search algorithm. Nucleic acids research **29** (2001) 4724–4735
16. Eddy, S.: RNABOB: a program to search for RNA secondary structure motifs in sequence databases (1996)
17. Graf, S., Strothmann, D., Kurtz, S., Steger, G.: Hypalib: a database of RNAs and RNA structural elements defined by hybrid patterns. Nucleic Acids Res. **29** (2001) 196–198
18. Strothmann, D., Gräf, S.A., Kurtz, S., Steger, G.: The syntax and semantics of a language for describing complex patterns in biological sequences. Technical report, Universität Bielefeld, Technische Fakultät, Arbeitsgruppe Praktische Informatik (2000)

19. Billoud, B., Kontic, M., Viari, A.: Palingol: a declarative programming language to describe nucleic acids' secondary structures and to scan sequence database. Nucleic Acids Res **24** (1996) 395–403
20. Meyer, F., Kurtz, S., Backofen, R., Will, S., Beckstette, M.: Structator: fast index-based search for RNA sequence-structure patterns. BMC Bioinformatics **12** (2011) 214
21. Pribnow, D.: Nucleotide sequence of an RNA polymerase binding site at an early t7 promoter. Proceedings of the National Academy of Sciences of the United States of America **72** (1975) 784–8
22. van Helden, J.: The Analysis of Regulatory Sequences. In: Multiple Aspects of DNA and RNA: from Biophysics to Bioinformatics: Lecture Notes of the Les Houches Summer School 2004. Gulf Professional Publishing (2005)
23. Parida, L.: Pattern Discovery in Bioinformatics: Theory & Algorithms. Chapman & Hall/CRC (2007)
24. Stormo, G.D., Schneider, T.D., Gold, L., Ehrenfeucht, A.: Use of the "perceptron" algorithm to distinguish translational initiation sites in E. coli. Nucleic Acids Res **10** (1982) 2997–3011
25. Schneider, T.D., Stormo, G.D., Gold, L., Ehrenfeucht, A.: Information content of binding sites on nucleotide sequences. Journal of molecular biology **188** (1986) 415–31
26. Schneider, T.: Information theory primer (1995)
27. Crooks, G.E., Hon, G., Chandonia, J.M., Brenner, S.E.: Weblogo: a sequence logo generator. Genome Res **14** (2004) 1188–1190
28. Kullback, S., Leibler, R.A.: On information and sufficiency. Ann. Math. Statistics **22** (1951) 79–86
29. Hertz, G.Z., Hartzell, 3rd, G., Stormo, G.D.: Identification of consensus patterns in unaligned DNA sequences known to be functionally related. Comput Appl Biosci **6** (1990) 81–92
30. Hertz, G.Z., Stormo, G.D.: Identifying DNA and protein patterns with statistically significant alignments of multiple sequences. Bioinformatics **15** (1999) 563–577
31. Stormo, G.D., Hartzell, 3rd, G.: Identifying protein-binding sites from unaligned DNA fragments. Proc Natl Acad Sci U S A **86** (1989) 1183–1187
32. Bailey, T.L., Elkan, C.: Fitting a mixture model by expectation maximization to discover motifs in biopolymers. Proc Int Conf Intell Syst Mol Biol **2** (1994) 28–36
33. Lawrence, C.E., Altschul, S.F., Boguski, M.S., Liu, J.S., Neuwald, A.F., Wootton, J.C.: Detecting subtle sequence signals: a Gibbs sampling strategy for multiple alignment. Science **262** (1993) 208–214
34. Neuwald, A.F., Liu, J.S., Lawrence, C.E.: Gibbs motif sampling: detection of bacterial outer membrane protein repeats. Protein Sci **4** (1995) 1618–1632
35. Neuwald, A.F., Liu, J.S., Lipman, D.J., Lawrence, C.E.: Extracting protein alignment models from the sequence database. Nucleic Acids Res **25** (1997) 1665–1677
36. Roth, F.P., Hughes, J.D., Estep, P.W., Church, G.M.: Finding DNA regulatory motifs within unaligned noncoding sequences clustered by whole-genome mRNA quantitation. Nat Biotechnol **16** (1998) 939–945
37. Thijs, G., Lescot, M., Marchal, K., Rombauts, S., De Moor, B., Rouzé, P., Moreau, Y.: A higher-order background model improves the detection of promoter regulatory elements by Gibbs sampling. Bioinformatics **17** (2001) 1113–1122
38. Liu, X., Brutlag, D.L., Liu, J.S.: Bioprospector: discovering conserved DNA motifs in upstream regulatory regions of co-expressed genes. Pac Symp Biocomput (2001) 127–138
39. Matys, V., Kel-Margoulis, O.V., Fricke, E., Liebich, I., Land, S., Barre-Dirrie, A., Reuter, I., Chekmenev, D., Krull, M., Hornischer, K., Voss, N., Stegmaier, P., Lewicki-Potapov, B., Saxel, H., Kel, A.E., Wingender, E.: TRANSFAC and its module TRANSCompel: transcriptional gene regulation in eukaryotes. Nucleic Acids Research **34** (2006) D108–D110
40. Sandelin, A., Alkema, W., Engström, P., Wasserman, W.W., Lenhard, B.: Jaspar: an open-access database for eukaryotic transcription factor binding profiles. Nucleic Acids Research **32** (2004) D91–D94
41. Taylor, W.R.: The classification of amino acid conservation. J Theor Biol **119** (1986) 205–218

42. Eddy, S.R.: Where did the BLOSUM62 alignment score matrix come from? Nat Biotechnol **22** (2004) 1035–1036
43. Needleman, S.B., Wunsch, C.D.: A general method applicable to the search for similarities in the amino acid sequence of two proteins. Journal of Molecular Biology **48** (1970) 443–453
44. Smith, T., Waterman, M.: Identification of common molecular subsequences. Journal of Molecular Biology **147** (1981) 195–197
45. Pearson, W.R., Lipman, D.J.: Improved tools for biological sequence comparison. Proc Natl Acad Sci U S A **85** (1988) 2444–2448
46. Altschul, S.F., Gish, W., Miller, W., Myers, E.W., Lipman, D.J.: A basic local alignment search tool. J. Mol. Biol. **215** (1990) 403–410
47. Thompson, J.D., Higgins, D.G., Gibson, T.J.: Clustal w: improving the sensitivity of progressive multiple sequence alignment through sequence weighting, position-specific gap penalties and weight matrix choice. Nucleic Acids Res **22** (1994) 4673–4680
48. Notredame, C., Higgins, D.G., Heringa, J.: T-coffee: A novel method for fast and accurate multiple sequence alignment. J Mol Biol **302** (2000) 205–217
49. Do, C.B., Mahabhashyam, M.S.P., Brudno, M., Batzoglou, S.: Probcons: Probabilistic consistency-based multiple sequence alignment. Genome Res **15** (2005) 330–340
50. Edgar, R.C.: Muscle: multiple sequence alignment with high accuracy and high throughput. Nucleic Acids Res **32** (2004) 1792–1797
51. Katoh, K., Misawa, K., Kuma, K.i., Miyata, T.: MAFFT: a novel method for rapid multiple sequence alignment based on fast Fourier transform. Nucleic Acids Res **30** (2002) 3059–3066
52. Morgenstern, B., Frech, K., Dress, A., Werner, T.: Dialign: finding local similarities by multiple sequence alignment. Bioinformatics **14** (1998) 290–294
53. Morgenstern, B.: Dialign 2: improvement of the segment-to-segment approach to multiple sequence alignment. Bioinformatics **15** (1999) 211–218
54. Eddy, S.R.: Profile hidden markov models. Bioinformatics **14** (1998) 755–763
55. Gribskov, M., McLachlan, A.D., Eisenberg, D.: Profile analysis: detection of distantly related proteins. Proceedings of the National Academy of Sciences of the United States of America **84** (1987) 4355–8
56. Krogh, A., Brown, M., Mian, I.S., Sjölander, K., Haussler, D.: Hidden Markov models in computational biology. applications to protein modeling. Journal of molecular biology **235** (1994) 1501–31
57. Baldi, P., Chauvin, Y., Hunkapiller, T., McClure, M.A.: Hidden Markov models of biological primary sequence information. Proceedings of the National Academy of Sciences of the United States of America **91** (1994) 1059–63
58. Rabiner, L.R.: A tutorial on hidden Markov models and selected applications in speech recognition. In: Proceedings of the IEEE. (1989) 257–286
59. Henikoff, J.G., Henikoff, S.: Using substitution probabilities to improve position-specific scoring matrices. Computer applications in the biosciences : CABIOS **12** (1996) 135–43
60. Claverie, J.M.: Some useful statistical properties of position-weight matrices. Comput Chem **18** (1994) 287–294
61. Sjölander, K., Karplus, K., Brown, M., Hughey, R., Krogh, A., Mian, I., Haussler, D.: Dirichlet mixtures: a method for improved detection of weak but significant protein sequence homology. Computer applications in the biosciences : CABIOS **12** (1996) 327–345
62. Brown, M., Hughey, R., Krogh, A., Mian, I.S., Sjölander, K., Haussler, D.: Using Dirichlet mixture priors to derive hidden Markov models for protein families. In Hunter, L., Searls, D.B., Shavlik, J.W., eds.: Proceedings of the 1st International Conference on Intelligent Systems for Molecular Biology, Bethesda, MD, USA, July 1993, AAAI (1993) 47–55
63. Hughey, R., Krogh, A.: Hidden Markov models for sequence analysis: extension and analysis of the basic method. Comput Appl Biosci **12** (1996) 95–107
64. Sonnhammer, E.L., Eddy, S.R., Durbin, R.: Pfam: a comprehensive database of protein domain families based on seed alignments. Proteins **28** (1997) 405–420
65. Finn, R.D., Bateman, A., Clements, J., Coggill, P., Eberhardt, R.Y., Eddy, S.R., Heger, A., Hetherington, K., Holm, L., Mistry, J., Sonnhammer, E.L.L., Tate, J., Punta, M.: Pfam: the protein families database. Nucleic Acids Res (2013)

66. Haft, D.H., Selengut, J.D., Richter, R.A., Harkins, D., Basu, M.K., Beck, E.: TIGRFAMS and genome properties in 2013. Nucleic Acids Res **41** (2013) D387–D395
67. Moult, J.: A decade of CASP: progress, bottlenecks and prognosis in protein structure prediction. Curr Opin Struct Biol **15** (2005) 285–289
68. Gough, J., Karplus, K., Hughey, R., Chothia, C.: Assignment of homology to genome sequences using a library of hidden Markov models that represent all proteins of known structure. J Mol Biol **313** (2001) 903–919
69. Altschul, S.F., Madden, T.L., Schäffer, A.A., Zhang, J., Zhang, Z., Miller, W., Lipman, D.J.: Gapped BLAST and PSI-BLAST: a new generation of protein database search programs. Nucleic Acids Res **25** (1997) 3389–3402
70. UniProt: Update on activities at the universal protein resource (UniProt) in 2013. Nucleic Acids Res **41** (2013) D43–D47
71. Pruitt, K.D., Tatusova, T., Maglott, D.R.: Ncbi reference sequence (refseq): a curated non-redundant sequence database of genomes, transcripts and proteins. Nucleic Acids Res **33** (2005) D501–D504
72. Karplus, K.: Hidden Markov models for detecting remote protein homologies. Bioinformatics **14** (1998) 846–865
73. Karplus, K., Karchin, R., Barrett, C., Tu, S., Cline, M., Diekhans, M., Grate, L., Casper, J., Hughey, R.: What is the value added by human intervention in protein structure prediction? Proteins **Suppl 5** (2001) 86–91
74. Karplus, K., Karchin, R., Draper, J., Casper, J., Mandel-Gutfreund, Y., Diekhans, M., Hughey, R.: Combining local-structure, fold-recognition, and new fold methods for protein structure prediction. Proteins **53 Suppl 6** (2003) 491–496
75. Eddy, S.R.: Accelerated profile HMM searches. PLoS Comput Biol **7** (2011) e1002195
76. Söding, J.: Protein homology detection by HMM-HMM comparison. Bioinformatics **21** (2005) 951–960
77. Remmert, M., Biegert, A., Hauser, A., Söding, J.: HHblits: lightning-fast iterative protein sequence searching by HMM-HMM alignment. Nat Methods **9** (2012) 173–175
78. Wheeler, T.J., Eddy, S.R.: nhmmer: DNA homology search with profile hmms. Bioinformatics **29** (2013) 2487–2489
79. Wheeler, T.J., Clements, J., Eddy, S.R., Hubley, R., Jones, T.A., Jurka, J., Smit, A.F.A., Finn, R.D.: Dfam: a database of repetitive DNA based on profile hidden markov models. Nucleic Acids Res **41** (2013) D70–D82
80. Eddy, S.R.: A memory-efficient dynamic programming algorithm for optimal alignment of a sequence to an RNA secondary structure. BMC Bioinformatics **3** (2002) 18
81. Sakakibara, Y., Brown, M., Hughey, R., Mian, I.S., Sjölander, K., Underwood, R.C., Haussler, D.: Recent methods for RNA modeling using stochastic context-free grammars. In: Proceedings of the Asilomar Conference on Combinatorial Pattern Matching, New York, NY, Springer-Verlag (1994) 289–306
82. Eddy, S.R., Durbin, R.: RNA sequence analysis using covariance models. Nucleic Acids Res **22** (1994) 2079–2088
83. Burge, S.W., Daub, J., Eberhardt, R., Tate, J., Barquist, L., Nawrocki, E.P., Eddy, S.R., Gardner, P.P., Bateman, A.: Rfam 11.0: 10 years of RNA families. Nucleic Acids Res **41** (2013) D226–D232
84. Nawrocki, E.P., Eddy, S.R.: Infernal 1.1: 100-fold faster RNA homology searches. Bioinformatics **29** (2013) 2933–2935
85. Uemura, Y., Hasegawa, A., Kobayashi, S., Yokomori, T.: Tree adjoining grammars for RNA structure prediction. Theoretical Computer Science **210** (1999) 277–303
86. Rivas, E., Eddy, S.: The language of RNA: a formal grammar that includes pseudoknots. Bioinformatics **16** (2000) 334
87. Cai, L., Malmberg, R.L., Wu, Y.: Stochastic modeling of RNA pseudoknotted structures: a grammatical approach. Bioinformatics **19 Suppl 1** (2003) i66–i73
88. Matsui, H., Sato, K., Sakakibara, Y.: Pair stochastic tree adjoining grammars for aligning and predicting pseudoknot RNA structures. Proc IEEE Comput Syst Bioinform Conf (2004) 290–299

89. Grundy, W.N., Bailey, T.L., Elkan, C.P., Baker, M.E.: Meta-meme: motif-based hidden Markov models of protein families. Comput Appl Biosci **13** (1997) 397–406
90. Jonassen, I. Collins, J., Higgins, D.: Finding flexible patterns in unaligned protein sequences. Protein Science **4** (1995) 1587–1595
91. Hulo, N., Bairoch, A., Bulliard, V., Cerutti, L., Cuche, B.A., de Castro, E., Lachaize, C., Langendijk-Genevaux, P.S., Sigrist, C.J.A.: The 20 years of PROSITE. Nucleic Acids Res **36** (2008) D245–D249
92. Yokomori, T., Ishida, N., Kobayashi, S.: Learning local languages and its application to protein α-chain identification. In: 27th Annual Hawaii International Conference on System Sciences (HICSS-27), January 4-7, 1994, Maui, Hawaii, USA, IEEE Computer Society (1994) 113–122
93. Yokomori, T., Kobayashi, S.: Learning local languages and their application to DNA sequence analysis. IEEE Transactions on Pattern Analysis and Machine Intelligence **20** (1998) 1067–1079
94. Garcia, P., Vidal, E., Oncina, J.: Learning locally testable languages in the strict sense. In: Proceedings of the International Conference on Algorithmic Learning Theory. (1990) 325–338
95. Garcia, P., Vidal, E.: Inference of k-testable languages in the strict sense and application to syntactic pattern recognition. IEEE Trans. Pattern Anal. Mach. Intell. **12** (1990) 920–925
96. Peris, P., López, D., Campos, M., Sempere, J.M.: Protein motif prediction by grammatical inference. In Sakakibara, Y., Kobayashi, S., Sato, K., Nishino, T., Tomita, E., eds.: Ig TM. Volume 4201 of Lecture Notes in Computer Science, Springer (2006) 175–187
97. Peris, P., López, D., Campos, M.: IGTM: An algorithm to predict transmembrane domains and topology in proteins. BMC Bioinformatics **9** (2008)
98. Garcia, P., Vidal, E., Casacuberta, F.: Local languages, the succesor method, and a step towards a general methodology for the inference of regular grammars. IEEE Trans. Pattern Anal. Mach. Intell. **9** (1987) 841–845
99. Oncina, J., Garcia, P.: Inferring regular languages in polynomial update time. In: Pattern Recognition and Image Analysis. (1992) 49–61
100. Lang, K.J. In: Random DFA's can be approximately learned from sparse uniform examples. Association for Computing Machinery (1992) 45–52
101. Lang, K.J., Pearlmutter, B.A., Price, R.A.: Results of the Abbadingo One DFA learning competition and a new evidence-driven state merging algorithm. In: Proceedings of the 4th International Colloquium on Grammatical Inference. ICGI '98, London, UK, Springer-Verlag (1998) 1–12
102. Coste, F., Kerbellec, G., Idmont, B., Fredouille, D., Delamarche, C.: Apprentissage d'automates par fusions de paires de fragments significativement similaires et premières expérimentations sur les protéines MIP. In: JOBIM. (2004)
103. Coste, F., Kerbellec, G.: A similar fragments merging approach to learn automata on proteins. In Gama, J., Camacho, R., Brazdil, P., Jorge, A., Torgo, L., eds.: ECML. Volume 3720 of Lecture Notes in Computer Science., Springer (2005) 522–529
104. Coste, F., Kerbellec, G.: Learning Automata on Protein Sequences. In Denise, A., Durrens, P., Robin, S., Rocha, E., de Daruvar, A., Groppi, A., eds.: JOBIM, Bordeaux, France (2006) 199–210
105. Kerbellec, G.: Apprentissage d'automates modélisant des familles de séquences protéiques. PhD thesis, Université de Rennes 1 (2008)
106. Bretaudeau, A., Coste, F., Humily, F., Garczarek, L., Corguillé, G.L., Six, C., Ratin, M., Collin, O., Schluchter, W.M., Partensky, F.: Cyanolyase: a database of phycobilin lyase sequences, motifs and functions. Nucleic Acids Research **41** (2013) 396–401
107. Burgos, A., Coste, F., Kerbellec, G.: Learning automata on protein sequences by partial multiple sequence alignment. (in preparation)
108. Coste, F., Fredouille, D.: What is the Search Space for the Inference of Non Deterministic, Unambiguous and Deterministic Automata? Rapport de recherche RR-4907, INRIA (2003)
109. Dyrka, W., Nebel, J.C.: A stochastic context free grammar based framework for analysis of protein sequences. BMC Bioinformatics **10** (2009) 323

110. Coste, F., Garet, G., Nicolas, J.: Local Substitutability for Sequence Generalization. In Heinz, J., de la Higuera, C., Oates, T., eds.: ICGI 2012. Volume 21 of JMLR Workshop and Conference Proceedings, University of Maryland, MIT Press (2012) 97–111
111. Clark, A., Eyraud, R.: Identification in the limit of substitutable context free languages. In Jain, S., Simon, H.U., Tomita, E., eds.: Proceedings of the 16th International Conference on Algorithmic Learning Theory, Springer-Verlag (2005) 283–296
112. Clark, A., Eyraud, R.: Polynomial identification in the limit of substitutable context-free languages. Journal of Machine Learning Research **8** (2007) 1725–1745
113. Yoshinaka, R.: Identification in the limit of k, l-substitutable context-free languages. In Clark, A., Coste, F., Miclet, L., eds.: ICGI. Volume 5278 of Lecture Notes in Computer Science., Springer (2008) 266–279
114. Harris, Z.: Distributional structure. Word **10** (1954) 146–162
115. Coste, F., Garet, G., Nicolas, J.: A bottom-up efficient algorithm learning substitutable languages from positive examples. In Clark, A., Kanazawa, M., Yoshinaka, R., eds.: ICGI 2014. Volume 34 of JMLR Workshop and Conference Proceedings. (2014) 49–63
116. Nevill-Manning, C.G., Witten, I.H.: Compression and explanation using hierarchical grammars. The Computer Journal **40** (1997) 103–116
117. Cherniavsky, N., Lander, R.: Grammar-based compression of DNA sequences. In: DIMACS Working Group on the Burrows-Wheeler Transform. (2004) 21
118. Lanctot, J.K., Li, M., Yang, E.H.: Estimating DNA sequence entropy. In: ACM-SIAM Symposium on Discrete Algorithms. (2000) 409–418
119. Apostolico, A., Lonardi, S.: Off-line compression by greedy textual substitution. Proceedings of the IEEE **88** (2000) 1733–1744
120. Apostolico, A., Lonardi, S.: Compression of biological sequences by greedy off-line textual substitution. In: Data Compression Conference. (2000) 143–153
121. Nevill-Manning, C., Witten, I.: On-line and off-line heuristics for inferring hierarchies of repetitions in sequences. In: Data Compression Conference, IEEE (2000) 1745–1755
122. Carrascosa, R., Coste, F., Gallé, M., López, G.G.I.: The smallest grammar problem as constituents choice and minimal grammar parsing. Algorithms **4** (2011) 262–284
123. Carrascosa, R., Coste, F., Gallé, M., López, G.G.I.: Searching for smallest grammars on large sequences and application to DNA. J. Discrete Algorithms **11** (2012) 62–72
124. Brejova, B., Vinar, T., Li, M.: Pattern Discovery: Methods and Software. In Krawetz, S.A., Womble, D.D., eds.: Introduction to Bioinformatics. Humana Press (2003) 491–522
125. Sakakibara, Y.: Grammatical inference in bioinformatics. IEEE Trans. Pattern Anal. Mach. Intell. **27** (2005) 1051–1062
126. Durbin, R., Eddy, S., Krogh, A., Mitchison, G.: Biological Sequence Analysis : Probabilistic Models of Proteins and Nucleic Acids. Cambridge University Press (1999)
127. Baldi, P., Brunak, S.: Bioinformatics: The Machine Learning Approach. 2nd edn. Cambridge: MIT Press (2001)
128. de la Higuera, C.: Grammatical Inference: Learning Automata and Grammars. Cambridge University Press, (2010)

Printed in the United States
By Bookmasters